World Wide Photos

CLASSICAL MECHANICS:
A Modern Perspective

CLASSICAL MECHANICS:
A Modern Perspective
Second Edition

Vernon Barger
University of Wisconsin, Madison

Martin Olsson
University of Wisconsin, Madison

McGRAW-HILL, INC.
New York   St. Louis   San Francisco   Auckland   Bogota´   Caracas   Lisbon
London   Madrid   Mexico City   Milan   Montreal   New Delhi
San Juan   Singapore   Sydney   Tokyo   Toronto

CLASSICAL MECHANICS:
A Modern Perspective

This book is printed on acid-free paper.

2 3 4 5 6 7 8 9 0   D O C   D O C   9 0 9 8 7 6 5

ISBN 0-07-003734-5

The editor was Jack Shira;
the production supervisor was Annette Mayeski.
The photo editor was Anne Manning.
R. R. Donnelley & Sons Company was printer and binder.

Library of Congress Catalog Card Number: 94-72897

Photo Credits

Front, 2-page spread, endpaper: Skydiving over Sydney, Australia (Associated Press photo). Wide World Photos.

Front, reverse side of right endpaper: Lunar module over the moon with earth in the background (NASA photo). Courtesy NASA.

Back, reverse of left endpaper: Jupiter and its Galilean moons (NASA Voyager photo). Courtesy Jet Propulsion Laboratory.

Back, left endpaper: Einstein ring due to gravitational lensing of a distant quasar (National Radio Astronomy Observatory photo).

Back, right endpaper: Rings of glowing gas encircling the 1987A supernova. (Hubble Space Telescope photo). Courtesy Space Telescope Science Institute.

Fig. 7.17: Richard Wainscoat, University of Hawaii, Institute for Astronomy.

Fig. 8.15: National Radio Astronomy Observatory.

About the Authors

Vernon D. Barger earned his B.S. and Ph.D. degrees at the Pennsylvania State University and joined the faculty of the University of Wisconsin-Madison in 1963, where he continues to teach and do research. He is currently Vilas Professor and Director of the Institute for Elementary Particle Physics Research. He has been a Visiting Professor at the University of Hawaii and the University of Durham, and a Visiting Scientist at the SLAC, CERN and Rutherford Laboratories. Dr. Barger's fellowships include Fellow of the American Physical Society, John Simon Guggenheim Memorial Foundation Fellow, and Senior Visiting Fellow of the British Science-Engineering Research Council. Barger has co-authored three other textbooks and has published more than 300 scientific articles.

Martin G. Olsson earned his B.S. degree at the California Institute of Technology and his Ph.D. degree at the University of Maryland. A member of the University of Wisconsin faculty since 1964, he has published over 100 research papers. With Barger he also coauthored the textbook *Classical Electricity and Magnetism: A Contemporary Perspective*. Olsson was the recipient of a University distinguised teaching award and has served as Chair of the Physics Department. He has held visiting positions at the University of Durham and CERN, Los Alamos and Rutherford Laboratories.

To Annetta, Victor, Charlene, Amy and Andrew
To Sallie, Marybeth, Nina and Anne

Contents

PREFACE

In the twenty-one years that have elapsed since the original edition was published, we have collected many ideas for improvements. In deciding which changes to make, we have continued with our original philosophy of a reasonably concise presentation that includes numerous applications of interest in the real world. By incorporating feedback from students in our classes, we have tried to make the textbook even more student friendly.

The original edition was designed for an intensive one semester course of 45 lectures and the present text preserves that option with the basic material contained in the first 8 chapters. A one-semester course may include Chapters 1, 2, 3.1–3.3, 3.7, 4, 5.1–5.4, 5.6, 6.1–6.5, 6.7–6.9, 7.1–7.7, 7.10, 8.1–8.2. Several new chapters are included to accommodate longer courses of two quarters or two semesters and to provide enrichment for students taking a one-semester course. Numerous new exercises have been added. Short answers to most exercises are given in an Appendix.

The major changes include the following:

• One of the salient features of the first edition was the introduction of Lagrangian methods at an early stage. In the new edition more Lagrangian material and examples are included which made it natural to devote a single chapter to an introduction to the Lagrangian approach. We have integrated a parallel track development of Lagrangian and Newtonian methods throughout the text.

• We updated the section on the Grand Tour of the outer planets in view of the spectacular success of the Voyager space mission. In the first edition, more than five years before the launch, we did not anticipate how truly revolutionary this odyssey would be.

• In the treatment of tops we now use the Euler angles and the Lagrangian to obtain the equations of motion.

• We have expanded the gravitation chapter to introduce the physical ideas that underlie general relativity and qualitatively explore its consequences.

• An area of exploding interest today is cosmology and we devote a new chapter to the Newtonian description of the universe as a whole. First we classify the possible universes consistent with Hubble's law and Newtonian dynamics; then we use the virial theorem together with astronomical observations to discuss the evidence that most of the matter in the universe is in the form of dark matter.

• A chapter on special relativity is added for curricula where relativity is taught in the mechanics course. A description is given of an experimental test of time dilation with round-the-world flights with atomic clocks.

• The years since the original edition saw the emergence of non-linear dynamics as a major area in physics. We give an introduction to this area by describing solutions to the Duffing equation for a damped and driven anharmonic oscillator. After considering approximate analytic solutions, we explore numerical solutions including the period-doubling route to chaos. This chapter may provide a convenient starting point for students who want to do an undergraduate thesis involving numerical studies of non-linear systems: it is at a somewhat higher level than the other chapters.

• We have deleted a few sections from the original edition in the interest of keeping a reasonable length. Numerous sections have been rewritten to make the derivations more understandable. Throughout the text we have made improvements in notation.

Many colleagues and students contributed greatly to the development of this new edition and we wish to thank them for their help and encouragement. In particular, we would like to express our appreciation to the following people. Throughout many drafts of the manuscript, Professor Charles Goebel generously gave us excellent advice and made substantial contributions to the contents. Amy Barger and Andrew Barger gave valuable student input on the manuscript and solved many of the exercises. Professor Micheal Berger provided input from his classroom experience with the book. Professors Art Code and Jacqueline Hewett were very helpful in providing photos. Collin Olson, James Ireland and Andrew Barger made computer-generated figures. Ed Stoeffhaas skillfully typeset the manuscript using TEX and created many of the new illustrations. Jack Shira, as editor of this series, was extremely helpful and supportive of our efforts to produce an improved textbook.

We have found classical mechanics to be an extremely interesting course to teach since it offers the opportunity for students to develop an appreciation for the physical explanation of diverse phenomena. We sincerely hope that students will enjoy using the book as much as we have enjoyed creating it!

Vernon Barger
Martin Olsson

CLASSICAL MECHANICS:
A MODERN PERSPECTIVE

Chapter 1

ONE-DIMENSIONAL MOTION

The formulation of classical mechanics represents a giant milestone in our intellectual and technological history, as the first mathematical abstraction of physical theory from empirical observation. This crowning achievement is rightly accorded to Isaac Newton (1642–1727), who modestly acknowledged that if he had seen further than others, "it is by standing upon the shoulders of Giants." However, the great physicist Pièrre Simon Laplace characterized Newton's work as the supreme exhibition of individual intellectual effort in the history of the human race.

Newton translated the interpretation of various physical observations into a compact mathematical theory. Three centuries of experience indicate that mechanical behavior in the everyday domain can be understood from Newton's theory. His simple hypotheses are now elevated to the exalted status of laws, and these are our point of embarkation into the subject.

1.1 Newtonian Theory

The Newtonian theory of mechanics is customarily stated in three laws. According to the first law, a particle continues in uniform motion (*i.e.*, in a straight line at constant velocity) unless a force acts on it. The first law is a fundamental observation that physics is simpler when viewed from a certain kind of coordinate system, called an *inertial frame*. One cannot define an inertial frame except by saying that it is a frame in which Newton's laws hold. However, once one finds (or imagines) such a frame, all other frames which move with respect to it at constant velocity, with no rotation, are also inertial frames. A coordinate system fixed on the surface of the earth is not an inertial frame because of the acceleration due to the rotation of the earth and the earth's motion around the sun. Nevertheless, for many purposes it is an adequate approximation to regard a coordinate frame fixed on the earth's surface as an inertial frame. Indeed, Newton himself discovered nature's true laws while riding on the earth!

The essence of Newton's theory is the second law, which states that *the time rate of change of momentum of a body is equal to the force acting on the particle.* For motion in one dimension, the second law is

$$F = \frac{dp}{dt} \tag{1.1}$$

where the momentum p is given by the product of (mass) × (velocity) for the particle

$$p = mv \tag{1.2}$$

The second law provides a definition of force. It is useful because experience has shown that the force on a body is related in a quantitative way to the presence of other bodies in its vicinity. Further, in many circumstances it is found that the force on a body can be expressed as a function of x, v, and t, and so (1.1) becomes

$$F(x, v, t) = \frac{dp}{dt} = m\frac{d^2x}{dt^2} \tag{1.3}$$

This differential equation is called the *equation of motion.* Here m is assumed to be constant. For the remainder of this book we use Newton's notation $\dot{x} = dx/dt$; $\ddot{x} = d^2x/dt^2$. Newton's second law is then

$$F(x, \dot{x}, t) = m\ddot{x} = ma \tag{1.4}$$

where $a = \ddot{x}$ is the acceleration. In the special case $F = 0$, integration of (1.1) gives $p = $ constant in accordance with the first law.

While Newton's laws apply to any situation in which one can specify the force, very few interesting physical problems lead to force laws amenable to simple mathematical solution. The fundamental force laws of gravitation and electromagnetism do have simple forms for which the second law of motion can often be solved exactly. The use of approximate empirical forms to approximate the true force laws of physical situations involving frictional and drag forces is one of the arts that will be taught in this book. However, in this modern age of computers, one can handle arbitrary force laws by the brute-force method of numerical integration.

The third law states that if body A experiences a force due to body B, then B experiences an equal but opposite force due to A. (One speaks of this as the force between the two bodies.) As a consequence, the rates of change of the momenta of particles A and B are equal but opposite, and therefore the total $p_A + p_B$ is constant. This law is extremely useful, for instance in the treatment of rigid-body motion, but its range of applicability is not as universal as the first two laws. The third law breaks down when the interaction between the particles is electromagnetic, because the electromagnetic field carries momentum.

It is a remarkable fact that macroscopic phenomena can be explained by such a simple set of mathematical laws. As we shall see, the mathematical solutions to some problems can be complex; nevertheless, the physical basis is just (1.1). Of course, there is still a great deal of physics to put into (1.1), namely, the laws of force for specific kinds of interactions.

1.2 Interactions

Using the planetary orbit data analysis by Kepler, Newton was able to show that all known planetary orbits could be accounted for by the following force law

$$F = -\frac{GM_1M_2}{r^2} \tag{1.5}$$

This states that force between masses M_1 and M_2 is proportional to the masses and inversely proportional to the square of the distance between them. The negative sign in (1.5) denotes an attractive force between the masses. The force acts along the line between the two masses and thus for non-rotational motion the problem is effectively one-dimensional. Newton proposed that this gravitational law was universal, the same force law applying between us and the earth as between celestial bodies (and more generally between any two masses). The universality of the gravitational law can be verified, and the proportionality constant G determined, by delicate experimental measurements of the force between masses in the laboratory. The value of G is

$$G = 6.672 \times 10^{-11}\,\mathrm{m}^3/(\mathrm{kg\,s})^2 \tag{1.6}$$

The dominant gravitational force on an object located on the surface of the earth is the attraction to the earth. The gravitational force between two spherically symmetric bodies is as if all the mass of each body were concentrated at its center, as Newton proved. We will give a proof of this assertion in Chapter 8. The earth is very nearly spherical so we can use the force law of (1.5). Thus for an object of mass m on the surface of earth, the force is

$$F = -m \frac{M_E G}{R_E^2} = -mg \tag{1.7}$$

where g is the gravitational acceleration,

$$g \simeq 9.8 \text{ m/s}^2 \tag{1.8}$$

Using the measured value of $R_E = 6,371$ km along with the measured values of g and G as given above, we may use (1.7) to deduce the mass of the earth to be

$$M_E = 5.97 \times 10^{24} \text{ kg} \tag{1.9}$$

Since the earth's radius is large, the gravitational force of an object anywhere in the biosphere is given to good accuracy by (1.7); even at the top of the atmosphere ($\approx$ 200 km up) the force has decreased by less than 10% from its value at the surface of the earth. Consequently, in many applications on earth, we can neglect the variation of the gravitational force with position.

The static Coulomb force between two charges e_1 and e_2 is similar in form to the gravitational-force law of (1.5),

$$F = k \frac{e_1 e_2}{r^2} \tag{1.10}$$

This force is attractive if the charges are of opposite sign and repulsive if the charges are of the same sign. The constant k depends on the system of electrical units; in *SI* units, $k = (4\pi\epsilon_0)^{-1} \simeq 9 \times 10^9 \text{N-m}^2/\text{C}^2$.

Another force with a wide range of application is the spring force or Hooke's law, which is expressed as

$$F = -kx \tag{1.11}$$

with $k > 0$. Here k is a spring constant which is dependent on the properties of the spring and x is the extension of the spring from its relaxed position. This particular force law is a very good approximation in many physical situations (*e.g.*, the stretching or bending of materials) which are initially in equilibrium.

Frictional forces prevent or damp motions. The static frictional force between two solid surfaces is

$$|F| \leq \mu_s N \tag{1.12}$$

The force F acts to prevent sliding motion. N is the perpendicular force (normal force) holding the surfaces together, and μ_s is a material-dependent coefficient. Equation (1.12) is an *approximate* formula for frictional forces which has been deduced from empirical observations. The frictional force which retards the motion of sliding objects is given by

$$F = \mu_k N \tag{1.13}$$

It is observed that this force is nearly independent of the velocity of the motion for velocities which are neither too small (where there is molecular adhesion) nor too large (where frictional heating becomes important). For a given pair of surfaces, the coefficient of kinetic friction μ_k is less than the coefficient of static friction μ_s.

Frictional laws to describe the motion of a solid through a fluid or a gas are often complicated by such effects as turbulence. However, for sufficiently small velocities, the approximate form

$$F = -bv \tag{1.14}$$

where b is a constant, holds. The drag coefficient b in (1.14) is proportional to the fluid viscosity. For a sphere of radius a moving slowly through a fluid of viscosity η the Stokes law of resistance is calculated to be

$$b_{\text{sphere}} = 6\pi a \eta \tag{1.15}$$

At higher, but still subsonic velocities, the drag law is

$$F = -cv^2 \tag{1.16}$$

For instance, the drag force on an airplane is remarkably well represented by a constant times the square of the velocity. The drag coefficient c for a body of cross-sectional area S moving through a fluid of density ρ is given by

$$c = \tfrac{1}{2}C_D S \rho \tag{1.17}$$

where C_D is a dimensionless factor related to the geometry of the body (about 0.4 for a sphere).

Externally imposed forces can take on a variety of forms. Of those depending explicitly on time, sinusoidally oscillating forces like

$$F = F_0 \cos \omega t \tag{1.18}$$

are frequently encountered in physical situations.

In a general case the forces can be position-, velocity-, and time-dependent,

$$F = F(x, v, t) \tag{1.19}$$

Among the most interesting and easily solved examples are those in which the forces depend on only one of the above three variables, as illustrated by the examples in the following three sections.

1.3 The Drag Racer: Frictional Force

A number of interesting engineering-type problems can be solved from straightforward application of Newton's laws. As an illustration, suppose we consider a drag racer that can achieve maximum possible acceleration when starting from rest. The external forces on the racer which must be taken into account are (1) gravity, (2) the normal forces supporting the racer at the wheels, and (3) the frictional forces which oppose the rotation of the powered rear wheels. A sketch indicating the various external forces is given in Fig. 1-1.

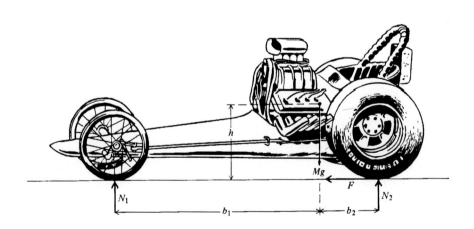

FIGURE 1-1. Forces on a drag racer.

Since the racer is in vertical equilibrium, the sum of the external vertical forces must vanish,

$$N_1 + N_2 - Mg = 0 \qquad (1.20)$$

Both N_1 and N_2 must be positive. For the horizontal motion we apply Newton's second law,

$$F = Ma \qquad (1.21)$$

The frictional force F is bounded by

$$F \leq \mu N_2 \qquad (1.22)$$

The maximum friction force occurs just as the racer tires begin to slip relative to the drag strip, because the coefficient of kinetic friction is smaller than the coefficient of static friction. For maximal initial acceleration we must have the maximum friction force $F = \mu N_2$. Referring back to (1.20), a maximal $N_2 = Mg$ is obtained when $N_1 = 0$, that is, when the back wheels completely support the racer. The greatest possible acceleration is then

$$a_{\max} = \frac{\mu (N_2)_{\max}}{M} = \mu g \qquad (1.23)$$

We see that the optimum acceleration is independent of the racer's mass. Under normal conditions the coefficient of friction μ between rubber and concrete is about unity. Thus a racer can achieve an acceleration of about 9.8 m/s^2. In actual design a small normal force N_1 on the front wheels is allowed for steering purposes.

The standard drag strip is ≈ 400 m (1/4 mi) in length. If we assume that the racer can maintain the maximum acceleration for the duration of a race and that the coefficient of friction is constant, we can calculate the final velocity and the elapsed time. The differential form of the second law is

$$F = Ma = M\frac{dv}{dt} = M\ddot{x} \qquad (1.24)$$

When the acceleration a is constant, a single integration

$$\int_{v_0}^{v} dv = a \int_{0}^{t} dt \qquad (1.25)$$

gives

$$v - v_0 = at \tag{1.26}$$

Using $dx = v dt$, a second integration

$$\int_{x_0}^{x} dx = \int_{0}^{t} (v_0 + at) dt \tag{1.27}$$

yields

$$x - x_0 = v_0 t + \tfrac{1}{2} a t^2 \tag{1.28}$$

We can eliminate t from (1.26) and (1.28) to obtain

$$v^2 = v_0^2 + 2a(x - x_0) \tag{1.29}$$

Substituting $a = 9.8 \, \text{m/s}^2$, $x = 0.40 \, \text{km}$, $x_0 = 0$ and $v_0 = 0$, we find $v = 89 \, \text{m/s}$ (or $320.4 \, \text{km/h}$)! The time elapsed, $t = v/a$, is about $9 \, \text{s}$. For comparison, the world drag-racing records (with a piston engine) as of 1992 are $v = 134.8 \, \text{m/s}$ ($485.3 \, \text{km/h}$) for velocity and $4.80 \, \text{s}$ for elapsed time. (These records were set in different races.) With tires that are several times wider than automobile tires and have treated surfaces, coefficients of friction considerably greater than $\mu = 1$ are realized in drag racing. The rubber laid down by previous racers in effect gives a rubber-rubber contact which also increases the coefficient of friction. Aerodynamic effects are important as well. The drag force from wind resistance reduces the speed of a racer, while a negative lift force on the back wheels can be produced by wind resistance against an up-tilted rear wing found on many racers, which increases the normal force, giving greater traction and allowing larger acceleration.

1.4 Sport Parachuting: Aerodynamic Drag

The sport of skydiving visually illustrates the effect of the viscous frictional force of (1.16). Immediately upon leaving the aircraft, the jumper accelerates downward due to the gravity force. As his velocity increases, the air resistance exerts a greater retarding force, and eventually approximately balances the pull of gravity. From this time onward the descent of the diver is at a uniform rate, called the *terminal velocity*. The terminal velocity in a spread-eagle position is roughly 120 mi/h. By assuming a vertical head-down position, the diver can decrease his cross sectional area (perpendicular to the direction of motion) thereby lowering the air resistance [smaller value of c in (1.16)], and increase his terminal velocity of descent. Eventually, of course, the diver opens his parachute. This dramatically increases the air resistance and correspondingly reduces his terminal velocity, to allow a soft impact with the ground.

To analyze the physics of skydiving, we shall assume that the motion is vertically downward and choose a coordinate system with $x = 0$ at the earth's surface and positive upward. In this coordinate frame, downward forces are negative. We approximate the external force on the diver as

$$F = -mg + cv^2 \tag{1.30}$$

The frictional force is positive, as required for an upward force. The terminal velocity is reached when the opposing gravity and frictional forces balance, giving $F = 0$. Under this condition, the terminal velocity is

$$v_t = \sqrt{\frac{mg}{c}} \tag{1.31}$$

To solve the differential equation of motion,

$$F = m\frac{dv}{dt} = -mg + cv^2 \tag{1.32}$$

we rearrange the factors and integrate

$$\int_0^v \frac{dv}{v_t^2 - v^2} = -\frac{g}{v_t^2} \int_0^t dt \tag{1.33}$$

In (1.32) the frictional coefficient c has been replaced by v_t from (1.31). We obtain

$$\frac{1}{2v_t} \ln\left(\frac{v_t + v}{v_t - v}\right) = -\frac{g}{v_t^2} t \tag{1.34}$$

which can be inverted to express v in terms of t,

$$v = -v_t \frac{1 - \exp(-2gt/v_t)}{1 + \exp(-2gt/v_t)} \tag{1.35}$$

At large times the decreasing exponentials go to zero rapidly and v approaches the terminal velocity,

$$v \to -v_t \tag{1.36}$$

Although the limiting velocity is exactly reached only at infinite time, it is approximately reached for times $t \gg v_t/2g$. A typical value for v_t

on a warm summer day is 54 m/s (194.4 km/h) for a 70 kg diver in a spread-eagle position. After a time

$$t = \frac{2v_t}{g} = \frac{2(54)}{9.8} = 11\,\text{s} \qquad (1.37)$$

the sky diver would be traveling about 52 m/s with his parachute unopened! The velocity of the diver as a function of time is plotted in Figs. 1-2 and 1-3. To calculate the distance the diver has fallen after a specific elapsed time, we integrate $dx = v\,dt$ using (1.35),

$$\int_h^x dx = -v_t \int_0^t \left(1 - \frac{2\exp(-2gt/v_t)}{1+\exp(-2gt/v_t)} \right) dt \qquad (1.38)$$

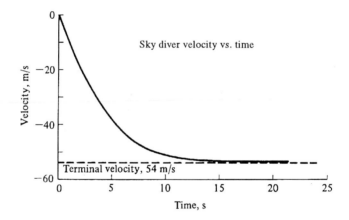

FIGURE 1-2. Velocity of a sky diver as a function of time for a terminal velocity of 54 m/s.

The result of the integration is

$$h - x = v_t \left[t - \frac{v_t}{g} \ln \left(\frac{2}{1+\exp(-2gt/v_t)} \right) \right] \qquad (1.39)$$

At time $t = 2v_t/g$, the diver has fallen a distance $(h - x)$, given by

$$h - x = \frac{v_t^2}{g} \left[2 - \ln \left(\frac{2}{1+e^{-4}} \right) \right] = \frac{(54)^2}{9.8}(2 - 0.7) = 385\,\text{m} \qquad (1.40)$$

Sky divers normally free-fall about 1,400 m (in 30 s) so much of the descent is at terminal velocity.

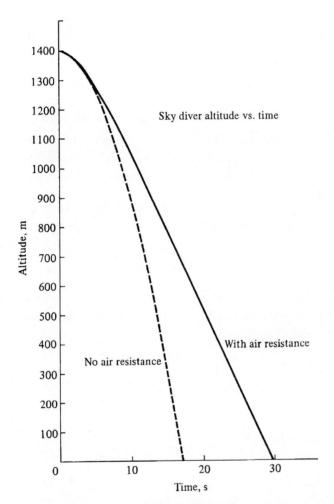

FIGURE 1-3. Altitude of a sky diver with unopened parachute as a function of time (for a terminal velocity of 54 m/s).

Finally, let us use the drag coefficient formula of (1.17) to estimate the free fall terminal velocity of a sky diver. By (1.31) and (1.17) we have

$$v_t = \sqrt{\frac{mg}{\frac{1}{2}C_D S\rho}} \tag{1.41}$$

Assuming that in a spread-eagle position $C_D \simeq 0.5$ and $S \simeq 1 \text{ m}^2$ and that the air density is 1 kg/m^3 we find for a 70 kg diver,

$$v_t = \sqrt{\frac{70(9.8)}{0.25(1)(1)}} = 52 \text{ m/s} \tag{1.42}$$

very near the actual value. The excellent agreement is fortuitous but the ability to make such estimates of the drag force is certainly useful.

1.5 Archery: Spring Force

The force exerted on an arrow by an archer's bow can be approximated by the spring force of (1.11). A 134 Newton bow with a 0.72 meter draw d has a spring constant k given by

$$k = \frac{|F|}{d} = \frac{134}{0.72} = 186 \, \text{kg/s}^2 \qquad (1.43)$$

After release of the bowstring, the motion of the arrow of mass m is described by the second law,

$$m\frac{dv}{dt} = -kx \, , \qquad x < 0 \qquad (1.44)$$

until it leaves the bowstring at $x = 0$. Here we neglect the mass of the bowstring. To integrate this differential equation for the velocity, we use the chain rule of differentiation

$$\frac{dv}{dt} = \frac{dv}{dx}\frac{dx}{dt} = \frac{dv}{dx}v \qquad (1.45)$$

Substituting into (1.44), rearranging factors, and integrating we obtain

$$m\int_0^v v \, dv = -k \int_{-d}^0 x \, dx \qquad (1.46)$$

or

$$\tfrac{1}{2}mv^2 = \tfrac{1}{2}kd^2 \qquad (1.47)$$

Thus the velocity of the arrow as it leaves the bowstring is given by

$$v = d\sqrt{\frac{k}{m}} \qquad (1.48)$$

The longer the draw and the stronger the bow, the higher the arrow velocity. For a typical target arrow, with weight $m = 23$ g, the velocity is

$$v = 0.72\sqrt{\frac{186}{23 \times 10^{-3}}} = 65 \, \text{m/s} \qquad (1.49)$$

This is almost double the maximum speed of a fastball thrown by a professional baseball player!

1.6   Methods of Solution

For the general motion of a particle in one dimension, the equation of motion is

$$m\ddot{x} = F(x, \dot{x}, t) \tag{1.50}$$

Since this is a second-order differential equation, the solution for x as a function of t involves two arbitrary constants. These constants can be fixed from physical conditions, such as the position and velocity at the initial time. In the examples of §1.3 to 1.5, we have introduced several techniques for solving (1.50). In the case where F depends on only one of the variables x, $\dot{x}$, or t, the formal solution of (1.50) is straightforward. We now run through the methods of solution to the differential equations of motion for these specific classes of force laws.

For a force that depends only on x, we may use the chain rule of (1.45), and integrate (1.50) to obtain

$$m \int^{v} v' dv' = \int^{x} F(x')\, dx' + C_1 \tag{1.51}$$

where C_1 is a constant of integration. Here we have used primes to denote the dummy variables of integration. The resulting expression for $v(x)$ is

$$v = \sqrt{\frac{2}{m}} \sqrt{\int^{x} F(x')\, dx' + C_1} \tag{1.52}$$

This method of solution was employed in the archery discussion of §1.5. The solution for $x(t)$ is found by substituting $v = \dot{x}$ in (1.52), rearranging factors so as to separate the variables, and integrating, to get

$$\int^{x} \frac{dx'}{\sqrt{\int^{x'} F(x'')dx'' + C_1}} = \sqrt{\frac{2}{m}} \int^{t} dt' + C_2 \tag{1.53}$$

The integration constants C_1 and C_2 can be fixed from the initial velocity and position.

With a velocity-dependent force we can integrate (1.50) as follows:

$$m \int^{v} \frac{dv'}{F(v')} = \int^{t} dt' + C_1 \tag{1.54}$$

We used this technique in the sky-diving analysis of §1.4. The result of the integration gives $v(t)$, which can then be integrated over t to find $x(t)$.

The solution of (1.50) for a time-dependent force $F(t)$ can be obtained from direct integration,

$$m \int^{v} dv' = \int^{t} F(t') \, dt' + C_1 \tag{1.55}$$

A second integration leads to the solution for $x(t)$,

$$m \int^{x} dx' = \int^{t} \left[\int^{t'} F(t'') dt'' + C_1 \right] dt' + C_2 \tag{1.56}$$

If the force law depends on more than one variable, the techniques for finding analytical solutions, when they exist, are more complicated.

For the forces involved in many physical problems, (1.50) cannot be solved in closed analytical form. However, we can then resort to numerical methods which can be evaluated using computers. To illustrate the numerical approach, we assume that the position x_0 and velocity v_0 are known at the initial time t_0. The acceleration a_0 then is given by (1.50) as

$$a_0 = \frac{F(x_0, v_0, t_0)}{m} \tag{1.57}$$

After a short time interval Δt,

$$\begin{aligned} t_1 &= t_0 + \Delta t \\ x_1 &= x_0 + v_0 \Delta t \\ v_1 &= v_0 + a_0 \Delta t \end{aligned} \tag{1.58}$$

where we have neglected the change in a and v over Δt. This approximation becomes more accurate as the time increment Δt is made smaller. From these new values of the variables, we can calculate the new acceleration using (1.50)

$$a_1 = \frac{F(x_1, v_1, t_1)}{m} \tag{1.59}$$

By repetition of this procedure n times, we can calculate x and v at time $t_n = t_0 + n \, \Delta t$

$$\begin{aligned} x_n &= x_{n-1} + v_{n-1} \, \Delta t \\ v_n &= v_{n-1} + a_{n-1} \, \Delta t \end{aligned} \tag{1.60}$$

We thereby obtain a complete numerical solution to the equation of motion. The solution becomes more accurate as the time increment Δt is made smaller. This illustrates that a unique solution to the differential equation of motion is always possible for any reasonable force law. For the numerical solution to a specific problem the use of more sophisticated numerical methods is usually desirable in order to increase the accuracy of the result for a given Δt.

1.7 Simple Harmonic Oscillator

Many common physical applications of the spring force involve oscillatory motion, such as vibrations of a mass attached to a spring. A system undergoing periodic steady-state motion under the action of a spring is called a *harmonic oscillator*. The motion is called *simple harmonic* when the restoring force is proportional to the displacement from an equilibrium position (for instance, proportional to the extension or compression of a spring). Any system in which there is a linear restoring force (such as AC circuits and certain servomechanisms) exhibits simple harmonic oscillations.

The equation of motion for a simple harmonic oscillator,

$$m\ddot{x} = -kx \tag{1.61}$$

with $k > 0$, can be solved by (1.52) and (1.53). However, we can cleverly construct the solution as follows. The functions $\cos \omega_0 t$ and $\sin \omega_0 t$ satisfy (1.61) if the angular frequency ω_0 is given by

$$\omega_0 = \sqrt{\frac{k}{m}} \tag{1.62}$$

The general solution to (1.61) is a linear superposition of $\cos \omega_0 t$ and $\sin \omega_0 t$ solutions

$$x(t) = A \cos \omega_0 t + B \sin \omega_0 t \tag{1.63}$$

where A and B are arbitrary constants. An equivalent form of the solution is

$$x(t) = a \cos(\omega_0 t + \alpha) \tag{1.64}$$

with constants related by

$$A = a \cos \alpha \qquad B = -a \sin \alpha \tag{1.65}$$

The constant a is called the *amplitude* of the motion, and α is called the *initial phase*. The initial conditions can be used to specify the arbitrary constants a and α. From (1.64) the velocity of the oscillator is

$$v(t) = -a\omega_0 \sin(\omega_0 t + \alpha) \tag{1.66}$$

The period τ of the motion is the time required for the system to undergo a complete oscillation and return to the initial values of x and v. The period for the oscillator is

$$\tau = \frac{2\pi}{\omega_0} \tag{1.67}$$

The frequency of the oscillator (number of oscillations per unit time) is

$$\nu = \frac{1}{\tau} = \frac{\omega_0}{2\pi} \tag{1.68}$$

We can illustrate our harmonic-oscillator solution with the bow-and-arrow example of §1.5. At $t = 0$ the bow is at full draw, $x = -d$, and the arrow velocity is zero. From (1.66) we find

$$\alpha = 0 \tag{1.69}$$

and from (1.64) we obtain

$$a = -d \tag{1.70}$$

The solution with proper boundary conditions is

$$x(t) = -d \cos \omega_0 t \tag{1.71}$$
$$v(t) = d\omega_0 \sin \omega_0 t \tag{1.72}$$

with $\omega_0 = \sqrt{k/m}$. As time increases from $t = 0$, x increases to zero at

$$t = \frac{1}{2}\left(\frac{\pi}{\omega_0}\right) \tag{1.73}$$

At this instant the arrow leaves the bowstring with velocity

$$v = d\omega_0 = d\sqrt{\frac{k}{m}} \tag{1.74}$$

which agrees with (1.48). For the bow described in §1.5 the arrow-propulsion time from (1.73) is

$$t = \frac{\pi}{2}\sqrt{\frac{m}{k}} = \frac{\pi}{2}\sqrt{\frac{23 \times 10^{-3}}{186}} \approx \frac{1}{60}s \tag{1.75}$$

In our archery example the simple-harmonic-force law does not apply beyond this time (one-fourth of the period τ), as illustrated in Figs. 1-4 and 1-5.

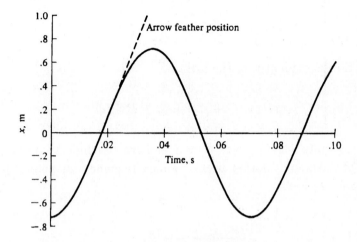

FIGURE 1-4. Displacement of a simple harmonic oscillator *vs.* time. The position of the feather end of the archer's arrow as a function of time is indicated by the dashed line after the arrow leaves the bow.

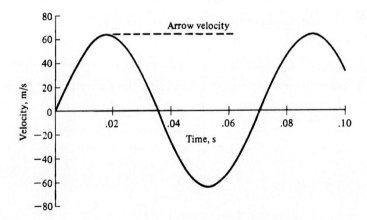

FIGURE 1-5. Velocity of a simple harmonic oscillator vs. time. The velocity of the arrow after it leaves the bow is indicated by the dashed line.

As another instructive example we consider the spring-mass system in Fig. 1-6. The spring, assumed massless, has a rest length ℓ. When the mass m is attached to the free end, the equation of motion in the absence of gravity is

$$m\ddot{x} = -k(x - \ell) \tag{1.76}$$

or

$$\ddot{x} + \omega_0^2 x = \omega_0^2 \ell, \qquad \omega_0^2 = \frac{k}{m} \tag{1.77}$$

The solution is evidently of the form

$$x(t) = C + A \cos \omega_0 t + B \sin \omega_0 t \tag{1.78}$$

Substitution into (1.77) yields $C = \ell$ and we conclude that the motion consists of harmonic motion with angular frequency ω_0 about the equilibrium point $x = \ell$.

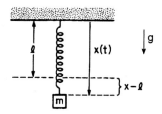

FIGURE 1-6. A mass m suspended by a spring of rest length ℓ undergoes vertical oscillations.

With gravity present, we must add mg to the forces acting on m and the equation of motion becomes

$$m\ddot{x} = -k(x - \ell) + mg \tag{1.79}$$

or

$$\ddot{x} + \omega_0^2 x = \omega_0^2 \ell + g \tag{1.80}$$

Comparing to the gravity-free case shows the equation of motion differs only by the constant on the right side. The solution to (1.80) is then (1.78) but with $C = \ell + \frac{mg}{k}$. The motion is again harmonic with angular frequency ω_0 except that the equilibrium point is $\ell + \frac{mg}{k}$. When the mass is at $\ell + \frac{mg}{k}$, the upward force due to the spring is $k(\frac{mg}{k}) = mg$, which just equals the weight force. The mass will remain at this position if released at rest there.

We conclude this section by solving the simple harmonic equation of motion (1.61) in a more systematic way. The equation of motion

$$\ddot{x} + \omega_0^2 x = 0 , \qquad \omega_0 = \sqrt{\frac{k}{m}} \tag{1.81}$$

is a linear differential equation with constant coefficients; such an equation always has a solution of the form

$$x(t) = e^{\lambda t} \tag{1.82}$$

With this substitution, (1.81) becomes

$$(\lambda^2 + \omega_0^2) e^{\lambda t} = 0 \tag{1.83}$$

which requires that $\lambda^2 = -\omega_0^2$. Thus (1.82) is a solution if $\lambda = i\omega_0$ or $\lambda = -i\omega_0$ and so the linear superposition

$$x(t) = c_1 e^{i\omega_0 t} + c_2 e^{-i\omega_0 t} \tag{1.84}$$

is a solution; here c_1 and c_2 are constants (generally complex). Since by appropriately choosing these constants we can fit any initial conditions x_0 and $\dot{x}_0$, (1.84) is the general solution. Using the identity $e^{\pm i\theta} = \cos\theta \pm i\sin\theta$ we can rewrite this general solution as

$$\begin{aligned}
x(t) &= c_1(\cos\omega_0 t + i\sin\omega_0 t) + c_2(\cos\omega_0 t - i\sin\omega_0 t) \\
&= (c_1 + c_2)\cos\omega_0 t + i(c_1 - c_2)\sin\omega_0 t
\end{aligned} \tag{1.85}$$

Since any physical quantity such as $x(t)$ must be real (no imaginary part) we must choose $c_2 = c_1^*$, where c_1^* is the complex conjugate of c_1. With $2\mathcal{Re}\, c_1 = A$ and $2\mathcal{Im}\, c_1 = -B$, we obtain the result in (1.63).

1.8   Damped Harmonic Motion

In almost all physical problems frictional forces play a role. For example, a harmonic oscillator that is subject to a damping force has an amplitude that decreases with time. For this reason, and also because a damped harmonic oscillator applies to such a broad range of physical phenomena, we treat its solution at some length. The form (1.14) chosen for the frictional force is linear in the velocity; the equation of motion is then linear in both x and its time derivatives and is solvable analytically.

The equation of motion of the damped harmonic oscillator is

$$m\ddot{x} = -kx - b\dot{x} \tag{1.86}$$

We define the dampling constant $\gamma = \frac{1}{2}(b/m)$ and the natural frequency $\omega_0 = \sqrt{k/m}$ to express (1.86) in the form

$$\ddot{x} + 2\gamma\dot{x} + \omega_0^2 x = 0 \tag{1.87}$$

Like the undamped harmonic oscillator equation of motion, (1.81), this is a linear differential equation with constant coefficients so again

$$x(t) = e^{\lambda t} \tag{1.88}$$

is a solution. Substituting this into (1.87) we find

$$(\lambda^2 + 2\gamma\lambda + \omega_0^2)e^{\lambda t} = 0 \tag{1.89}$$

which is satisfied only if the term in parentheses vanishes. Solving the quadratic equation, the possible values of λ are

$$\lambda = -\gamma \pm \sqrt{\gamma^2 - {\omega_0}^2} \equiv -\gamma \pm \Omega \tag{1.90}$$

where we have defined

$$\Omega \equiv \sqrt{\gamma^2 - \omega_0^2} \tag{1.91}$$

The qualitative nature of the solution depends on the relative magnitude of the frictional coefficient γ and the natural frequency ω_0. We distinguish the three cases:

$$
\begin{array}{lll}
\text{I.} & \gamma > \omega_0\,, & \Omega \text{ real} \\
\text{II.} & \gamma = \omega_0\,, & \Omega \text{ zero} \\
\text{III.} & \gamma < \omega_0\,, & \Omega \text{ imaginary}
\end{array} \tag{1.92}
$$

For $\Omega \neq 0$ the general solution is a superposition of $e^{\lambda t}$ terms with both possible values of λ. In case I the solution is

$$
\begin{aligned}
\text{I.} \qquad x(t) &= c_1 e^{-(\gamma-\Omega)t} + c_2 e^{-(\gamma+\Omega)t} \\
&= e^{-\gamma t}(c_1 e^{\Omega t} + c_2 e^{-\Omega t})
\end{aligned} \tag{1.93}
$$

If $\Omega = 0$ the two terms in (1.93) have the same t-dependence. Then, since the expression depends only on the one constant $c_1 + c_2$, (1.93) is not the general solution of the second order differential equation. Treating the $\Omega = 0$ case as a limit $\Omega \to 0$ we can expand the exponentials

$$e^{\pm\Omega t} = 1 \pm \Omega t + \mathcal{O}(\Omega^2) \tag{1.94}$$

and group the terms in (1.93) as

$$x(t) = e^{-\gamma t}\left[(c_1 + c_2) + (c_1 - c_2)\Omega t\right] + \mathcal{O}(\Omega^2) \tag{1.95}$$

Then defining $C = c_1 + c_2$ and $D = (c_1 - c_2)\Omega$, the solution for $\Omega = 0$ is

II. $$x(t) = e^{-\gamma t}[C + Dt] \tag{1.96}$$

In case III, we express

$$\Omega = \sqrt{\gamma^2 - \omega_0^2} = i\sqrt{\omega_0^2 - \gamma^2} = i\Omega' \tag{1.97}$$

in terms of the real quantity

$$\Omega' = \sqrt{\omega_0^2 - \gamma^2} \tag{1.98}$$

Then the form of the solution (1.93) is

III. $$x(t) = e^{-\gamma t}\left(c_1 e^{i\Omega' t} + c_2 e^{-i\Omega' t}\right) \tag{1.99}$$

For $x(t)$ to be real the constants c_1 and c_2 must be complex conjugates, $c_2 = c_1^*$. Thus the solution can be expressed in the form

$$x(t) = 2e^{-\gamma t}\mathcal{Re}\left(c_1 e^{i\Omega' t}\right) \tag{1.100}$$

Writing c_1 in polar form as $c_1 = \frac{1}{2}ae^{i\alpha}$ where a and α are real, we obtain

III. $$x(t) = ae^{-\gamma t}\cos(\Omega' t + \alpha) \tag{1.101}$$

The two constants which appear in the above solutions can be related to the initial conditions $x(0) \equiv x_0$ and $\dot{x}(0) \equiv v_0$ at time $t = 0$. After solving for the constants from the initial conditions, the solutions are of the forms

$$\text{I. } x(t) = \frac{1}{2}\left[x_0 + \frac{(v_0 + \gamma x_0)}{\Omega}\right]e^{-(\gamma - \Omega)t} + \frac{1}{2}\left[x_0 - \frac{(v_0 + \gamma x_0)}{\Omega}\right]e^{-(\gamma + \Omega)t}$$

(1.102)

$$\text{II.} \qquad\qquad x(t) = e^{-\gamma t}\left[x_0 + (v_0 + \gamma x_0)t\right] \qquad\qquad (1.103)$$

$$\text{III.} \qquad\qquad x(t) = ae^{-\gamma t}\cos(\Omega' t + \alpha) \qquad\qquad (1.104)$$

with $a = \left(\omega_0^2 x_0^2 + 2\gamma v_0 x_0 + v_0^2\right)^{1/2}/\Omega'$ and $\tan\alpha = -(v_0 + \gamma x_0)/x_0\Omega'$.

In all three cases the amplitude of the displacement decays exponentially with time, although in II the exponential factor is multiplied by a linear function of t. At large times the rates of falloff are characterized by the exponentials:

I. $e^{-(\gamma - \Omega)t}$ $\gamma > \omega_0$ (overdamped)

II. $e^{-\gamma t} * $ (linear function of t) $\gamma = \omega_0$ (critically damped)

III. $e^{-\gamma t} * $ (sinusoidal function of t) $\gamma < \omega_0$ (underdamped)

(1.105)

Illustrations of the time dependences for the three cases are given in Fig. 1-7 for the initial conditions $x = x_0$, $v_0 = 0$. An exception to the above rates of decrease occurs when the initial conditions are such that the coefficient of the $e^{-(\gamma - \Omega)t}$ term of solution I vanishes. In that circumstance, the mass returns to rest like $e^{-(\gamma + \Omega)t}$.

There are endless applications of damped harmonic oscillators. The pneumatic spring return on a door represents an everyday situation where solution II is the ideal. Upon releasing the door with no initial velocity, we want it to close as rapidly as possible without slamming. Equations (1.105) indicate that solution II should be selected; the spring-tube system should be designed with $\gamma = \omega_0$. Solution III might close the door faster, due to the vanishing of the cosine factor in (1.104), but this would let the door slam! On the other hand, solution III describes physical systems that undergo damped periodic oscillations.

The behavior of simple electric circuits is determined by a differential equation which has the same mathematical form as the damped harmonic

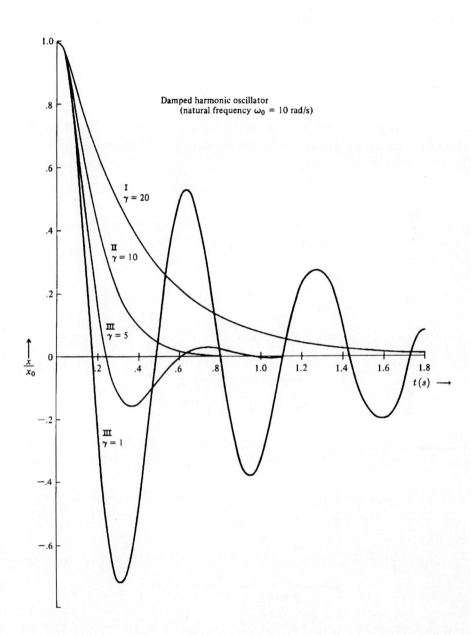

FIGURE 1-7. Time dependence of the displacement of a damped harmonic oscillator for the initial conditions $x = x_0$, $v = 0$. The natural frequency of the oscillator is $\omega_0 = 10$ rad/s. Results for various strengths of the damping constant γ are illustrated.

oscillator.   As an example we consider the circuit of Fig. 1-8 with an inductor L, resistor R, and capacitor C in series.   When the switch is closed, the sum of the voltage drops across the elements of the circuit must add up to zero. This leads to the differential equation

$$L\frac{di}{dt} + Ri + \frac{q}{C} = 0 \qquad (1.106)$$

where $i(t)$ is the current flowing in the circuit and $q(t)$ is the charge on one of the capacitor plates. Since $i = dq/dt = \dot{q}$, the circuit equation can be written as

$$L\ddot{q} + R\dot{q} + \frac{q}{C} = 0 \qquad (1.107)$$

This equation has the form of the damped-harmonic-oscillator equation (1.86) with the following correspondences:

$$
\begin{aligned}
x &\to q & \gamma &= \frac{b}{2m} \to \frac{R}{2L} \\
m &\to L & \omega_0 &= \sqrt{\frac{k}{m}} \to \sqrt{\frac{1}{LC}} \\
b &\to R & \Omega &= \sqrt{\gamma^2 - \omega_0^2} \to \sqrt{\left(\frac{R}{2L}\right)^2 - \frac{1}{LC}} \\
k &\to \frac{1}{C} & \Omega' &= \sqrt{\omega_0^2 - \gamma^2} \to \sqrt{\frac{1}{LC} - \left(\frac{R}{2L}\right)^2}
\end{aligned}
\qquad (1.108)
$$

Since it is often far easier to connect circuit elements than to build and test a mechanical system, this analogy has been of considerable practical importance.

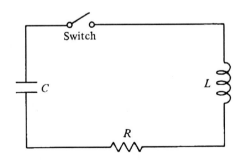

FIGURE 1-8. Simple L, R, C electric series circuit.

If the circuit in Fig. 1-8 is in a static state, when the switch is closed at time $t = 0$ the initial conditions are

$$q(t = 0) = q_0 = CV_0$$

$$i(0) = \dot{q}(t = 0) = 0$$

(1.109)

where V_0 is the voltage across the capacitor. By reference to (1.102)–(1.104) the solutions for the charge as a function of time are

I.

$$\frac{R}{2L} > \sqrt{\frac{1}{LC}} \qquad (\gamma > \omega_0 , \text{ overdamped})$$

(1.110)

$$q(t) = q_0 \left[\left(1 + \frac{\gamma}{\Omega} \right) e^{-(\gamma - \Omega)t} + \left(1 - \frac{\gamma}{\Omega} \right) e^{-(\gamma + \Omega)t} \right]$$

II.

$$\frac{R}{2L} = \sqrt{\frac{1}{LC}} \qquad (\gamma = \omega_0 , \text{ critically damped})$$

(1.111)

$$q(t) = q_0 (1 + \gamma t) e^{-\gamma t}$$

III.

$$\frac{R}{2L} < \sqrt{\frac{1}{LC}} \qquad (\gamma < \omega_0 , \text{ underdamped})$$

(1.112)

$$q(t) = (\Omega')^{-1} \omega_0 q_0 e^{-\gamma t} \cos \left(\Omega' t - \arctan \frac{\gamma}{\Omega'} \right)$$

For a circuit with a voltage source, as in Fig. 1-9, the sum of the voltage drops around the circle must equal 0. Thus the differential equation for the circuit in Fig. 1-9 is

$$L \frac{di}{dt} + Ri + \frac{q}{C} = V(t)$$

(1.113)

where $V(t)$ is the voltage of the generator. This differential equation is of the form of the equation of motion for a damped harmonic oscillator subjected to an external force, a topic which we take up in the following section.

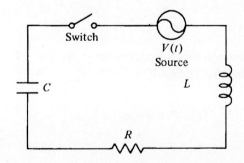

FIGURE 1-9. Series L, R, C circuit with a voltage generator $V(t)$.

1.9 Damped Oscillator With Driving Force: Resonance

Numerous physical systems can be described in terms of a damped harmonic oscillator driven by an external force that oscillates sinusoidally with time as

$$F(t) = F_0 \cos \omega t = mf \cos \omega t \tag{1.114}$$

where we have introduced $f = F_0/m$ for later convenience. The equation of motion in (1.87) gets modified to

$$\ddot{x} + 2\gamma \dot{x} + {\omega_0}^2 x = f \cos \omega t \tag{1.115}$$

A *particular solution* to this inhomogeneous linear differential equation is most readily obtained by using complex numbers and solving for a related equation with a complex driving force. For this purpose we introduce

$$\begin{aligned} z &= x + iy \\ e^{i\omega t} &= \cos \omega t + i \sin \omega t \end{aligned} \tag{1.116}$$

and observe that the real part of

$$\ddot{z} + 2\gamma \dot{z} + {\omega_0}^2 z = f e^{i\omega t} \tag{1.117}$$

is identical with (1.115). This latter form is more convenient to solve. Once we find the solution for z, the physical displacement x is obtained from $x = \mathcal{R}e\, z$. Note that if the left-hand side of (1.117) were not linear in z this method would not work. Since the first and second derivatives of $e^{i\omega t}$ are $i\omega e^{i\omega t}$ and $-\omega^2 e^{i\omega t}$, there is a solution with the time dependence $e^{i\omega t}$. Thus, as a possible solution to (1.117), we try

$$z = \frac{f}{R} e^{i\omega t} \tag{1.118}$$

where $1/R$ is a time-independent response factor. The differential equation is satisfied by this choice if

$$\left[(i\omega)^2 + 2\gamma(i\omega) + \omega_0^2 \right] \frac{f}{R} = f \tag{1.119}$$

or

$$R = \omega_0^2 - \omega^2 + 2i\gamma\omega \tag{1.120}$$

The complex factor R can be written in polar form

$$R = re^{i\theta} \tag{1.121}$$

where

$$r^2 = |R|^2 = \left(\omega_0^2 - \omega^2\right)^2 + 4\gamma^2\omega^2 \tag{1.122}$$

and

$$\tan\theta = \left(\frac{\mathcal{I}mR}{\mathcal{R}eR}\right) = \frac{2\gamma\omega}{\omega_0^2 - \omega^2} \tag{1.123}$$

The angle θ lies between 0 and π. Using (1.116), (1.118), and (1.121), we arrive at the desired solution to (1.115)

$$x = \mathcal{R}ez = \mathcal{R}e\left(\frac{f}{r}e^{i(\omega t - \theta)}\right) \tag{1.124}$$

or

$$x(t) = \frac{f}{r}\cos(\omega t - \theta) \tag{1.125}$$

The response $x(t)$ to the force $mf\cos\omega t$ is thus proportional to $1/r$. The response oscillates with a phase $(\omega t - \theta)$ that lags the oscillations of the force by a phase angle θ.

Both r and θ depend on the relative size of the driving frequency ω and natural frequency ω_0. For small damping $\gamma \ll \omega_0$ and values of ω near to ω_0, we can make the following approximations in (1.121) and (1.122):

$$r^2 = (\omega_0 - \omega)^2(\omega_0 + \omega)^2 + 4\gamma^2\omega^2 \approx 4\omega_0^2\left[(\omega_0 - \omega)^2 + \gamma^2\right] \tag{1.126}$$

$$\tan\theta = \frac{2\gamma\omega}{(\omega_0 - \omega)(\omega_0 + \omega)} \approx \frac{\gamma}{\omega_0 - \omega} \tag{1.127}$$

From these approximate expressions, we see that r^2 has a minimum when the driving force is at the natural frequency of the oscillator, $\omega = \omega_0$. The large response $x(t)$ produced by a driving frequency in the vicinity $\omega = \omega_0$ is called a *resonance*. The magnitude r_m of r at the resonance frequency $\omega = \omega_0$ is governed by the size of the frictional coefficient γ.

$$r_m = 2\omega_0\gamma \tag{1.128}$$

The width of the resonance is defined as the difference of the two values of ω at which r^2 is twice its minimum value. From (1.126) and (1.128) these values are $\omega = \omega_0 \pm \gamma$ and thus the width is 2γ. The resonance becomes narrower and the maximum displacement x larger as friction is made smaller. Plots of $[r(\omega_0)/r(\omega)]^2$ and $\theta(\omega)$ are shown in Figs. 1-10 and 1-11. The phase lag θ is 90° at resonance. At small frequencies ω, the phase lag tends to zero, and far above resonance it approaches 180°, as is evident from (1.123) or (1.120). Resonance phenomena analogous to that discussed here play an extremely important role in all branches of physics and engineering.

The solution we have been discussing is known as a *particular solution* since there are no integration constants. This particular solution is often called the *steady state* solution. It is not the most general solution since it does not match a general initial state of the oscillator. The general solution to the forced-oscillator differential equation is obtained by adding to the particular solution in (1.125) the general solution of the homogeneous equation (*i.e.*, the oscillation equation with no driving force). The result for the underdamped case is

$$x(t) = ae^{-\gamma t} \cos (\Omega' t + \alpha)$$
$$+ \frac{f}{\left[\left(\omega_0^2 - \omega^2 \right)^2 + 4\gamma^2 \omega^2 \right]^{\frac{1}{2}}} \cos \left(\omega t - \arctan \frac{2\gamma\omega}{\omega_0^2 - \omega^2} \right) \quad (1.129)$$

The sum satisfies (1.117) and contains two arbitrary constants, a and α. The initial conditions determine these constants. The term with the decaying exponential is called a transient—it vanishes at large times. The force-dependent term describes the steady-state oscillatory motion of the harmonic system.

Any periodic force can be Fourier-analyzed into an infinite series of $\cos(n\omega t + \phi_n)$ terms

$$F(t) = m \sum_n f_n \cos(n\omega t + \phi_n) \quad (1.130)$$

where F_n and ϕ_n are constants and the period is $2\pi/\omega$. The solution (1.129) for a driving force $F_0 \cos \omega t$ can be used for a force $F_n \cos(n\omega t + \phi_n)$. Then the solution for a superposition of driving frequencies in (1.130) can be obtained as a summation over solutions with driving frequencies $n\omega$.

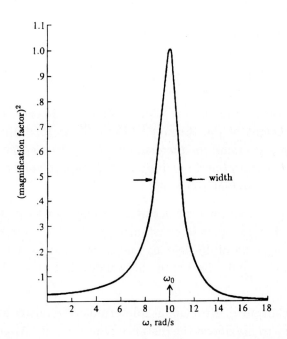

FIGURE 1-10. Square of the magnification factor, $\left[r(\omega = \omega_0)/r(\omega)\right]^2$, as a function of driving frequency ω for forced oscillator of natural frequency $\omega_0 = 10$ and damping constant $\gamma = 1$.

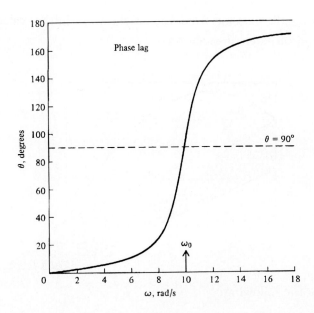

FIGURE 1-11. Phase lag θ as a function of driving frequency ω for the forced oscillator.

PROBLEMS

1.2 Interactions

1-1. An athlete can throw a javelin 60 m from a standing position. If he can run 100 m at constant velocity in 10 s, how far could he hope to throw the javelin while running? Neglect air resistance and the height of the thrower in the interest of simplicity. (*Hint: derive an expression for the distance R in terms of the initial angle θ to the horizontal and maximize R.*) Compare your answer with a world-class throw of 105 m for the javelin.

1-2. A world class shotputter can put a 7.26 kg shot a distance of 22 m. Assume that the shot is constantly accelerated over a distance of 2 m at an angle of 45 degrees and is released at a height of 2 m above the ground. Estimate the weight that this athlete can lift with one hand.

1-3. For the shotput of Problem 1-2 determine the initial angle θ of the trajectory to maximize the distance R of the put. Approximate the value of v_0 by that obtained in Problem 1-2. A photographic study found that expert athletes have learned by trial and error to release the shotput at this optimum angle.

1-4. A projectile is shot from the origin with initial velocity v_0 and inclination angle θ as shown.

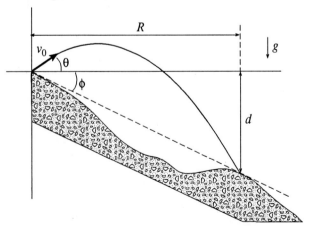

Show the following:

a) The range R (maximum horizontal distance), v_0, θ and the drop d

are related by

$$R \sin 2\theta + d(1 + \cos 2\theta) = R^2/R_0$$

where

$$R_0 \equiv v_0^2/g$$

b) The condition for maximum range R_m is

$$\tan 2\theta_m = R_m/d$$

[Note that if $d = 0$ then $\theta_m = 45°$.]

c) If the land falls off with a constant slope angle ϕ (*i.e.*, $d = R_m \tan \phi$) then the maximum range angle θ_m and ϕ are related by

$$2\theta_m + \phi = 90°$$

[Note that if $\phi = 0$ then $\theta_m = 45°$.]

d) The maximum range is given by

$$R_0^2 + 2dR_0 = R_m^2$$

[Note that if $d = 0$ then $R_m = R_0$.]

e) The optimum angle, maximum range, slope to impact angle ϕ, and the elevation drop d satisfy the triangle relation

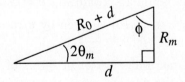

1-5. A perfectly flexible cable has length l. Initially, the cable is at rest, with a length x_0 of it hanging vertically over the edge of a table. Neglecting friction, compute the length hanging over the edge after a time t. Assume that the sections of the cable remain straight during the motion.

1-6. A particle of mass m, initially at rest, moves on a horizontal line subject to a force $F(t) = ae^{-bt}$. Find its position and velocity as a function of time.

1.4 Drag Force

1-7. A boat is slowed by a drag force $F(v)$. Its velocity decreases according to the formula

$$v = c^2(t - t_1)^2$$

where c is a constant and t_1 is the time at which it stops. Find the force $F(v)$ as a function of v.

1-8. A mass m sliding horizontally is subject to a viscous drag force. For an initial velocity v_0 (at $x = t = 0$) and a retarding force $F = -b\dot{x}$ find the velocity as a function of distance, $v(x)$, and show that the mass moves a finite distance before coming to rest. For the same initial conditions and a retarding force $F = -c\dot{x}^2$ find $v(x)$ and $x(t)$, and show that the mass never comes to rest.

1-9. Integrate the equation of motion in (1.32) to directly find the velocity as a function of distance fallen for a sky diver in free fall. At what free-fall distance does the velocity reach two-thirds of the terminal velocity? Assume that $v_t = 54\,\text{m/s}$.

1-10. A diver of mass m begins a descent from a 10 meter diving board with zero initial velocity.

a) Calculate the velocity v_0 on impact with the water and the approximate elapsed time from dive until impact.

b) Set up the equation of motion for vertical descent of the diver through the water, assuming that the buoyancy force balances the gravity force underwater and the drag force is given by (1.16). Solve for the velocity v as a function of the depth x under water and impose the boundary condition $v = v_0$ at $x = 0$.

c) If the constant c in (1.16) is given by $c/m = 0.4\ (\text{meter})^{-1}$, estimate the depth at which $v = \frac{1}{10} v_0$.

d) Solve for the time under water in terms of the depth. How long does it take for the diver to reach the bottom of a 5 m deep pool?

1-11. A ball of mass m is thrown vertically upward with initial velocity v_i. If the air resistance is proportional to v^2 and the terminal velocity is v_t, show that the ball returns to its initial position with velocity v_f given by

$$\frac{1}{v_f^2} = \frac{1}{v_i^2} + \frac{1}{v_t^2}$$

1-12. A bicyclist is able to pedal at a maximum speed v_0 on the level with no wind. If there is a wind force w parallel to the biker's path the biker will slow down or speed up. The air resistance is proportional to the square of the air speed. The biker's power output is equal to the applied force times the ground velocity. Assume the power output is constant and that there are no other power losses. Find an equation that relates v to v_0 and w. Compute the velocity numerically for $v_0 = 15$ m/s in the cases of a head wind $w = 5$ m/s and a tail wind $w = 5$ m/s.

1-13. A drag racer experiences a retarding force due to wind resistance that is proportional to the square of the racer's velocity. Assuming that the racer is designed for optimum acceleration, set up the equation of motion and derive a relation between v and t. Also derive a relation between v and x. Eliminate the coefficient of friction and solve the resulting equation numerically for the terminal velocity that can reproduce the 1988 world record of $v = 129.1$ m/s, $t = 4.99$ s for $x = 0.4$ km. Then determine the coefficient of friction.

1.5 Spring Force

1-14. A massless spring of rest length ℓ and spring constant k has a mass m attached to one end. The system is set on a table with the mass vertically above the spring as shown.

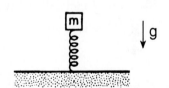

a) What is the new equilibrium height of the mass above the table?

b) The spring is compressed a distance c below the new equilibrium point and released. Find the motion of the mass assuming the free end of the spring remains in contact with the table.

c) Find the critical compression for which the spring will break contact with the table.

1-15. An archer using the equipment described in § 1.5 aims horizontally at a target 50 m away.

a) How far below the aiming point will the arrow strike? (Neglect air resistance.)

b) At what angle should the arrow be released so as to hit the target?

c) What would be the maximum possible flight distance on level ground? (Neglect air resistance and the height of the archer.)

d) Suppose that the arrow is released at a height of 1.6 m above the ground (typical shoulder-height of a person) at the angle found in part b) above. Calculate the horizontal distance at which the arrow would hit the ground.

1.7 Simple Harmonic Oscillator

1-16. Solve the damped unforced oscillator by the following method. Define a new variable y by

$$x = e^{\beta t} y$$

Substitute into the equation of motion (1.87) to find the equation satisfied by $y(t)$. Choose β such that the coefficient of $\dot{y}$ vanishes and solve in the underdamped, critical, and overdamped cases.

1-17. Show that the underdamped oscillator solution (1.104) can be expressed as $x(t) = x_0 e^{-\gamma t} \left[\cos \Omega' t + \left(\frac{v_0 + \gamma x_0}{x_0 \Omega'} \right) \sin \Omega' t \right]$ and demonstrate by direct calculation that $x(0) = x_0$ and $\dot{x}(0) = v_0$.

1.8 Forced Oscillator With Damping

1-18. Show by direct substitution that $x = r^{-1} f \cos(\omega t - \theta)$ satisfies the forced damped oscillator equation of motion

$$\ddot{x} + 2\gamma \dot{x} + \omega_0^2 x = f \cos \omega t$$

and that r and θ are the same as in (1.122) and (1.123).

1-19. An electric motor of mass 100 kg is supported by vertical springs which compress by 1 mm when the motor is installed. If the motor's armature is not properly balanced, for what revolutions/minute would a resonance occur?

1-20. Find the initial conditions such that an underdamped harmonic oscillator will immediately begin steady-state motion under the time-dependent force $F = mf \cos \omega t$.

1-21. Find the steady-state solution for a damped harmonic oscillator driven by the force

$$F(t) = mf \sin \omega t$$

1-22. An AM radio station transmits a signal consisting of a carrier wave at frequency $\nu_c = 10^6$ Hz whose amplitude is modulated at frequency $\nu_m = 10^4$ Hz, as illustrated in the accompanying figure.

This means that many oscillations of the carrier occur while the modulation only changes slightly. A rudimentary radio receiver circuit is shown schematically in the accompanying figure. The incident radio waves induce an oscillating voltage $V_0 \cos \omega_c t$ in the antenna, where $\omega_c = 2\pi\nu_c$ and $V_0 \simeq 1\,\text{mV}$.

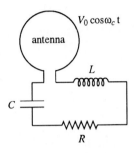
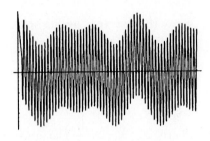

a) Given the capacitance $C = 300$ pico-farads and resistance $R = 5$ ohms find the proper inductance L to create a resonance with the incident wave. *Hint: at first assume the resistance has little effect on the resonant frequency and then verify that this is a good approximation.*

b) Compute the damping constant and verify that the transients die out much faster than the modulation varies. This insures that the receiver will faithfully amplify the incident signal.

c) Compute the maximum voltage across the capacitor in terms of the above value of V_0. This is the voltage amplification of the circuit.

d) An adjacent station in carrier frequency is 20 kHz higher. If our receiver is tuned to the original 10^6 Hz how much will the adjacent stations's carrier be amplified?

1-23. A critically damped oscillator with $\omega_0 = 1$ rad/sec is acted upon by a driving force F_{driver}

a) Find a particular solution for $F_{\text{driver}} = mfe^t$.

b) Find a particular solution for $F_{\text{driver}} = mfe^{-t}$. *Hint: Try $x = At^n e^{-t}$ for $n = 0, 1, 2$.*

c) Using the preceding results obtain the general solution for $F_{\text{driver}} = mf \cosh t$ with initial conditions $x(0) = \dot{x}(0) = 0$.

1-24. An undamped harmonic oscillator with natural frequency ω_0 is subjected to a driving force $F(t) = ae^{-bt}$. The oscillator starts from rest at the origin $(x = 0)$ at time $t = 0$. Find the solution of the equation of motion which satisfies the specified initial condition.

1-25. Find the average power dissipated per driving period by the frictional force of a sinusoidally driven harmonic oscillator in steady state. (Recall that power = force $\times$ velocity.) Show that maximum dissipation occurs at $\omega = \omega_0$ and evaluate this maximum.

1-26. A sawtooth wave (see accompanying figure) can be decomposed into an infinite sum of cosines

$$f(t) = \sum_{n=0}^{\infty} \frac{1}{(2n + 1)^2} \cos(\omega_n t),$$

where $\omega_n = (2n + 1)\omega$ and ω is the "angular frequency" of the sawtooth. Find the steady-state motion of an oscillator driven by this force per unit mass

$$\ddot{x} + 2\gamma\dot{x} + \omega_0^2 x = f(t).$$

Hint: find the solution for a given ω_n and use the principle of superposition.

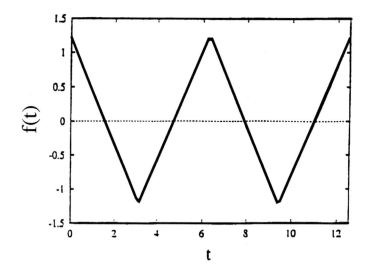

Chapter 2

ENERGY CONSERVATION

There are three important conservation laws of mechanics: energy, momentum, and angular momentum. The three laws can be derived from Newtonian theory. However, their range of validity is much broader, extending even to the domain of relativistic elementary particles, although slightly changed in form. In their ramifications in all branches of science, these conservation laws have exceptionally far-reaching consequences. In this chapter we discuss energy conservation and then in later chapters we take up in turn momentum and angular momentum conservation.

2.1 Potential Energy

To derive the energy-conservation law in the case of one-dimensional motion, we start with the second law of motion for a body of mass m

$$\frac{d}{dt}\left(mv\right) = F(x, v, t) \tag{2.1}$$

and multiply by v. Since $v\, dv/dt = \frac{1}{2} d(v^2)/dt$ we obtain the equation

$$\frac{d}{dt}\left(\tfrac{1}{2}mv^2\right) = F(x, v, t)v \tag{2.2}$$

Substituting $v = dx/dt$ on the right-hand side and integrating with respect to t gives

$$\tfrac{1}{2}mv^2(t_2) - \tfrac{1}{2}mv^2(t_1) = \int_{t_1}^{t_2} F\big(x(t), v(t), t\big)\frac{dx}{dt}dt = \int_{x_1}^{x_2} F\big(x, v(x), t(x)\big)dx \tag{2.3}$$

The left-hand side is the difference at two times of the familiar expression for the kinetic energy

$$K = \tfrac{1}{2}mv^2 \tag{2.4}$$

Equation (2.3) is the *Work-Energy theorem*

$$K_2 - K_1 = \Delta K = \text{Work} = \int_{x_1}^{x_2} F\big(x, v(x), t(x)\big)\, dx \tag{2.5}$$

This theorem states that the work done by the force acting on m equals the change in kinetic energy.

Some forces depend only on position, and then the integrand of (2.5) is a function only of x and the integral does not depend on the particular motion $x(t)$. In such cases it is valuable to define the potential energy

$$V(x) \equiv - \int_{x_s}^{x} F(x')dx' \tag{2.6}$$

where x_s is an arbitrary but fixed reference point. The right-hand side of (2.5) can be expressed in terms of V as

$$\int_{x_1}^{x_2} F(x)dx = \int_{x_1}^{x_s} F(x)dx + \int_{x_s}^{x_2} F(x)dx = V(x_1) - V(x_2) \tag{2.7}$$

for any x_s, even one which is outside the range x_1 to x_2. Using this in (2.5) yields

$$K_2 + V(x_2) = K_1 + V(x_1) \equiv E \tag{2.8}$$

The quantity E is known as the total energy of the body. Since this E is independent of coordinate, the energy is constant in time; *i.e.*, the energy is conserved. If F has an explicit dependence on either v or t, there is no conserved energy of the form (2.8). This does not mean that the energy of the universe is not conserved, but only that energy is transferred between the mechanical form (2.8) and other forms such as thermal energy (microscopic motion of molecules).

The expression for the energy in (2.8) can be simply written as

$$E = K + V(x) = \tfrac{1}{2}mv^2(x) + V(x) \tag{2.9}$$

for any value of the coordinate x. The term *potential energy* means that V is a form of energy which potentially may appear as kinetic energy.

By differentiating (2.6), we can solve for the force in terms of the potential energy:

$$F(x) = -\frac{dV(x)}{dx} \tag{2.10}$$

A force that is derivable in this way from a potential energy is called a *conservative force*. In one-dimensional motion, any force which is a function only of position is a conservative force. The effect of changing from a reference point x_s to a new reference point x'_s is just to change $V(x)$ in (2.6) by a constant, independent of x. The force is unaffected by a change in reference point since it is calculated from the derivative of the potential energy. Because the motion of the particle is determined by the force, all measurable quantities are independent of x_s; hence x_s can be chosen arbitrarily.

Since the energy is a constant of the motion for a conservative force, we can use (2.9) to determine v as a function of x, given E. If the energy is known for a conservative system, it can be used as one of the initial conditions. The conservative nature of spring and gravitational forces allows the use of energy conservation. The potential energy of the spring force is

$$V(x) = -\int_0^x (-kx')dx' = \tfrac{1}{2}kx^2 \tag{2.11}$$

where we have chosen $x_s = 0$. Equation (1.47) of the archery example is a special case of (2.11) with this potential energy for $x < 0$. The total energy of an oscillator can be calculated from (2.11) using the solutions for x and v from (1.64) and (1.66). We find

$$E = \tfrac{1}{2}mv^2 + \tfrac{1}{2}kx^2 = m\big[-\omega a \, \sin(\omega t + \alpha)\big]^2 + \tfrac{1}{2}k\big[a \, \cos(\omega t + \alpha)\big]^2 \tag{2.12}$$

which simplifies to

$$E = \tfrac{1}{2}ka^2 \tag{2.13}$$

where we have used $\omega^2 = k/m$. The energy is proportional to the square of the maximum displacement a, which is called the amplitude. At the turning points of the motion, $x = \pm a$, the energy is entirely potential energy. At $x = 0$, the kinetic energy is greatest; see Fig. 2-1.

2.2 Gravitational Escape

The gravitational potential energy due to the earth's attraction on a mass m at a distance $x \geq R_E$ from the center of the earth is

$$V(x) = -\int_\infty^x \left(-\frac{GmM_E}{x'^2}\right) dx' = -\frac{GmM_E}{x} \tag{2.14}$$

We have chosen x_s so that the potential energy vanishes at infinite distance. We may express the gravitational constant G in terms of the

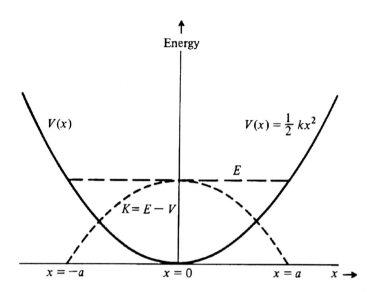

FIGURE 2-1. Potential and kinetic energies for motion under the spring force.

gravitational acceleration on the surface of the earth using (1.7),

$$GM_E = gR_E^2 \tag{2.15}$$

to obtain

$$V(x) = -\frac{mgR_E^2}{x}, \qquad x \geq R_E \tag{2.16}$$

We can use (2.16) in (2.9) to calculate the minimum velocity needed by a rocket at the earth's surface to go to $x = \infty$, that is, to "to escape from the earth's gravitational attraction" (see Fig. 2-2). From (2.9) the velocity at some position x is

$$v(x) = \pm\sqrt{\frac{2}{m}\left(E + \frac{mgR_E^2}{x}\right)} \tag{2.17}$$

For the velocity to be always real as x increases to ∞, $E \geq 0$ is required. The minimum velocity for escape from the earth's surface is consequently obtained by putting $E = 0$, $x = R_E$ in (2.17), yielding

$$\begin{aligned}
v_{\text{esc}} &= \sqrt{2gR_E} \\
&= \sqrt{2(9.8)(6.371 \times 10^6)} \text{ m/s} \\
&= 11.2 \text{ km/s} \quad (40{,}200 \text{ km/h})
\end{aligned} \tag{2.18}$$

The escape velocity is independent of the mass of the rocket. To get to the moon, a spacecraft launched from the earth needs a velocity nearly equal to the escape velocity.

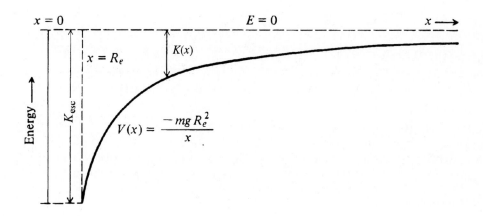

FIGURE 2-2.   Gravitational potential energy due to the earth and the minimum kinetic energy K_{esc} needed for escape from the earth's gravitational attraction.

2.3   Small Oscillations

For a general potential energy the velocity can be calculated from (2.9) to be

$$v(x) = \pm\sqrt{\frac{2}{m}[E - V(x)]} \tag{2.19}$$

This expression determines only the magnitude of the velocity. The sign depends on the previous history of the motion. Since the velocity must be real, the accessible region is

$$V(x) \leq E \tag{2.20}$$

The positions at which $V(x) = E$ are turning points, where the velocity goes through zero and changes sign, *i.e.*, the particle comes to rest and reverses its direction of motion. The qualitative nature of the motion of a particle can be described using (2.19); see Fig. 2-3.

For the potential energy sketched in Fig. 2-4, at the total energy indicated by the dashed horizontal line there are three turning points,

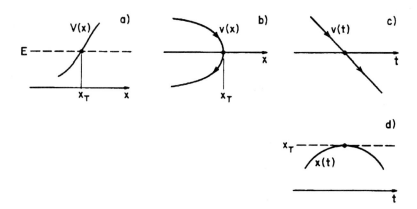

FIGURE 2-3. Behavior of a) the potential energy $V(x)$, b) the velocity $v(x)$, c) the velocity $v(t)$, and d) the position $x(t)$, near a turning point x_T of the motion. We illustrate here the case of a particle initially moving with a velocity in the positive x direction.

x_1, x_2, x_3. The regions $0 \le x < x_1$ and $x_2 < x < x_3$ are forbidden by (2.20). The motion $x(t)$ of a particle in the region $x_1 \le x < x_2$ will be oscillatory, $i.e.$, the particle will move back and forth between x_1 and x_2. The sign of the velocity in this region changes at the turning points as in (2.19). Finally, a particle approaching x_3 from infinity will slow down, reverse its motion at $x = x_3$, and go back out toward infinity.

The motion of a particle in the potential valley, $x_1 \le x \le x_2$, is particularly simple if the maximum displacements from the minimum potential energy at $x = x_e$ are small. In such a case, we can approximate the potential by a few terms of a series expansion about $x = x_e$:

$$V(x) = V(x_e) + (x - x_e)\left[\frac{dV(x)}{dx}\right]_{x=x_e} + \tfrac{1}{2}(x - x_e)^2\left[\frac{d^2V(x)}{dx^2}\right]_{x=x_e} + \cdots$$

$$\text{(2.21)}$$

The derivative dV/dx vanishes at a minimum. Since the second derivative of $V(x)$ is positive at a minimum of $V(x)$, a particle at $x = x_e$ is in stable equilibrium, so for small displacements the potential energy can be approximated by

$$V(x) \cong V(x_e) + \tfrac{1}{2}k(x - x_e)^2 \qquad\qquad (2.22)$$

where

$$k \equiv \left[\frac{d^2V(x)}{dx^2}\right]_{x=x_e} \ge 0 \qquad\qquad (2.23)$$

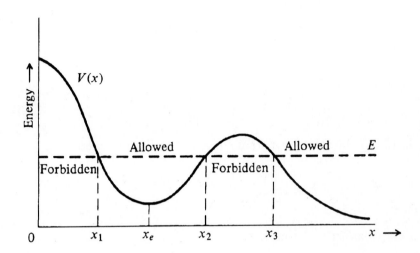

FIGURE 2-4. Allowed and forbidden regions for motion of a particle with energy E for a potential energy $V(x)$.

The constant term $V(x_e)$ can be dropped since it has no consequences for the physical motion. If we make a change of variable to the displacement from equilibrium, we see that the potential energy in (2.22) is that of a simple harmonic oscillator, (2.11), with x replaced by $x - x_e$. Small oscillations in any system can be approximately treated in terms of simple harmonic motion. The expansion about a potential-energy minimum as in (2.22) also provides justification for Hooke's law on the springlike elastic deformation in solids.

In the case discussed above, the effective spring constant k was positive and x_e was a *stable equilibrium point*. If instead, $V(x_e)$ were a local maximum, $F(x_e)$ would still vanish since $\left.\frac{dV}{dx}\right|_{x_e} = 0$, but then $\left.\frac{d^2V}{dx^2}\right|_{x_e}$ would be negative. Since the effective spring constant k would be negative under a small displacement from x_e, the force would be directed away from x_e. In this instance x_e is called an *unstable equilibrium point*.

As an illustration, we find an approximate solution for the motion of a particle of mass m in the potential energy

$$V(x) = \frac{-g^2}{x} + \frac{h^2}{x^2} \tag{2.24}$$

At the equilibrium position,

$$\left[\frac{dV(x)}{dx}\right]_{x=x_e} = \frac{g^2}{x_e{}^2} - \frac{2h^2}{x_e{}^3} = 0 \tag{2.25}$$

which gives

$$x_e = \frac{2h^2}{g^2} \tag{2.26}$$

From (2.23) the spring constant for small oscillations about x_e is

$$k = \left[\frac{d^2V(x)}{dx^2}\right]_{x=x_e} = \frac{-2g^2}{x_e{}^3} + \frac{6h^2}{x_e{}^4} \tag{2.27}$$

or upon substitution from (2.26),

$$k = \frac{g^8}{8h^6} \tag{2.28}$$

which is positive so that x_e is a point of stable equilibrium. The approximate solution to the equation of motion from (2.28), (1.62) and (1.64) is

$$x(t) - \frac{2h^2}{g^2} = a\cos\left[\left(\frac{g^4}{h^3\sqrt{8m}}\right)t - \alpha\right] \tag{2.29}$$

where a and α are arbitrary constants to be determined from the initial conditions.

2.4 Three-Dimensional Motion: Vector Notation

In three dimensions the position of a particle of mass m can be specified by its cartesian coordinates (x, y, z). Newton's second law can then be stated as the three equations

$$\begin{aligned} m\ddot{x} &= F_x \\ m\ddot{y} &= F_y \\ m\ddot{z} &= F_z \end{aligned} \tag{2.30}$$

where (F_x, F_y, F_z) are called the x, y, z components of the force of the particle. If one chooses to use a different cartesian coordinate system, which is translated and rotated with respect to the original system, the equations of motion must have the same form. In the new coordinate system the equations of motion are

$$m\ddot{x}' = F_{x'}$$
$$m\ddot{y}' = F_{y'} \tag{2.31}$$
$$m\ddot{z}' = F_{z'}$$

where (x', y', z') are the coordinates of the particle in the new coordinate frame. Each of (2.31) is a linear combination of (2.30). As an example, suppose the new coordinate system has the same origin as the original system but is rotated by an angle ϕ around the z axis, as illustrated in Fig. 2-5. The coordinates of the particle in the two frames are related by

$$x' = x \cos \phi + y \sin \phi$$
$$y' = -x \sin \phi + y \cos \phi \tag{2.32}$$
$$z' = z$$

By time-differentiating, we see that analogous relations hold for velocities and accelerations; *e.g.*,

$$\ddot{x}' = \ddot{x} \cos \phi + \ddot{y} \sin \phi$$
$$\ddot{y}' = -\ddot{x} \sin \phi + \ddot{y} \cos \phi \tag{2.33}$$
$$\ddot{z}' = \ddot{z}$$

Substituting (2.30) into (2.33), we obtain

$$m\ddot{x}' = F_x \cos \phi + F_y \sin \phi$$
$$m\ddot{y}' = -F_x \sin \phi + F_y \cos \phi \tag{2.34}$$
$$m\ddot{z}' = F_z$$

When we identify

$$F_{x'} = F_x \cos \phi + F_y \sin \phi$$
$$F_{y'} = -F_x \sin \phi + F_y \cos \phi \tag{2.35}$$
$$F_{z'} = F_z$$

the set of all three new equations is equivalent to the old set; the new equations are just linear combinations of the old equations. For instance, $m\ddot{x}' = F_{x'}$ is just $\cos \phi$ times the equation $m\ddot{x} = F_x$ plus $\sin \phi$ times the equation $m\ddot{y} = F_y$. Notice that $F_{x'}, F_{y'}, F_{z'}$ are related to F_x, F_y, F_z in the same way as x', y', z' are related to x, y, z [(2.32)].

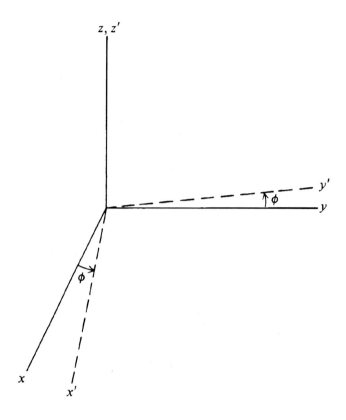

FIGURE 2-5. Two coordinate systems related by a rotation by an angle ϕ about the z axis.

To symbolize the above state of affairs and at the same time realize a great simplification in notation, we introduce the concept of a vector. A *vector* is a set of three quantities (in a three-dimensional coordinate system) whose components in differently oriented (*i.e.*, rotated) coordinate systems are related in the same way as the set of coordinates (x, y, z). Symbolically, we denote a vector with components (a_x, a_y, a_z) by **a**. Examples of vectors which we have already encountered are the position vector $\mathbf{r} \equiv (x, y, z)$, the velocity vector $\mathbf{v} = \dot{\mathbf{r}} = (\dot{x}, \dot{y}, \dot{z})$, the acceleration vector $\ddot{\mathbf{r}} = (\ddot{x}, \ddot{y}, \ddot{z})$, and the force vector $\mathbf{F} \equiv (F_x, F_y, F_z)$. The basic idea of a vector is that it is a quantity with components that change in a specific way when the coordinate system is changed [*e.g.*, (2.32)].

In vector notation Newton's second law can be written as a single equation

$$m\ddot{\mathbf{r}} = \mathbf{F} \tag{2.36}$$

This is shorthand for the set of (2.30) or (2.31). An advantage of vector notation is that no specific reference frame is necessary in this statement of the laws of motion. Vector notation is also often useful in manipulation and solving the equations of motion.

The distance of a point (x, y, z) from the origin of the coordinate system is $\sqrt{x^2 + y^2 + z^2}$. This distance is independent of the rotational orientation of the coordinate system,

$$\sqrt{x'^2 + y'^2 + z'^2} = \sqrt{x^2 + y^2 + z^2} \tag{2.37}$$

as can easily be checked for the transformation in (2.32). The above quantity is called the length, or magnitude, of $\mathbf{r}$ and is denoted by

$$|\mathbf{r}| \equiv r \equiv \sqrt{x^2 + y^2 + z^2} \tag{2.38}$$

For a general vector $\mathbf{a} = (a_x, a_y, a_z)$, the length, or magnitude, is similarly defined as

$$|\mathbf{a}| \equiv a \equiv \sqrt{a_x{}^2 + a_y{}^2 + a_z{}^2} \tag{2.39}$$

Since by the definition of a vector given above the components of $\mathbf{a}$ transform under rotations of the coordinate system in the same way as the components of $\mathbf{r}$, the length of $\mathbf{a}$ is independent of the orientation of the coordinate frame. A quantity, such as $|\mathbf{a}|$, that is independent of frame orientation is called a *scalar*, to distinguish it from a quantity such as F_x, which is the component of a vector, and therefore is different in different cartesian coordinate systems [see (2.35)].

If vectors are multiplied by scalars and added together by the rule

$$\alpha\mathbf{a} + \beta\mathbf{b} = (\alpha a_x + \beta b_x, \; \alpha a_y + \beta b_y, \; \alpha a_z + \beta b_z) \tag{2.40}$$

the resulting quantity is again a vector because its components transform under coordinate-system rotations according to the definition of a vector. Since any linear combination of vectors is a vector, many new vectors can be generated from the position vector $\mathbf{r}$. For instance, the relative coordinate of two particles $\mathbf{r}_2 - \mathbf{r}_1 = (x_2 - x_1,\ y_2 - y_1,\ z_2 - z_1)$ is a vector, as is the change of coordinate of a particle between two times:

$$\Delta\mathbf{r} = \mathbf{r}(t + \Delta t) - \mathbf{r}(t) = \left[x(t + \Delta t) - x(t),\ y(t + \Delta t) - y(t),\ z(t + \Delta t) - z(t) \right] \tag{2.41}$$

It follows that the velocity,

$$\mathbf{v} = \dot{\mathbf{r}} = \lim_{\Delta t \to 0} \frac{\Delta\mathbf{r}}{\Delta t} \tag{2.42}$$

and the acceleration,

$$\mathbf{a} = \ddot{\mathbf{r}} = \lim_{\Delta t \to 0} \frac{\mathbf{v}(t + \Delta t) - \mathbf{v}(t)}{\Delta t} \tag{2.43}$$

are vectors. We note that all vectors which are constructed from the difference of two position vectors (such as $\mathbf{r}_2 - \mathbf{r}_1$, $\Delta\mathbf{r}$, $\dot{\mathbf{r}}$, $\ddot{\mathbf{r}}$) are unchanged by a shift in origin of the coordinate frame. Under a change in origin, all position vectors $\mathbf{r}$ are replaced by $\mathbf{r}' = \mathbf{r} + \mathbf{s}$, where $\mathbf{s}$ is a constant vector. It follows that the vectors formed from differences of two position vectors are independent of $\mathbf{s}$. Since the acceleration is unchanged by a shift in origin, the force vector must also share this property in order for (2.36) to hold in translated frames. The position vector $\mathbf{r}$ is the only vector which depends upon the origin of the coordinate system, and therefore is sometimes said not to be a true vector.

The geometrical representation of a vector as a directed line segment, or "arrow," is a powerful intuitive tool. We represent the position vector $\mathbf{r} = (x,\ y,\ z)$ by an arrow drawn from the origin to the point $(x,\ y,\ z)$, as illustrated in Fig. 2-6. The length of $\mathbf{r}$ is then just the length of the arrow. The components of $\mathbf{r}$ are the coordinates of the orthogonal projections of the arrow's point onto the coordinate axes. We can also represent an arbitrary vector $\mathbf{a}$ by an arrow, since under rotations of the coordinate frame the components of $\mathbf{a}$ transform the same way as the components of $\mathbf{r}$. The length of the arrow is proportional to the magnitude of the vector, and the projections of the arrow on the coordinate axes are proportional to the components of the vector, as illustrated in Fig. 2-7. The location of the arrow is arbitrary (so long as the arrow represents a "true" vector, not the position vector) and may be chosen for convenience. For instance, arrows representing the velocity, acceleration, or force on a particle may be attached to the point representing the position of the particle. The addition of vectors is represented by the head-to-tail construction illustrated in Fig. 2-8.

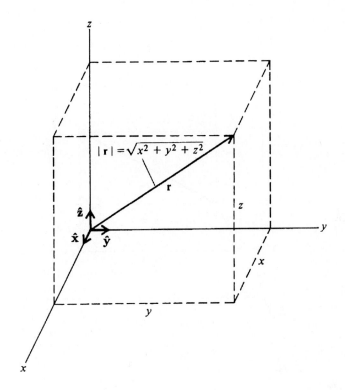

FIGURE 2-6. Position vector **r** and coordinate-system unit vectors $\hat{\mathbf{x}}$, $\hat{\mathbf{y}}$, $\hat{\mathbf{z}}$.

The *dot product* of two vectors is defined as

$$\mathbf{a} \cdot \mathbf{b} \equiv a_x b_x + a_y b_y + a_z b_z \tag{2.44}$$

The dot product is a *scalar* (*i.e.*, independent of the frame orientation), as we can readily demonstrate from the identity

$$\mathbf{a} \cdot \mathbf{b} = a_x b_x + a_y b_y + a_z b_z = \tfrac{1}{2}\Big[(a_x + b_x)^2 - a_x^2 - b_x^2 + (a_y + b_y)^2$$
$$- a_y^2 - b_y^2 + (a_z + b_z)^2 - a_z^2 - b_z^2\Big]$$
$$= \tfrac{1}{2}\left(|\mathbf{a} + \mathbf{b}|^2 - |\mathbf{a}|^2 - |\mathbf{b}|^2\right)$$

$$\tag{2.45}$$

Since the vector magnitudes $|\mathbf{a}|$, $|\mathbf{b}|$, $|\mathbf{a} + \mathbf{b}|$ are scalars, it follows that the dot product is a scalar. From the defining (2.44), we further observe that

$$\mathbf{a} \cdot \mathbf{a} = |\mathbf{a}|^2 \equiv a^2$$
$$\mathbf{a} \cdot \mathbf{b} = \mathbf{b} \cdot \mathbf{a} \tag{2.46}$$
$$(\mathbf{a} + \mathbf{b}) \cdot \mathbf{c} = \mathbf{a} \cdot \mathbf{c} + \mathbf{b} \cdot \mathbf{c}$$

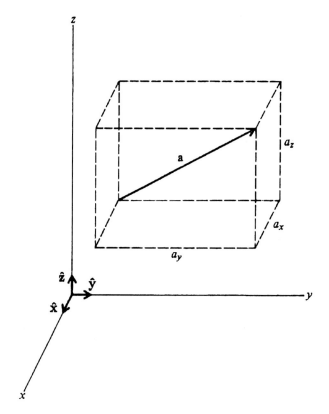

FIGURE 2-7. Arrow representation of an arbitrary vector **a**.

The magnitude of the vector $\mathbf{a} + \mathbf{b}$ is given in terms of the dot product $\mathbf{a} \cdot \mathbf{b}$ by

$$|\mathbf{a} + \mathbf{b}|^2 = (\mathbf{a} + \mathbf{b}) \cdot (\mathbf{a} + \mathbf{b}) = a^2 + b^2 + 2\mathbf{a} \cdot \mathbf{b} \qquad (2.47)$$

Furthermore, inasmuch as the vectors **a**, **b**, and $\mathbf{a} + \mathbf{b}$ form a triangle as illustrated in Fig. 2-8, the opposite side $|\mathbf{a} + \mathbf{b}|$ of the triangle is related by trigonometry to the adjacent sides $|\mathbf{a}|$ and $|\mathbf{b}|$ by

$$|\mathbf{a} + \mathbf{b}|^2 = a^2 + b^2 + 2ab \cos \theta \qquad (2.48)$$

where θ is the angle between the arrows representing **a** and **b**. Equating the above two formulas for $|\mathbf{a} + \mathbf{b}|^2$, we deduce the following result for the dot product:

$$\mathbf{a} \cdot \mathbf{b} = ab \cos \theta \qquad (2.49)$$

Thus the dot product represents the product of the length of one vector times the orthogonal projection of the other vector on it, as indicated in

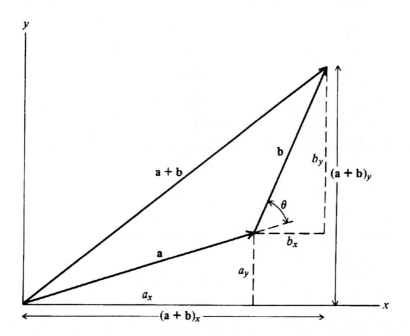

FIGURE 2-8. Head-to-tail construction of the addition of two vectors **a** and **b**. (For convenience of illustration the x, y-coordinate axes are taken to lie in the plane defined by **a** and **b**.)

Fig. 2-9. If $\mathbf{a} \cdot \mathbf{b} = 0$, even though $a \neq 0$ and $b \neq 0$, the angle between the vectors is 90° and the vectors are said to be *orthogonal*.

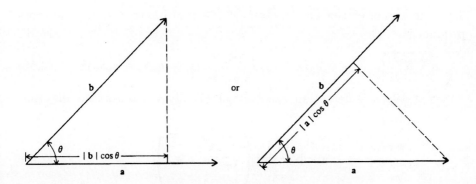

FIGURE 2-9. Geometrical illustration of the dot product.

It is useful to define a set of coordinate-axis vectors $\hat{\mathbf{x}}$, $\hat{\mathbf{y}}$, $\hat{\mathbf{z}}$ of unit length $|\hat{\mathbf{x}}| = |\hat{\mathbf{y}}| = |\hat{\mathbf{z}}| = 1$ which are directed along the x, y, z axes of the coordinate system, as in Fig. 2-6. The components of these orthogonal unit vectors are

$$\hat{\mathbf{x}} = (1, 0, 0)$$
$$\hat{\mathbf{y}} = (0, 1, 0) \qquad (2.50)$$
$$\hat{\mathbf{z}} = (0, 0, 1)$$

From this definition the dot products of unit vectors with each other are

$$\hat{\mathbf{x}} \cdot \hat{\mathbf{x}} = \hat{\mathbf{y}} \cdot \hat{\mathbf{y}} = \hat{\mathbf{z}} \cdot \hat{\mathbf{z}} = 1$$
$$\hat{\mathbf{x}} \cdot \hat{\mathbf{y}} = \hat{\mathbf{x}} \cdot \hat{\mathbf{z}} = \hat{\mathbf{y}} \cdot \hat{\mathbf{z}} = 0 \qquad (2.51)$$

In a given frame a general vector $\mathbf{a}$ can be represented in terms of the unit vectors of the frame as

$$\mathbf{a} = a_x \hat{\mathbf{x}} + a_y \hat{\mathbf{y}} + a_z \hat{\mathbf{z}} \qquad (2.52)$$

The sum of two vectors can be expressed as

$$\mathbf{a} + \mathbf{b} = (a_x + b_x)\hat{\mathbf{x}} + (a_y + b_y)\hat{\mathbf{y}} + (a_z + b_z)\hat{\mathbf{z}} \qquad (2.53)$$

Another type of product of two vectors of considerable importance is the *cross product*, written $\mathbf{a} \times \mathbf{b}$. The cross product has three components, defined by

$$(\mathbf{a} \times \mathbf{b})_x = a_y b_z - a_z b_y$$
$$(\mathbf{a} \times \mathbf{b})_y = a_z b_x - a_x b_z \qquad (2.54)$$
$$(\mathbf{a} \times \mathbf{b})_z = a_x b_y - a_y b_x$$

Thus, in terms of the unit vectors of the coordinate system, the cross product is

$$\mathbf{a} \times \mathbf{b} = (a_y b_z - a_z b_y)\hat{\mathbf{x}} + (a_z b_x - a_x b_z)\hat{\mathbf{y}} + (a_x b_y - a_y b_x)\hat{\mathbf{z}} \qquad (2.55)$$

Alternatively, the definition can be symbolically written as the determinant

$$\mathbf{a} \times \mathbf{b} = \det \begin{pmatrix} \hat{\mathbf{x}} & \hat{\mathbf{y}} & \hat{\mathbf{z}} \\ a_x & a_y & a_z \\ b_x & b_y & b_z \end{pmatrix} \qquad (2.56)$$

From the symmetry properties of the determinant or directly from (2.55), we note that

$$\mathbf{a} \times \mathbf{b} = -\mathbf{b} \times \mathbf{a} \tag{2.57}$$

The cross product of a vector with itself vanishes:

$$\mathbf{a} \times \mathbf{a} = 0 \tag{2.58}$$

The cross product transforms like an ordinary vector under rotations of the coordinate system. For instance, consider the transformation equation (2.32). The vectors $\mathbf{a}$ and $\mathbf{b}$ transform in the same way as the position vector $\mathbf{r}$; that is,

$$
\begin{aligned}
a_{x'} &= a_x \cos \phi + a_y \sin \phi & b_{x'} &= b_x \cos \phi + b_y \sin \phi \\
a_{y'} &= -a_x \sin \phi + a_y \cos \phi & b_{y'} &= -b_x \sin \phi + b_y \cos \phi \\
a_{z'} &= a_z & b_{z'} &= b_z
\end{aligned} \tag{2.59}
$$

The components of $\mathbf{a} \times \mathbf{b}$ in the rotated frame are then found to be

$$
\begin{aligned}
(\mathbf{a} \times \mathbf{b})_{x'} &= (a_{y'} b_{z'} - a_{z'} b_{y'}) \\
&= (a_y b_z - a_z b_y) \cos \phi + (a_z b_x - a_x b_z) \sin \phi \\
&= (\mathbf{a} \times \mathbf{b})_x \cos \phi + (\mathbf{a} \times \mathbf{b})_y \sin \phi \\
(\mathbf{a} \times \mathbf{b})_{y'} &= -(\mathbf{a} \times \mathbf{b})_x \sin \phi + (\mathbf{a} \times \mathbf{b})_y \cos \phi \\
(\mathbf{a} \times \mathbf{b})_{z'} &= (\mathbf{a} \times \mathbf{b})_z
\end{aligned} \tag{2.60}
$$

which corresponds to the transformation of (x, y, z) in (2.32). For this reason the cross product is sometimes called the vector product. However, the cross product behaves differently from ordinary vectors under *inversion* of the coordinate axes (that is, $x' = -x$, $y' = -y$, $z' = -z$). We have

$$\mathbf{r}' = -\mathbf{r} \qquad \mathbf{a}' = -\mathbf{a} \qquad (\mathbf{a} \times \mathbf{b})' = (\mathbf{a} \times \mathbf{b}) \tag{2.61}$$

A three-component quantity such as $(\mathbf{a} \times \mathbf{b})$, which behaves like a vector under rotation of the coordinate axes but does not change sign under inversion, is called an *axial vector*.

The dot product of the vector $\mathbf{a}$ with $\mathbf{a} \times \mathbf{b}$ is zero, as we show by use of (2.44) and (2.54),

$$\mathbf{a} \cdot (\mathbf{a} \times \mathbf{b}) = a_x(a_y b_z - a_z b_y) + a_y(a_z b_x - a_x b_z) + a_z(a_x b_y - a_y b_x)$$
$$= 0$$

$$(2.62)$$

Equivalently

$$\mathbf{b} \cdot (\mathbf{a} \times \mathbf{b}) = 0 \qquad (2.63)$$

since $\mathbf{a}$ and $\mathbf{b}$ are arbitrary vectors, and we can rename $\mathbf{a} \leftrightarrow \mathbf{b}$. Thus the cross-product vector $\mathbf{a} \times \mathbf{b}$ is orthogonal to the vectors $\mathbf{a}$ and $\mathbf{b}$. The arrow representing $\mathbf{a} \times \mathbf{b}$ must therefore be perpendicular to the plane defined by the arrows of $\mathbf{a}$ and $\mathbf{b}$. By the definition in (2.55), the direction of $\mathbf{a} \times \mathbf{b}$ is the direction in which a right-hand screw moves when it turns from $\mathbf{a}$ toward $\mathbf{b}$, as indicated in Fig. 2-10. The square of the magnitude of the cross product

$$|\mathbf{a} \times \mathbf{b}|^2 = (a_y b_z - a_z b_y)^2 + (a_z b_x - a_x b_z)^2 + (a_x b_y - a_y b_x)^2 \qquad (2.64)$$

can be rewritten

$$|\mathbf{a} \times \mathbf{b}|^2 = (a_x^2 + a_y^2 + a_z^2)(b_x^2 + b_y^2 + b_z^2) - (a_x b_x + a_y b_y + a_z b_z)^2$$
$$= a^2 b^2 - |\mathbf{a} \cdot \mathbf{b}|^2$$

$$(2.65)$$

Since a, b and $\mathbf{a} \cdot \mathbf{b}$ are scalars under rotations, the length of $\mathbf{a} \times \mathbf{b}$ is also a scalar. By substitution of (2.49), we obtain

$$|\mathbf{a} \times \mathbf{b}|^2 = a^2 b^2 (1 - \cos^2 \theta) \qquad (2.66)$$

and so

$$|\mathbf{a} \times \mathbf{b}| = ab |\sin \theta| \qquad (2.67)$$

where θ is the angle between the arrows representing $\mathbf{a}$ and $\mathbf{b}$. The length of $\mathbf{a} \times \mathbf{b}$ is just the area of the parallelogram, with the arrows $\mathbf{a}$ and $\mathbf{b}$ as sides.

The cross products of the unit vectors $\hat{\mathbf{x}}$, $\hat{\mathbf{y}}$, $\hat{\mathbf{z}}$ of (2.50) are found [from (2.55)] to be

$$
\begin{array}{lll}
\hat{\mathbf{x}} \times \hat{\mathbf{y}} = \hat{\mathbf{z}} & \hat{\mathbf{y}} \times \hat{\mathbf{x}} = -\hat{\mathbf{z}} & \hat{\mathbf{x}} \times \hat{\mathbf{x}} = 0 \\
\hat{\mathbf{y}} \times \hat{\mathbf{z}} = \hat{\mathbf{x}} & \hat{\mathbf{z}} \times \hat{\mathbf{y}} = -\hat{\mathbf{x}} & \hat{\mathbf{y}} \times \hat{\mathbf{y}} = 0 \\
\hat{\mathbf{z}} \times \hat{\mathbf{x}} = \hat{\mathbf{y}} & \hat{\mathbf{x}} \times \hat{\mathbf{z}} = -\hat{\mathbf{y}} & \hat{\mathbf{z}} \times \hat{\mathbf{z}} = 0
\end{array}
\qquad (2.68)
$$

A new kind of scalar can be formed by taking the dot product of a vector $\mathbf{a}$ with an axial vector $(\mathbf{b} \times \mathbf{c})$. This scalar, $\mathbf{a} \cdot (\mathbf{b} \times \mathbf{c})$ is called the

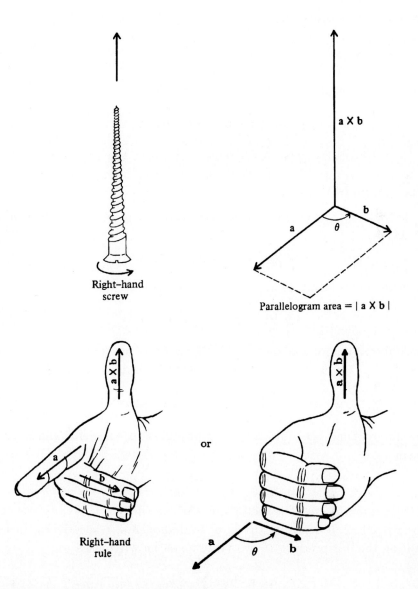

FIGURE 2-10. Geometrical illustration of the cross product.

triple product. From (2.44) and (2.56) the triple product can be written as a determinant of the vector components:

$$\mathbf{a} \cdot (\mathbf{b} \times \mathbf{c}) = \det \begin{pmatrix} a_x & a_y & a_z \\ b_x & b_y & b_z \\ c_x & c_y & c_z \end{pmatrix} \tag{2.69}$$

The symmetry properties of the determinant under interchange of rows imply that

$$\mathbf{a} \cdot (\mathbf{b} \times \mathbf{c}) = \mathbf{c} \cdot (\mathbf{a} \times \mathbf{b}) = \mathbf{b} \cdot (\mathbf{c} \times \mathbf{a}) \tag{2.70}$$

This interchangeability of the dot and cross products,

$$\mathbf{a} \cdot (\mathbf{b} \times \mathbf{c}) = (\mathbf{a} \times \mathbf{b}) \cdot \mathbf{c} \tag{2.71}$$

is a useful property of vector algebra. Although the triple product is a scalar under rotations it is called a *pseudoscalar* because it changes sign under coordinate inversion.

The repeated cross product of three vectors $\mathbf{a} \times (\mathbf{b} \times \mathbf{c})$ can be worked out to

$$\mathbf{a} \times (\mathbf{b} \times \mathbf{c}) = \mathbf{b}(\mathbf{a} \cdot \mathbf{c}) - \mathbf{c}(\mathbf{a} \cdot \mathbf{b}) \tag{2.72}$$

When the cross products are carried out in a different order, the result is

$$(\mathbf{a} \times \mathbf{b}) \times \mathbf{c} = \mathbf{b}(\mathbf{a} \cdot \mathbf{c}) - \mathbf{a}(\mathbf{b} \cdot \mathbf{c}) \tag{2.73}$$

A useful formula for the dot product of two cross products can be derived from (2.73). We take the dot product of (2.73) with a vector $\mathbf{d}$,

$$(\mathbf{a} \times \mathbf{b}) \times \mathbf{c} \cdot \mathbf{d} = (\mathbf{a} \cdot \mathbf{c})(\mathbf{b} \cdot \mathbf{d}) - (\mathbf{a} \cdot \mathbf{d})(\mathbf{b} \cdot \mathbf{c}) \tag{2.74}$$

then interchange the dot and cross products on the left-hand side to obtain

$$(\mathbf{a} \times \mathbf{b}) \cdot (\mathbf{c} \times \mathbf{d}) = (\mathbf{a} \cdot \mathbf{c})(\mathbf{b} \cdot \mathbf{d}) - (\mathbf{a} \cdot \mathbf{d})(\mathbf{b} \cdot \mathbf{c}) \tag{2.75}$$

The components of a vector are often labeled $\mathbf{a} = (a_1, a_2, a_3)$, the subscripts 1, 2, 3 denoting the x, y, z components, respectively. In this notation the dot product of two vectors can be written as

$$\mathbf{a} \cdot \mathbf{b} = \sum_i a_i b_i \tag{2.76}$$

where the summation is over $i = 1, 2, 3$. As a convenient shorthand notation, we shall often omit the $\sum_i$ symbol and simply write

$$\mathbf{a} \cdot \mathbf{b} = a_i b_i \tag{2.77}$$

where a summation over the repeated vector component index i is implied. This is known as the *summation convention*.

From the components a_i and b_i of two vectors **a** and **b**, we can form $3 \times 3 = 9$ products $a_i b_j$. We denote these nine components by the symbol T_{ij}:

$$\mathrm{T}_{ij} = a_i b_j \tag{2.78}$$

In vector notation we regard the nine quantities as components of

$$\mathbb{T} = \mathbf{ab} \tag{2.79}$$

with no dot or cross between the vectors **a** and **b**. This is sometimes called the *direct* or *outer product* of the vectors **a** and **b**. Any such quantity whose nine elements in one coordinate system transform to those in a rotated coordinate system in the same way as a product of vector components transform is called a *tensor* (more precisely, a *tensor of second rank*). Any linear combination of tensors is also a tensor. A general tensor can always be written as a linear combination of outer products. The sum of the diagonal elements $(i = j)$ of the tensor $\mathbb{T} = \mathbf{ab}$,

$$\mathrm{T}_{11} + \mathrm{T}_{22} + \mathrm{T}_{33} = a_1 b_1 + a_2 b_2 + a_3 b_3 \tag{2.80}$$

is just the dot product $\mathbf{a} \cdot \mathbf{b}$. The components of the cross product $\mathbf{a} \times \mathbf{b}$ are constructed from the off-diagonal elements $(i \neq j)$ of this tensor.

If we make a dot product of the tensor $(\mathbf{ab})$ with a vector **c**, we get a vector

$$\begin{aligned} (\mathbf{ab}) \cdot \mathbf{c} &= \mathbf{a}(\mathbf{b} \cdot \mathbf{c}) \\ \mathbf{c} \cdot (\mathbf{ab}) &= \mathbf{b}(\mathbf{c} \cdot \mathbf{a}) \end{aligned} \tag{2.81}$$

because $(\mathbf{b} \cdot \mathbf{c})$ and $(\mathbf{c} \cdot \mathbf{a})$ are scalars, and a vector multiplied by a scalar is a vector. Hence, for a general tensor $\mathbb{T}$, the dot products $\mathbb{T} \cdot \mathbf{c}$ and $\mathbf{c} \cdot \mathbb{T}$ are vectors. In terms of components,

$$\begin{aligned} (\mathbb{T} \cdot \mathbf{c})_i &= \mathrm{T}_{ij} c_j \\ (\mathbf{c} \cdot \mathbb{T})_i &= c_j \mathrm{T}_{ji} \end{aligned} \tag{2.82}$$

with a summation over the index j implied. The most important use of a tensor is to relate two vectors in this way. The unit tensor $\mathbb{I}$, with the property that

$$\mathbf{a} \cdot \mathbb{I} = \mathbb{I} \cdot \mathbf{a} = \mathbf{a} \tag{2.83}$$

for any vector **a**, is given in terms of unit vectors by

$$\mathbb{I} = \hat{\mathbf{x}}\hat{\mathbf{x}} + \hat{\mathbf{y}}\hat{\mathbf{y}} + \hat{\mathbf{z}}\hat{\mathbf{z}} \tag{2.84}$$

The components of a second-rank tensor are often written in a 3×3 matrix array as

$$\mathbb{T} = \begin{pmatrix} T_{11} & T_{12} & T_{13} \\ T_{21} & T_{22} & T_{23} \\ T_{31} & T_{32} & T_{33} \end{pmatrix} \tag{2.85}$$

and a vector $\mathbf{c}$ is represented by a column vector

$$\mathbf{c} = \begin{pmatrix} c_1 \\ c_2 \\ c_3 \end{pmatrix} \tag{2.86}$$

or a row vector

$$\mathbf{c} = (c_1, \, c_2, \, c_3) \tag{2.87}$$

The dot product $\mathbb{T} \cdot \mathbf{c}$ can then be worked out by matrix multiplication:

$$\mathbb{T} \cdot \mathbf{c} = \begin{pmatrix} T_{11} & T_{12} & T_{13} \\ T_{21} & T_{22} & T_{23} \\ T_{31} & T_{32} & T_{33} \end{pmatrix} \begin{pmatrix} c_1 \\ c_2 \\ c_3 \end{pmatrix} = \begin{pmatrix} T_{11}c_1 + T_{12}c_2 + T_{13}c_3 \\ T_{21}c_1 + T_{22}c_2 + T_{22}c_3 \\ T_{31}c_1 + T_{32}c_2 + T_{33}c_3 \end{pmatrix}$$
$$\tag{2.88}$$

Similarly, to evaluate $\mathbf{c} \cdot \mathbb{T}$ we use the row vector form of $\mathbf{c}$. Tensor methods are important in the treatment of rigid body dynamics, as discussed in Chapter 6.

2.5 Conservative Forces in Three Dimensions

We want to find the conditions on the force $\mathbf{F}$ for which energy conservation methods apply in three dimensions. With vector notation, Newton's laws of motion can compactly be expressed as

$$\frac{d}{dt}(m\mathbf{v}) = \mathbf{F}(\mathbf{r}, \mathbf{v}, t) \tag{2.89}$$

The appearance of the vectors $\mathbf{r}$ and $\mathbf{v}$ in the argument of $\mathbf{F}$ indicates that each component of $\mathbf{F}$ can depend on all the components of $\mathbf{r}$ and $\mathbf{v}$. [For example, $F_x(x, \, y, \, z, \, v_x, \, v_y, \, v_z, \, t).$]

In analogy to our derivation in (2.3) to (2.9) of energy conservation in one-dimensional motion, we take the dot product with $\mathbf{v}$ of both sides of (2.89) to obtain

$$\mathbf{v}\cdot\frac{d}{dt}(m\mathbf{v}) = \mathbf{F}(\mathbf{r},\mathbf{v},t)\cdot\mathbf{v} \tag{2.90}$$

or equivalently,

$$d\left(\tfrac{1}{2}m\mathbf{v}\cdot\mathbf{v}\right) = \mathbf{F}(\mathbf{r},\mathbf{v},t)\cdot d\mathbf{r} \tag{2.91}$$

From (2.44), the dot product $\mathbf{v}\cdot\mathbf{v}$ is

$$\mathbf{v}\cdot\mathbf{v} = v^2 = v_x^2 + v_y^2 + v_z^2 \tag{2.92}$$

Thus the differential on the left-hand side of (2.91) is the kinetic energy for three-dimensional motion. Integrating, we obtain the *work-energy theorem* in three dimensions:

$$\Delta K = K_2 - K_1 = \int_{\mathbf{r}_1}^{\mathbf{r}_2} \mathbf{F}(\mathbf{r},\mathbf{v},t)\cdot d\mathbf{r} = \text{Work} \tag{2.93}$$

An integral of the above form is called a *line integral.* Using the definitions of the dot product it can be expressed as

$$\int_{\mathbf{r}_1}^{\mathbf{r}_2} \mathbf{F}\cdot d\mathbf{r} = \int_{x_1}^{x_2} F_x\big(\mathbf{r}(x),\mathbf{v}(x),t(x)\big)dx + \int_{y_1}^{y_2} F_y\big(\mathbf{r}(y),\mathbf{v}(y),t(y)\big)dy$$
$$+ \int_{z_1}^{z_2} F_z\big(\mathbf{r}(z),\mathbf{v}(z),t(z)\big)dz \tag{2.94}$$

where in the integral over dx the line along which the integral is carried out is described by the functions $y = y(x)$ and $z = z(x)$, and similarly for the integrals over dy and dz.

As in the one-dimensional case, we define a potential energy $V(\mathbf{r})$ by the line integral

$$V(\mathbf{r}) = -\int_{\mathbf{r}_s}^{\mathbf{r}} \mathbf{F}(\mathbf{r}')\cdot d\mathbf{r}' \tag{2.95}$$

This line integral is illustrated in Fig. 2-11. By the same reasoning as in the one-dimensional case, a necessary condition that the potential energy is a unique function of coordinate is that the force be a function of

coordinate only. But it is also necessary that the value of the integral in
(2.95) be independent of the integration path. Assuming this, we obtain
the energy conservation condition as before:

$$E = \tfrac{1}{2}mv^2 + V(\mathbf{r}) = \text{constant} \qquad (2.96)$$

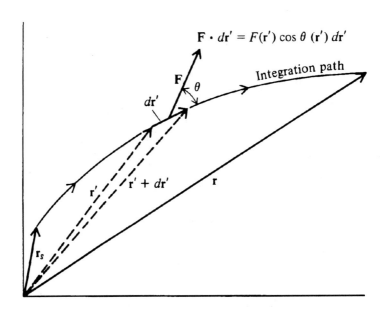

FIGURE 2-11. Geometrical interpretation of the line integral $\int_{\mathbf{r}_s}^{\mathbf{r}} \mathbf{F}(\mathbf{r}') \cdot d\mathbf{r}'$, where $\mathbf{r}'$
is the integration variable and $\mathbf{r}_s$, $\mathbf{r}$ are the limits of integration. The projection angle
θ generally varies along the path.

Before proceeding further, we investigate the condition on the force
for the above line integral to be path-independent. To find this condition
on $\mathbf{F}$, we first consider integration paths which include two adjacent sides
of an infinitesimal rectangle in the y, z plane, as shown in Fig. 2-12. We
locate a corner of the rectangle at the point (y_s, z_s) and calculate $V(\mathbf{r})$
at the diagonal corner $(y_s + dy,\ z_s + dz)$ by two different paths:

$$\text{Path I}: \quad (y_s, z_s) \to (y_s, z_s + dz) \to (y_s + dy, z_s + dz)$$

$$\text{Path II}: \quad (y_s, z_s) \to (y_s + dy, z_s) \to (y_s + dy, z_s + dz)$$

The value of $V(\mathbf{r})$ calculated from Path I is

$$V(\mathbf{r}) = -F_z(x_s, y_s, z_s)dz - F_y(x_s, y_s, z_s + dz)dy \qquad (2.97)$$

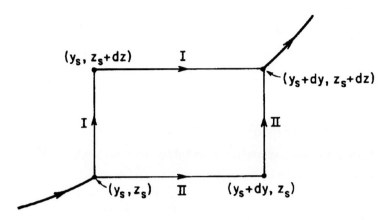

FIGURE 2-12. Integration path in the y, z plane including two alternative routes around an infinitesimal rectangle.

The corresponding result from Path II is

$$V(\mathbf{r}) = -F_y(x_s, y_s, z_s)dy - F_z(x_s, y_s + dy, z_s)dz \qquad (2.98)$$

Demanding that the same $V(\mathbf{r})$ results from both integration paths yields by subtraction

$$\begin{aligned}
&\left[F_y(x_s, y_s, z_s+dz) - F_y(x_s, y_s, z_s)\right]dy \\
&-\left[F_z(x_s, y_s+dy, z_s) - F_z(x_s, y_s, z_s)\right]dz = 0
\end{aligned} \qquad (2.99)$$

The quantities in brackets are immediately recognizable in terms of partial derivatives as

$$\left[\frac{\partial F_y}{\partial z}(x_s, y_s, z_s)\right] dz$$

and

$$\left[\frac{\partial F_z}{\partial y}(x_s, y_s, z_s)\right] dy$$

Canceling the factor $dy\, dz$ in (2.99) gives the condition,

$$\frac{\partial F_y}{\partial z} - \frac{\partial F_z}{\partial y} = 0 \qquad (2.100)$$

for the force $\mathbf{F}$ to be conservative. This condition must hold for any choice of (x_s, y_s, z_s) on the curve. To derive the above condition on an energy-conserving force, we have used an integration path in the y, z plane. If instead we integrate along a differential rectangle in the x, y plane, we get

$$\frac{\partial F_x}{\partial y} - \frac{\partial F_y}{\partial x} = 0 \tag{2.101}$$

and similarly for a rectangle in the x, z plane.

At this point it is convenient to introduce the vector differentiation operator ∇, defined as

$$\nabla = \hat{\mathbf{x}}\frac{\partial}{\partial x} + \hat{\mathbf{y}}\frac{\partial}{\partial y} + \hat{\mathbf{z}}\frac{\partial}{\partial z} \tag{2.102}$$

This is called the *del operator*, also known as the *gradient* or *grad*. In vector notation the requirement in (2.100) can then be written

$$(\nabla \times \mathbf{F})_x = 0 \tag{2.103}$$

where $\nabla \times \mathbf{F}$ is called the *curl* of $\mathbf{F}$, sometimes written curl $\mathbf{F}$. To verify this assertion we recall the cross-product definition from (2.54):

$$(\nabla \times \mathbf{F})_x = \nabla_y F_z - \nabla_z F_y = \frac{\partial F_z}{\partial y} - \frac{\partial F_y}{\partial z} \tag{2.104}$$

In general, the requirement for a force to be conservative (*i.e.*, derivable from a path-independent potential) is

$$\nabla \times \mathbf{F} = 0 \tag{2.105}$$

where from (2.55) with $\mathbf{a} = \nabla$ and $\mathbf{b} = \mathbf{F}$ we obtain the complete expansion for $\nabla \times \mathbf{F}$ in cartesian coordinates:

$$\nabla \times \mathbf{F} = \left(\frac{\partial F_z}{\partial y} - \frac{\partial F_y}{\partial z}\right)\hat{\mathbf{x}} + \left(\frac{\partial F_x}{\partial z} - \frac{\partial F_z}{\partial x}\right)\hat{\mathbf{y}} + \left(\frac{\partial F_y}{\partial x} - \frac{\partial F_x}{\partial y}\right)\hat{\mathbf{z}} \tag{2.106}$$

To summarize the preceding discussion, we have shown that if $\nabla \times \mathbf{F} = 0$ then $\int_{\mathbf{r}_s}^{\mathbf{r}} \mathbf{F}(\mathbf{r}') \cdot d\mathbf{r}'$ is path-independent and there is therefore a unique potential energy.

Conversely, if the potential-energy function $V(\mathbf{r})$ exists, then from (2.95) we can express $\mathbf{F}(\mathbf{r})$ in terms of it. The differential of (2.95) reads

$$dV(\mathbf{r}) = -\mathbf{F}(\mathbf{r}) \cdot d\mathbf{r} \qquad (2.107)$$

Comparing the right-hand side with the total differential dV

$$dV = \frac{\partial V}{\partial x}\, dx + \frac{\partial V}{\partial y}\, dy + \frac{\partial V}{\partial z}\, dz = (\boldsymbol{\nabla} V) \cdot d\mathbf{r} \qquad (2.108)$$

we see

$$\mathbf{F}(\mathbf{r}) = -\boldsymbol{\nabla} V(\mathbf{r}) \qquad (2.109)$$

In component form we can write

$$F_k(\mathbf{r}) = -\nabla_k V(\mathbf{r}) = -\frac{\partial V(\mathbf{r})}{\partial x_k} \qquad (2.110)$$

Forming the curl of this $\mathbf{F}$, we find

$$\boldsymbol{\nabla} \times \mathbf{F} = -\boldsymbol{\nabla} \times \boldsymbol{\nabla} V(\mathbf{r}) = 0 \qquad (2.111)$$

since $\boldsymbol{\nabla} \times \boldsymbol{\nabla} = 0$. Hence, there is a potential energy if and only if $\boldsymbol{\nabla} \times \mathbf{F} = 0$.

Among the most important physical examples of conservative forces are central forces. The magnitude of a central force at each point depends only on the distance from a certain point, the force center, and the direction of the force is radial to the force center, *i.e.*, towards or away from it. The gravitational and Coulomb forces are both of this type. If the force center is at the origin of the coordinate system, the force field has the form

$$\mathbf{F}(\mathbf{r}) = F(r)\hat{\mathbf{r}} \qquad (2.112)$$

where $\hat{\mathbf{r}} = \mathbf{r}/r$. If $F(r) < 0$ the force is towards the center and is called *attractive* (at that value of r), while if $F(r) > 0$ the force is *repulsive*. More generally, the center can be at any point $\mathbf{r}_0$ and the expression for the force looks like (2.112) with $\mathbf{r}$ replaced by $\mathbf{r} - \mathbf{r}_0$ (and r by $|\mathbf{r} - \mathbf{r}_0|$). The superposition of several such central forces with arbitrary centers is also conservative.

To prove explicitly that central forces are conservative, it suffices to take the center at the origin, (2.112). Using cartesian components

$$F_x = \frac{x}{r} F(r), \qquad F_y = \frac{y}{r} F(r), \qquad F_z = \frac{z}{r} F(r) \qquad (2.113)$$

we construct dV according to (2.107)

$$dV = -(F_x dx + F_y dy + F_z dz)$$
$$= -\frac{F(r)}{r}(x\,dx + y\,dy + z\,dz) = -F(r)dr \qquad (2.114)$$

In the last step we have used the differential of

$$r^2 = x^2 + y^2 + z^2 \qquad (2.115)$$

Since the right-hand side of (2.114) depends only on the radial coordinate r (not on θ or ϕ), its integral is path-independent. This establishes the conservative nature of a central force; equivalently we could have directly shown that $\nabla \times \mathbf{F} = 0$.

From (2.114) we obtain the central potential energy from the force law as

$$V(r) = -\int_{r_s}^{r} F(r')dr' \qquad (2.116)$$

The above formula could have been found directly from the line integral (2.95). For instance, from the gravitational force law

$$\mathbf{F} = -\frac{GMm}{r^2}\hat{\mathbf{r}} \qquad (2.117)$$

the gravitational potential energy due to a mass M at $\mathbf{r} = 0$ is

$$V(\mathbf{r}) = -\int_{\infty}^{\mathbf{r}} \left(-\frac{GMm}{r'^2}\hat{\mathbf{r}}'\right) \cdot d\mathbf{r}' \qquad (2.118)$$

and with $d\mathbf{r}' = \hat{\mathbf{r}}'dr' + \hat{\boldsymbol{\theta}}'r'd\theta'$ we have $\hat{\mathbf{r}}' \cdot d\mathbf{r}' = dr'$ and therefore

$$V(\mathbf{r}) = \int_{\infty}^{\mathbf{r}} \frac{GMm}{r'^2} dr' = -\frac{GMm}{r} \qquad (2.119)$$

Only the magnitude of the velocity enters into the three dimensional energy conservation law and hence the launch direction of a rocket is arbitrary as long as the rocket does not hit the earth. If $v \geq v_{\text{esc}}$, the spacecraft will not return.

2.6 Motion in a Plane

For the analysis of the mechanical motion of some systems cartesian coordinates are not the most convenient choice. For example, some kinds
of motion in a plane can frequently be described more simply in terms of
polar coordinates (r, θ) than (x, y). Since Newton's equations of motion
do not have the same form in polar coordinates, we cannot just substitute
$\ddot{r}$ and $\ddot{\theta}$ for $\ddot{x}$ and $\ddot{y}$ in Newton's equations; Newton's equations have the
same form only in different *cartesian* coordinate systems. Therefore we
must do some algebra to express Newton's equations in terms of polar
coordinates.

In cartesian coordinates we have

$$m\ddot{x} = F_x \qquad\qquad m\ddot{y} = F_y \qquad\qquad (2.120)$$

which is written in vector form as

$$m\ddot{\mathbf{r}} = \mathbf{F} \qquad\qquad (2.121)$$

with

$$\mathbf{r} = \hat{\mathbf{x}}x + \hat{\mathbf{y}}y \qquad \text{and} \qquad \mathbf{F} = \hat{\mathbf{x}}F_x + \hat{\mathbf{y}}F_y \qquad (2.122)$$

In polar coordinates the vector $\mathbf{r}$ is given by

$$\mathbf{r} = \hat{\mathbf{x}}r\cos\theta + \hat{\mathbf{y}}r\sin\theta \equiv \hat{\mathbf{r}}r \qquad\qquad (2.123)$$

The unit vectors $\hat{\mathbf{x}}$ and $\hat{\mathbf{y}}$ in the cartesian system do not change with
time. The differential $d\mathbf{r}$ is thus

$$
\begin{aligned}
d\mathbf{r} &= \hat{\mathbf{x}}(\cos\theta\, dr - r\sin\theta\, d\theta) + \hat{\mathbf{y}}(\sin\theta\, dr + r\cos\theta\, d\theta) \\
&= (\hat{\mathbf{x}}\cos\theta + \hat{\mathbf{y}}\sin\theta)dr + (-\hat{\mathbf{x}}\sin\theta + \hat{\mathbf{y}}\cos\theta)r\, d\theta \qquad (2.124) \\
&= \hat{\mathbf{r}}\, dr + \hat{\boldsymbol{\theta}}r\, d\theta
\end{aligned}
$$

where in the last form we have defined the unit vectors $\hat{\mathbf{r}}$ and $\hat{\boldsymbol{\theta}}$ to be
along the direction of $d\mathbf{r}$ when only r or θ, respectively, are increased:

$$
\begin{aligned}
\hat{\mathbf{r}} &= \hat{\mathbf{x}}\cos\theta + \hat{\mathbf{y}}\sin\theta \\
\hat{\boldsymbol{\theta}} &= -\hat{\mathbf{x}}\sin\theta + \hat{\mathbf{y}}\cos\theta
\end{aligned}
\qquad\qquad (2.125)
$$

[see Fig. 2-13 for a geometrical representation.] By direct calculation it
is seen that $\hat{\mathbf{r}} \cdot \hat{\mathbf{r}} = \hat{\boldsymbol{\theta}} \cdot \hat{\boldsymbol{\theta}} = 1$.

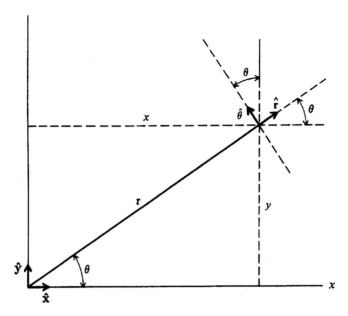

FIGURE 2-13. Polar variables and unit vectors for motion in a plane.

Dividing $d\mathbf{r}$ by dt we obtain

$$\mathbf{v} = \dot{\mathbf{r}} = \hat{\mathbf{r}}\dot{r} + \hat{\boldsymbol{\theta}}r\dot{\theta} \tag{2.126}$$

Differentiating $\mathbf{v}$ with respect to time, we find

$$\dot{\mathbf{v}} = \hat{\mathbf{r}}\ddot{r} + \dot{\hat{\mathbf{r}}}\dot{r} + \hat{\boldsymbol{\theta}}r\ddot{\theta} + (\dot{\hat{\boldsymbol{\theta}}}r + \hat{\boldsymbol{\theta}}\dot{r})\dot{\theta} \tag{2.127}$$

The derivatives of $\hat{\mathbf{r}}$ and $\hat{\boldsymbol{\theta}}$ are found from (2.125) to be

$$\dot{\hat{\mathbf{r}}} = \dot{\theta}\hat{\boldsymbol{\theta}}$$
$$\dot{\hat{\boldsymbol{\theta}}} = -\dot{\theta}\hat{\mathbf{r}} \tag{2.128}$$

Substituting these results for $\dot{\hat{\mathbf{r}}}$ and $\dot{\hat{\boldsymbol{\theta}}}$ into the expressions for $\mathbf{v}$ and $\dot{\mathbf{v}}$ above, we arrive at

$$\mathbf{v} = \dot{\mathbf{r}} = \dot{r}\hat{\mathbf{r}} + \dot{\theta}r\hat{\boldsymbol{\theta}}$$
$$\mathbf{a} = \dot{\mathbf{v}} = \hat{\mathbf{r}}(\ddot{r} - r\dot{\theta}^2) + \hat{\boldsymbol{\theta}}(r\ddot{\theta} + 2\dot{r}\dot{\theta}) \tag{2.129}$$

In polar coordinates we write $\mathbf{F} = \hat{\mathbf{r}}F_r + \hat{\boldsymbol{\theta}}F_\theta$, so Newton's law $m\ddot{\mathbf{r}} = \mathbf{F}$ is

$$m(\ddot{r} - r\dot{\theta}^2) = F_r$$
$$m(r\ddot{\theta} + 2\dot{r}\dot{\theta}) = F_\theta \tag{2.130}$$

Notice the difference between the left-hand side of these equations and the cartesian equations (2.30). In other coordinate systems the structure of the equations in motion can be even more complicated, and the derivation of results similar to (2.130) correspondingly more difficult. In the next chapter we will see that the derivation of the equations of motion is greatly simplified using Lagrange's method.

2.7 Simple Pendulum

The plane pendulum is a familiar system of historical importance whose motion cannot be described in terms of elementary functions. Nevertheless, we can easily find the equations of motion and an approximate solution for small oscillations.

The simple plane pendulum consists of a point mass m at the end of a weightless rod or string of constant length ℓ which swings back and forth in a vertical plane. We take the origin of the coordinate system at the pivot point, with x positive down and y positive to the right. The two forces acting upon the mass m are gravity and the tension in the rod T_r, as shown in Fig. 2-14. If T_r is positive, the force on m is radially inward. In terms of polar coordinates the force components are

$$F_r = mg\cos\theta - T_r$$
$$F_\theta = -mg\sin\theta \tag{2.131}$$

Newton's law in polar coordinates, (2.130), with $r = \ell$, together with the above forces gives

$$-m\ell\dot{\theta}^2 = mg\cos\theta - T_r \tag{2.132}$$
$$m\ell\ddot{\theta} = -mg\sin\theta \tag{2.133}$$

The first equation can be ignored if we are not interested in the value of $T_r(t)$, but only in the motion $\theta(t)$. After solving the second equation for $\theta(t)$, the first equation then gives $T_r(\theta)$. The second equation, (2.133), can be written as

$$\ddot{\theta} + \omega_0^2\sin\theta = 0 \tag{2.134}$$

where $\omega_0 = \sqrt{g/\ell}$.

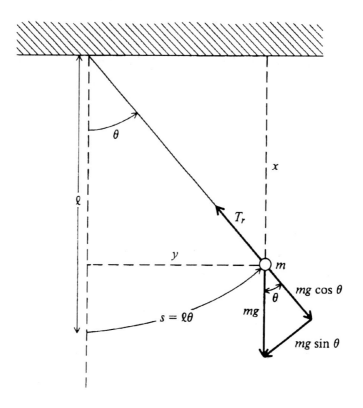

FIGURE 2-14. Simple plane pendulum.

For small oscillations, $|\theta| \ll 1$, we can approximate $\sin \theta \approx \theta$, with θ in radians, to obtain

$$\ddot{\theta} + \omega_0^2 \theta = 0 \tag{2.135}$$

This is readily recognizable as simple harmonic motion in θ with angular frequency ω_0 [see (1.61) through (1.64)]. The general solution of (2.135) is

$$\theta = a \cos(\omega_0 t + \alpha) \tag{2.136}$$

where the arbitrary constants a and α are to be fixed by the initial conditions. The period of small oscillations,

$$\tau = \frac{2\pi}{\omega_0} = 2\pi \sqrt{\frac{\ell}{g}} \tag{2.137}$$

is independent of a and α. This approximately amplitude-independent feature of the period of motion, called *isochronism*, is incorporated in the pendulum clock.

To solve for the motion exactly, without a small displacement approximation, we can use the energy method. Using the chain rule,

$$\ddot\theta = \frac{d\dot\theta}{d\theta}\frac{d\theta}{dt} = \frac{\dot\theta d\dot\theta}{d\theta} = \frac{d}{d\theta}\left(\frac{\dot\theta^2}{2}\right) \tag{2.138}$$

we integrate (2.134)

$$\int_0^{\dot\theta} d\left(\frac{\dot\theta^2}{2}\right) = -\frac{g}{\ell}\int_{\theta_0}^{\theta}\sin\theta d\theta \tag{2.139}$$

where θ_0 is the angle at which $\dot\theta = 0$ (θ_0 is the maximum angle of the motion). The evaluation of the integrals gives

$$\dot\theta^2 = \frac{2g}{\ell}(\cos\theta - \cos\theta_0) \tag{2.140}$$

If we multiply by $\frac{1}{2}m\ell^2$, we can recognize this as the statement of energy conservation.

Before finding the motion, we make the observation that we can use (2.140) to eliminate $\dot\theta$ from the formula (2.132) for the tension

$$T_r = mg\cos\theta + m\ell\dot\theta^2 \tag{2.141}$$

to get

$$T_r = 3mg\cos\theta - 2mg\cos\theta_0 \tag{2.142}$$

Equation (2.140) gives the angular velocity to be

$$\frac{d\theta}{dt} = \pm\sqrt{\frac{2g}{\ell}}\sqrt{\cos\theta - \cos\theta_0} \tag{2.143}$$

This differential equation is separable; it can be written as

$$\frac{d\theta}{\sqrt{\cos\theta - \cos\theta_0}} = \pm\sqrt{\frac{2g}{\ell}}\,dt \tag{2.144}$$

and integrated

$$\sqrt{\frac{2g}{\ell}}\int_0^t dt' = -\int_{\theta_0}^{\theta}\frac{d\theta'}{\sqrt{\cos\theta' - \cos\theta_0}} \tag{2.145}$$

The minus sign is required if θ_0 is positive (noting that $\theta < \theta_0$ and the θ

integral is negative). This integral can be cast into a standard form by substitution of the identity

$$\cos \theta = 1 - 2 \sin^2 \frac{\theta}{2} \tag{2.146}$$

in (2.145)

$$2\sqrt{\frac{g}{\ell}}\, t = -\int_{\theta_0}^{\theta} \frac{d\theta'}{\sqrt{\sin^2(\theta_0/2) - \sin^2(\theta'/2)}} \tag{2.147}$$

If we now introduce a new variable

$$\sin \beta' = \frac{\sin(\theta'/2)}{\sin(\theta_0/2)} \tag{2.148}$$

the solution in (2.147) becomes

$$\sqrt{\frac{g}{\ell}}\, t = \int_{\beta}^{\pi/2} \frac{d\beta'}{\sqrt{1 - \sin^2(\theta_0/2)\sin^2 \beta'}} \tag{2.149}$$

Setting $\theta = 0$, hence $\beta = 0$, gives a quarter period, so the period is given by

$$\tau = 4\sqrt{\frac{\ell}{g}} \int_0^{\pi/2} \frac{d\beta}{\sqrt{1 - \sin^2(\theta_0/2)\sin^2 \beta}} \tag{2.150}$$

In terms of the simple harmonic period $\tau_0 = 2\pi\sqrt{\frac{\ell}{g}}$,

$$\frac{\tau}{\tau_0} = \frac{2}{\pi} \int_0^{\pi/2} \frac{d\beta}{\sqrt{1 - \sin^2 \frac{\theta_0}{2}\sin^2 \beta}} \tag{2.151}$$

This integral is known as the complete elliptic integral of the first kind. It cannot be evaluated in closed form, but numerical evaluations are available in tabular form or from computer software. To find an approximation to the period for small angular displacements, the integrand can be expanded by power series and then integrated term by term

$$\frac{\tau}{\tau_0} = \frac{2}{\pi} \int_0^{\pi/2} d\beta \left[1 + \frac{1}{2} \sin^2 \left(\frac{\theta_0}{2} \right) \sin^2 \beta + \cdots \right]$$

$$= \frac{2}{\pi} \left[\beta + \frac{1}{4} \sin^2 \left(\frac{\theta_0}{2} \right) \left(\beta - \frac{\sin 2\beta}{2} \right) + \cdots \right]_0^{\pi/2} \qquad (2.152)$$

$$= \left[1 + \frac{1}{4} \sin^2 \left(\frac{\theta_0}{2} \right) \right] + \cdots$$

Again using the approximation of θ_0 small and $\sin^2(\theta_0/2) \approx \theta_0^2/4$, we find

$$\tau = \tau_0 \left(1 + \frac{\theta_0^2}{16} + \cdots \right) \qquad (2.153)$$

The period is increased over the simple harmonic period. The fractional lengthening of the period is

$$\frac{\tau - \tau_0}{\tau_0} = \frac{\theta_0^2}{16} \qquad (2.154)$$

For a 30° maximal displacement, the fractional lengthening of the period

$$\frac{\tau - \tau_0}{\tau_0} = \frac{1}{16} \left(\frac{30°}{57.3°} \right)^2 = 0.017 \qquad (2.155)$$

is less than 2 percent. For a pendulum clock of period 1 s and $\theta_0 = 5°$, the amplitude θ_0 must be regulated to $\pm 3°$ if the clock is to be accurate to 1 min/day. If the clock is desired to have an accuracy of 1 s/day, θ_0 must be regulated to 0.06 degrees.

2.8   Coupled Harmonic Oscillators

In physical problems that can be approximated by several small oscillations there is usually a coupling between the oscillators. As a specific example, we investigate the motion of two simple pendulums whose bobs are connected by a spring, as indicated in Fig. 2-15.

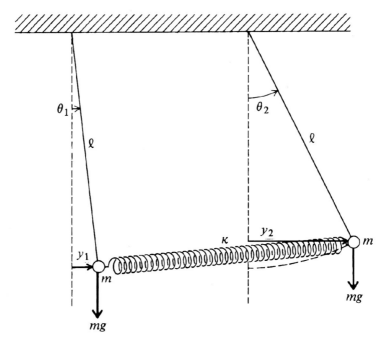

FIGURE 2-15. Two simple pendulums coupled by a spring.

For small angular displacements the equation of motion of a single isolated pendulum is given by (2.135).

$$\ddot{\theta} + \omega_0^2 \theta = 0 \qquad (2.156)$$

This equation can be alternatively expressed in terms of the x and y coordinates of the pendulum bob. For $\theta \ll 1$, we have

$$\begin{aligned} x &= \ell \cos\theta \approx \ell \\ y &= \ell \sin\theta \approx \ell\theta \end{aligned} \qquad (2.157)$$

and (2.135) becomes

$$\ddot{y} + \omega_0^2 y = 0 \qquad (2.158)$$

In this approximation the pendulum executes simple harmonic motion in the horizontal direction. The pendulum spring system of Fig. 2-15 is therefore equivalent for small displacements to the three-spring system of Fig. 2-16, with spring constants $k = m\omega_0^2 = mg/\ell$ for the outer springs and κ for the inner spring.

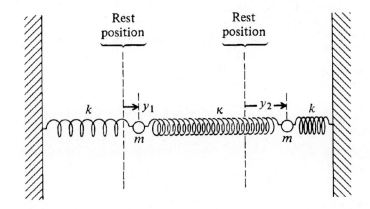

FIGURE 2-16. Equivalent three-spring system for the coupled-pendulum system of Fig. 2-15.

The equations of motion can be obtained by considering the forces on each mass separately. First fix y_2 at $y_2 = 0$ and imagine a positive y_1 displacement. The restoring forces on m_1 are $-ky_1$ due to the extension of left spring and $-\kappa y_1$ due to the compression of the right spring. Now include a positive y_2 displacement and consider its effect on m_1. The stretching of the middle spring gives a positive force κy_2 on m_1. Combining these forces the equation of motion for m_1 is

$$m\ddot{y}_1 = -ky_1 - \kappa y_1 + \kappa y_2 \qquad (2.159)$$

Note that the restoring force due to the middle spring depends only on the difference $y_1 - y_2$. A similar exercise applied to the right-hand mass yields

$$m\ddot{y}_2 = -ky_2 - \kappa y_2 + \kappa y_1 \qquad (2.160)$$

The differential equations of motion for the system are thus

$$m\ddot{y}_1 = -ky_1 - \kappa(y_1 - y_2)$$
$$m\ddot{y}_2 = -ky_2 + \kappa(y_1 - y_2) \qquad (2.161)$$

To solve these differential equations, we look for linear combinations of y_1 and y_2 that yield differential equations of simple harmonic form. Later in this Section we discuss the solution to coupled equations in some generality. In the present case it suffices to take the sum and the difference of the equations to uncouple them. If we add the equations, we find

$$m(\ddot{y}_1 + \ddot{y}_2) = -k(y_1 + y_2) \tag{2.162}$$

and if we subtract

$$m(\ddot{y}_1 - \ddot{y}_2) = -(k + 2\kappa)(y_1 - y_2) \tag{2.163}$$

The solutions of these two uncoupled equations are found directly from (1.64) to be

$$
\begin{aligned}
y_1 + y_2 &= a_+ \cos(\omega_+ t + \alpha_+) \\
y_1 - y_2 &= a_- \cos(\omega_- t + \alpha_-)
\end{aligned}
\tag{2.164}
$$

where the angular frequencies are

$$
\omega_+ = \sqrt{\frac{k}{m}} = \sqrt{\frac{g}{\ell}}
$$

$$
\omega_- = \sqrt{\frac{k + 2\kappa}{m}} = \sqrt{\frac{g}{\ell} + \frac{2\kappa}{m}}
\tag{2.165}
$$

The combinations $(y_1 + y_2)$ and $(y_1 - y_2)$ oscillate independently, simple harmonically, and are called the *normal modes*. The motion of the pendulum bobs is in general a superposition of the two normal modes of vibration and from (2.164) we have

$$
\begin{aligned}
y_1 &= \tfrac{1}{2}a_+ \cos(\omega_+ t + \alpha_+) + \tfrac{1}{2}a_- \cos(\omega_- t + \alpha_-) \\
y_2 &= \tfrac{1}{2}a_+ \cos(\omega_+ t + \alpha_+) - \tfrac{1}{2}a_- \cos(\omega_- t + \alpha_-)
\end{aligned}
\tag{2.166}
$$

The four constants a_+, a_-, α_+, and α_- in (2.166) are to be determined by the initial conditions. If only the amplitude of one normal mode is excited, that is, only a_+ or a_- is non-zero, the bobs swing in phase with frequency ω_+ or out of phase with frequency ω_-, as illustrated in Fig. 2-17. We note that ω_+ does not depend on κ since the middle spring is never stretched.

In the weak coupling limit $\kappa \ll k$, the coupling between the two pendulums causes a gradual interchange of energy between the two oscillators. To demonstrate this we suppose that both bobs are initially at rest and the motion of the system is started by displacing the first bob

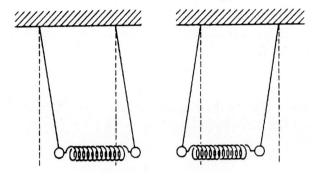

In phase motion: frequency ω_+

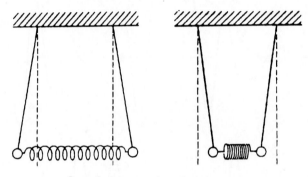

Out of phase motion: frequency ω_-

FIGURE 2-17. Normal modes for the coupled-pendulum system.

by a distance a. When these initial conditions are imposed on (2.166), we obtain

$$y_1 = \frac{a}{2}(\cos \omega_+ t + \cos \omega_- t)$$

$$y_2 = \frac{a}{2}(\cos \omega_+ t - \cos \omega_- t)$$

(2.167)

From trigonometric identities for the sum and difference of cosine functions, (2.167) can be written as

$$y_1 = a \cos\left(\frac{\omega_- + \omega_+}{2}t\right) \cos\left(\frac{\omega_- - \omega_+}{2}t\right)$$

$$y_2 = a \sin\left(\frac{\omega_- + \omega_+}{2}t\right) \sin\left(\frac{\omega_- - \omega_+}{2}t\right)$$

(2.168)

At time $t = \pi/(\omega_- - \omega_+)$, the first pendulum has come to rest and all the energy has been transferred through the coupling to the second oscillator. For the weak coupling limit, $\omega_- - \omega_+ \ll \omega_+$, the last factors in (2.168) are slowly varying functions of time. These slowly varying factors constitute an envelope for the rapidly oscillating sinusoidal factors of argument $[(\omega_- + \omega_+)/2]t$, as illustrated in Fig. 2-18. This phenomenon is known as *beats*. The beat frequency is $(\omega_- - \omega_+)/2$, and the period of the envelope of the amplitude is $2\pi/(\omega_- - \omega_+)$.

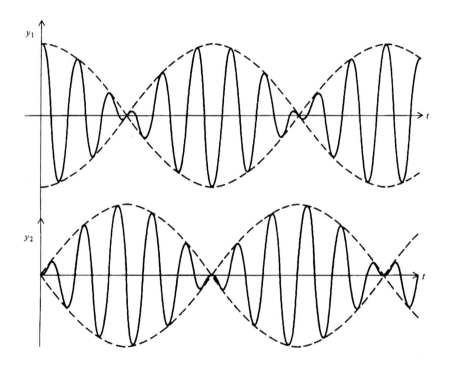

FIGURE 2-18. Beats and envelope exhibited by the coordinates of two weakly coupled oscillators.

We solved the coupled differential equations (2.161) by adding and subtracting to obtain uncoupled equations. We now discuss an alternative method of solution which can be more straightforwardly generalized to more complex coupled systems with different masses and spring constants. Denoting

$$\omega_0^2 \equiv \frac{k}{m} \qquad \text{and} \qquad \Delta^2 \equiv \frac{\kappa}{m} \qquad (2.169)$$

the equations (2.161) become

$$\ddot{y}_1 + (\omega_0^2 + \Delta^2)y_1 - \Delta^2 y_2 = 0$$
$$\ddot{y}_2 + (\omega_0^2 + \Delta^2)y_2 - \Delta^2 y_1 = 0 \tag{2.170}$$

This coupled set of differential equations is homogeneous and linear with constant coefficients. It has complex solutions of the form

$$y_1 = C_1 e^{i\omega t}$$
$$y_2 = C_2 e^{i\omega t} \tag{2.171}$$

where C_1 and C_2 are constants that are, in general, complex. The physical solutions are the real parts of these complex y_1, y_2 solutions. Substitution of (2.171) into (2.170) gives a pair of coupled linear equations for C_1 and C_2 that can be written in matrix form as

$$\begin{pmatrix} -\omega^2 + \omega_0^2 + \Delta^2 & -\Delta^2 \\ -\Delta^2 & -\omega^2 + \omega_0^2 + \Delta^2 \end{pmatrix} \begin{pmatrix} C_1 \\ C_2 \end{pmatrix} = 0 \tag{2.172}$$

For a non-trivial solution the determinant of the matrix must vanish, giving

$$\left(-\omega^2 + \omega_0^2 + \Delta^2\right)^2 = \Delta^4 \tag{2.173}$$

Solving the quadratic equation for ω^2, in this simple case by taking the square root, we find two solutions, $\omega^2 = \omega_+^2$ and $\omega^2 = \omega_-^2$, where

$$\omega_+^2 = \omega_0^2$$
$$\omega_-^2 = \omega_0^2 + 2\Delta^2 \tag{2.174}$$

Then solving for C_2/C_1 from (2.172) with these ω^2 values gives

$$C_2/C_1 = +1 \quad \text{for} \quad \omega^2 = \omega_+^2$$
$$C_2/C_1 = -1 \quad \text{for} \quad \omega^2 = \omega_-^2 \tag{2.175}$$

We shall parameterize the complex constant C_1 for solutions $\omega_\pm$ by

$$(C_1)_\pm = \tfrac{1}{2}a_\pm e^{i\alpha_\pm} \tag{2.176}$$

where $a_\pm$ and $\alpha_\pm$ are real. Then the most general motion is given by the linear superpositions

$$y_1 = \tfrac{1}{2}a_+ e^{i(\omega_+ t + \alpha_+)} + \tfrac{1}{2}a_- e^{i(\omega_- t + \alpha_-)}$$
$$y_2 = \tfrac{1}{2}a_+ e^{i(\omega_+ t + \alpha_+)} - \tfrac{1}{2}a_- e^{i(\omega_- t + \alpha_-)} \tag{2.177}$$

The general physical solution obtained by taking the real part of the superposition is the same as (2.166). If only one mode is excited, by a particular choice of initial conditions, the system will oscillate with a single frequency—the normal mode frequency ω_+ or ω_-.

Equation (2.172) is an example of an eigenvalue problem with matrix equation

$$\begin{pmatrix} \omega_0^2 + \Delta^2 & -\Delta^2 \\ -\Delta^2 & \omega_0^2 + \Delta^2 \end{pmatrix} \begin{pmatrix} C_1 \\ C_2 \end{pmatrix} = \omega^2 \begin{pmatrix} C_1 \\ C_2 \end{pmatrix} \tag{2.178}$$

The eigenvalues are $\omega^2 = \omega_+^2$ and $\omega^2 = \omega_-^2$. The vectors $\begin{pmatrix} C_1 \\ C_2 \end{pmatrix}_+$ and $\begin{pmatrix} C_1 \\ C_2 \end{pmatrix}_-$ are known as eigenvectors. Two eigenvectors corresponding to different eigenvalues are orthogonal, or normal. The modes of motion corresponding to ω_+^2 and ω_-^2 are accordingly called normal modes.

PROBLEMS

2.1 Potential Energy

2-1. The potential energy of a mass element dm at a height z above the earth's surface is $dV = (dm)gz$. Compute the potential energy in a pyramid of height h, square base $b \times b$, and mass density ρ. The Great Pyramid of Khufu is 147 m high and has a base of 234×234 m. Estimate its potential energy using $\rho = 2.5$ g/cm^3 for the density of its material. If an average worker lifted 50 kg through a distance of 1 m each minute of a 10 hr work day, estimate the person-years of labor expended in the construction of the Great Pyramid. This ignores friction and the considerable effort required to quarry and transport the stone.

2-2. The Turkish bow of the 15th and 16th centuries greatly outperformed western bows. The draw force $F(x)$ of the Turkish bow versus the bowstring displacement x (for x negative) is approximately represented by a quadrant of the ellipse

$$\left(\frac{F(x)}{F_{max}} \right)^2 + \left(\frac{x+d}{d} \right)^2 = 1$$

Calculate the work done by the bow in accelerating the arrow, taking $F_{max} = 360$ N, $d = 0.7$ m, and arrow mass $m = 34$ g. Assuming that all of the work ends up as arrow kinetic energy, determine the maximum range R of the arrow. (The actual range is about 430 m.) Compare with the range for a bow that acts like a simple spring force with the same F_{max} and d.

2.2 Gravitational Escape

2-3. From the radius and mass ratios

$$R\,(\text{moon})/R\,(\text{earth}) \simeq 1/3.66$$
$$M\,(\text{moon})/M\,(\text{earth}) \simeq 1/81.6$$

show that the gravitational acceleration on the moon and earth are related by

$$g\,(\text{moon})/g\,(\text{earth}) \simeq 1/6$$

Find the escape velocity from the surface of the moon.

2-4. A projectile is fired from the surface of the earth to the moon. Neglecting the orbital motion of the moon, what is the minimum velocity of impact on the surface of the moon? Take into account the gravitational pull of both the moon and the earth.

2-5. An iron meteor enters the earth's atmosphere at the escape velocity. Compute the kinetic energy per molecule and compare with the rough vaporization energy of $1\,\text{eV/molecule}$.

2.3 Small Oscillations

2-6. A particle of mass m moves under the action of a force

$$F = -F_0 \sinh\,(ax) = -\frac{F_0}{2}\,(e^{ax} - e^{-ax})$$

where $a > 0$. Sketch the potential energy, discuss the motion, and solve for the frequency of small oscillation if there exists a point of stability.

2-7. A particle moves subject to the potential energy

$$V\,(x) = V_0\left(\frac{a}{x} + \frac{x}{a}\right)$$

where V_0 and a are positive. Locate any equilibrium points, determine which are stable and obtain the frequency of small oscillations about those points.

2-8. Estimate the spring constant in units of eV/Å^2 for the hydrogen (H_2) molecule from the potential energy curve shown below, where r is the distance between protons. From the spring constant and the "reduced mass" $m = \frac{1}{2}m_{\text{proton}}$, compute the vibrational frequency ν. This frequency corresponds to infrared light.

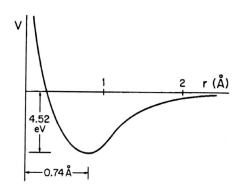

2.4 Three Dimensional Motion: Vector Notation

2-9. Given the vectors

$$\mathbf{A} = 2\hat{\mathbf{x}} + 3\hat{\mathbf{y}} + 4\hat{\mathbf{z}} \qquad \mathbf{B} = 3\hat{\mathbf{x}} + 2\hat{\mathbf{y}} - 2\hat{\mathbf{z}}$$

find

a) $A = |\mathbf{A}|$ and $B = |\mathbf{B}|$,

b) $\mathbf{A} \cdot \mathbf{B}$ and the angle θ between $\mathbf{A}$ and $\mathbf{B}$,

c) $\mathbf{A} \times \mathbf{B}$ and the angle θ between $\mathbf{A}$ and $\mathbf{B}$.

From b) and c) deduce the consistent choice of the angle θ.

2-10. A force field is given by

$$F_x = kyz \sin kxy$$
$$F_y = kxz \sin kxy$$
$$F_z = - \cos kxy$$

a) Evaluate $\nabla \times \mathbf{F}$ to show that $\mathbf{F}$ is conservative.

b) If the reference potential energy at $(x = 0,\ y = 0,\ z = 0)$ is zero, compute the potential energy at the point $(x = 1.0,\ y = 1.0,\ z = 1.0)$. Use any convenient path, such as along the axes.

c) Using a different path, compute the potential energy at the same point to check path independence.

2-11. Consider the following force:

$$\mathbf{F} = -K(x - z)^2 (\hat{\mathbf{x}} - \hat{\mathbf{z}})$$

a) Show that it is conservative.

b) Find the potential energy $V(\mathbf{r})$ assuming $V(\mathbf{0}) = 0$.

c) Calculate $\nabla V(\mathbf{r})$ to verify that it gives $\mathbf{F}$ correctly.

2-12. For a central force $\mathbf{F}(\mathbf{r}) = F(r)\hat{\mathbf{r}}$ show directly that $\nabla \times \mathbf{F} = 0$ for $r \neq 0$.

2-13. Determine whether or not the force $\mathbf{F} = \mathbf{r} \times \mathbf{a}$ (where $\mathbf{a}$ is a constant vector) leads to a conservative potential energy. Compute $\int \mathbf{F} \cdot d\mathbf{r}$ around a circle of radius R in the x, y plane centered at $\mathbf{r} = 0$.

2-14. Show that a force consisting of a superposition of N central forces with centers at $\mathbf{r} = \mathbf{r}_k$, $k = 1$ to N, is also a conservative force. *Hint: the $\mathbf{r}_k$ are constant vectors so $\nabla_{\mathbf{r}} = \nabla_{\mathbf{r}-\mathbf{r}_k}$. Use problem 2-12.*

2-15. Show that the ∇ operator can be expressed in spherical coordinates as

$$\nabla = \hat{\mathbf{r}}\frac{\partial}{\partial r} + \frac{\hat{\boldsymbol{\theta}}}{r}\frac{\partial}{\partial \theta} + \frac{\hat{\boldsymbol{\phi}}}{r\sin\theta}\frac{\partial}{\partial \phi}$$

where $(\hat{\mathbf{r}}, \hat{\boldsymbol{\theta}}, \hat{\boldsymbol{\phi}})$ are perpendicular unit vectors in the direction of increasing $(r, \theta, \phi,)$. (*Hint: Use $df = d\mathbf{r} \cdot \nabla f$ where $d\mathbf{r}$ is given by $d\mathbf{r} = \hat{\mathbf{r}}dr + \hat{\boldsymbol{\theta}}rd\theta + \hat{\boldsymbol{\phi}}r\sin\theta d\phi$ and f is an arbitrary scalar function. Express df in terms of partial derivatives.*) Show that the ∇ operator in cylindrical coordinates (ρ, ϕ, z) is

$$\nabla = \hat{\boldsymbol{\rho}}\frac{\partial}{\partial \rho} + \frac{\hat{\boldsymbol{\phi}}}{\rho}\frac{\partial}{\partial \phi} + \hat{\mathbf{z}}\frac{\partial}{\partial z}$$

2.6 Motion in a Plane

2-16. The bob of a pendulum moves in a horizontal circle as illustrated. Find the angular frequency of the circular motion in terms of the angle θ and the length ℓ of the string. This is known as a conical pendulum

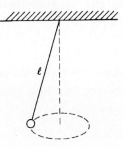

2.7 Simple Pendulum

2-17. A hemispherical thin glass goblet of radius $R = 5\,\text{cm}$ will withstand a perpendicular force of up to $2\,\text{N}$. If a 100-g steel ball is released from rest at the lip of the goblet and allowed to slide down the inside, at what point on the goblet will the ball break through? Neglect the radius of the ball.

2-18. A mass m is attached at one end of a massless rigid rod of length ℓ, and the rod is suspended at its other end by a frictionless pivot, as illustrated. The rod is released from rest at an angle $\alpha_0 < \pi/2$ with the vertical. At what angle α does the force in the rod change from compression to tension?

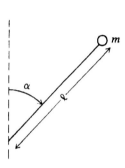

2-19. A ball of mass m is suspended by a string of length ℓ. For what ranges of the total energy will the string remain taut when the ball swings in an arc in a vertical plane? Choose the lowest point on the arc as the reference point for the potential energy.

2-20. A physics professor holds a bowling ball suspended as a pendulum. The ball is initially 1.9 m above the floor, the pendulum wire is 7 m in length and the ceiling height is 7.5 m. The bowling ball has diameter 0.15 m, mass 15 kg; the drag parameter is $C_D = 0.4$ and the air density is $1\ \text{kg/m}^3$. The professor gently releases the ball just in front of her nose and confidently expects that it will return short of its original position.

a) Estimate the work done by friction over one period, using (1.17) for the drag coefficient. Approximate the motion by that of a simple pendulum and use (2.140) in your calculation of this work.

b) Using the work-energy theorem, estimate the change in height when the ball swings back to the professor and by the given geometry find how close the pendulum returns to its release point.

2.8 Coupled Harmonic Oscillators

2-21. A mass $2m$ is suspended from a fixed support by a spring with spring constant $2k$. A second mass m is suspended from the first mass by a spring of constant k. Find the equation of motion for this coupled system and determine the frequencies of oscillation of normal modes. Neglect the masses of the springs. *Hint: It is easiest to choose the coordinates of the two masses at their equilibrium positions.*

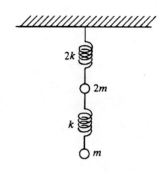

2-22. A mass m is suspended from a support by a spring with spring constant $m\omega_1^2$. A second mass m is suspended from the first by a spring with spring constant $m\omega_2^2$. A vertical harmonic force $F_0 \cos \omega t$ is applied to the upper mass. Find the steady-state motion for each mass. Examine what happens when $\omega = \omega_2$.

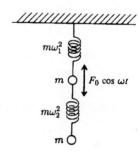

Chapter 3

LAGRANGIAN METHOD

The form that Newton's equations of motion take depends on the coordinate system used. For instance, the equations in a polar system are different from those in a cartesian system. The Lagrangian method is a reformulation which makes it simple to write the equations of motion in any coordinate system. In addition, it provides a straightforward and systematic way to handle constraints and to identify conserved quantities. The Lagrangian method allows an attack on many problems whose equations of motion would not otherwise be easy to find. Lagrange's equations (and the related Hamilton's equations) are of fundamental importance in classical mechanics and quantum mechanics.

3.1 Lagrange Equations

For a system consisting of N particles moving in three dimensions a total of $3N$ cartesian coordinates are required. The first particle's coordinates are labeled $\mathbf{r}_1$, the second $\mathbf{r}_2$, and so on up to $\mathbf{r}_N$. There is a Newton's equation for each of these coordinates. As a first step in the Lagrange approach we choose a new set of coordinates $q_1, q_2, \ldots q_{3N}$ called *general coordinates*, collectively denoted by $\{q_j\}$, to describe the configuration of the system. These coordinates do not necessarily have the dimensions of distance; in fact they are often angles. Newton's equations can be expressed in terms of the new coordinates by everywhere substituting for each cartesian coordinate its expression in terms of the new coordinates. These expressions relate the values of the new coordinates to the corresponding cartesian coordinate values which describe the same configuration of the system.

$$
\begin{aligned}
\mathbf{r}_1 &= \mathbf{r}_1(q_1, q_2, \ldots q_{3N}; \, t) \\
\mathbf{r}_2 &= \mathbf{r}_2(q_1, q_2, \ldots q_{3N}; \, t) \\
&\vdots \\
\mathbf{r}_N &= \mathbf{r}_N(q_1, q_2, \ldots q_{3N}; \, t)
\end{aligned}
\tag{3.1}
$$

Note that the expressions (or coordinate transformations) may be differ-

ent at different times. We use the common physics shorthand notation
that the same symbol **r** is used for the function **r**(q) and its value **r**. A
simple specific example of the transformation (3.1) might be the choice
of spherical polar coordinates (r, θ, ϕ) as general coordinates. For one
particle, (3.1) becomes

$$x = r \sin \theta \cos \phi$$
$$y = r \sin \theta \sin \phi \qquad (3.2)$$
$$z = r \cos \theta$$

If the particle moves on the surface of a sphere of radius ℓ centered at the
origin we may set $r = \ell$ in (3.2) and only two of the general coordinates,
θ and ϕ, will vary in time. A relation of this type is called a *constraint*.

The equations of motion which result directly from the substitutions
of (3.1) in Newton's equations are usually a mess. A much nicer set
of equations, both because they exhibit explicitly the simplifications of
symmetries and constraints, and because they are easier to write down,
are Lagrange's equations. They are not the same as Newton's but are
equivalent; in fact, each Lagrange equation is a linear combination of
Newton's equations, and *vice versa*.

3.2 Lagrange's Equations in One Dimension

We will first derive the Lagrange equation for one particle moving in one
dimension. With this as a guide we can then extend the derivation to a
system with an arbitrary number of degrees of freedom. The derivation
is purely mathematical and involves formal manipulations with partial
and total derivatives.

We introduce a general coordinate $q(t)$ expressed in terms of x by

$$q(t) = q[x(t), t] \qquad (3.3)$$

or inversely, as in (3.1),

$$x(t) = x[q(t), t] \qquad (3.4)$$

An explicit dependence on t in the transformation allows for the possibility that the q- and x-coordinates are related differently at different times.
The velocity $\dot{x} = dx/dt$ can be expressed in terms of the general velocity

$\dot{q}$ by chain differentiation

$$\dot{x} = \frac{\partial x}{\partial q}\dot{q} + \frac{\partial x}{\partial t} \tag{3.5}$$

where $\partial x/\partial q$ and $\partial x/\partial t$ are by (3.4) functions of q and t.

The momentum $p = m\dot{x}$ of the particle can be written in terms of the kinetic energy $K(\dot{x}) = \frac{1}{2}m\dot{x}^2$ as

$$p = \frac{dK}{d\dot{x}} \tag{3.6}$$

We introduce a new momentum $p(t)$ called the *general momentum* by a formula analogous to (3.6)

$$p(t) = \frac{\partial K}{\partial \dot{q}}(\dot{q}, q, t) \tag{3.7}$$

where $K(q, \dot{q}, t)$ means $\frac{1}{2}m\dot{x}^2$ with $\dot{x}$ expressed by (3.5). By chain differentiation the general momentum p is related to the ordinary momentum p

$$p = \frac{dK}{d\dot{x}}\frac{\partial \dot{x}}{\partial \dot{q}} = p\frac{\partial \dot{x}}{\partial \dot{q}} \tag{3.8}$$

By use of (3.4) and (3.5), the partial derivative $\partial \dot{x}/\partial \dot{q}$ (q held fixed) simplifies to

$$\frac{\partial \dot{x}}{\partial \dot{q}} = \frac{\partial x}{\partial q} \tag{3.9}$$

Therefore we have

$$p = p\frac{\partial x}{\partial q} \tag{3.10}$$

Newton's equation of motion is

$$\dot{p} = F(x, \dot{x}, t) \tag{3.11}$$

The corresponding Lagrange equation of motion has $\dot{p}$ instead of $\dot{p}$ on the left-hand side; it is derived by differentiating both sides of (3.10) with respect to t

$$\dot{p} = \dot{p}\frac{\partial x}{\partial q} + p\frac{d}{dt}\left(\frac{\partial x}{\partial q}\right) \tag{3.12}$$

To simplify the second term, we interchange the order of differentiation,

$$\frac{d}{dt}\left(\frac{\partial x}{\partial q}\right) = \frac{\partial \dot{x}}{\partial q} \tag{3.13}$$

We briefly digress to justify the result in (3.13). Proceeding just as in Eq. (3.5), the total time derivative of $\partial x / \partial q$ is

$$\frac{d}{dt}\left(\frac{\partial x}{\partial q}\right) = \frac{\partial}{\partial q}\left(\frac{\partial x}{\partial q}\right)\dot{q} + \frac{\partial}{\partial t}\left(\frac{\partial x}{\partial q}\right) \tag{3.14}$$

An important point must now be addressed. In the Lagrangian formalism we regard q and $\dot{q}$ as independent variables in the sense that $\partial \dot{q}/\partial q \equiv 0$. Other quantities such as the general momentum defined in (3.7) are *derived* since they ultimately depend on $q, \dot{q}$ and t. Differentiating (3.5) with respect to q (and treating $\dot{q}$ as an independent variable) we obtain

$$\frac{\partial \dot{x}}{\partial q} = \frac{\partial}{\partial q}\left(\frac{\partial x}{\partial q}\right)\dot{q} + \frac{\partial}{\partial q}\left(\frac{\partial x}{\partial t}\right) \tag{3.15}$$

Since the right-hand sides of (3.14) and (3.15) are identical, (3.13) follows.

Returning to the derivation, we multiply (3.13) by p and replace p on the right-hand side by the expression (3.6) to obtain

$$p\frac{d}{dt}\left(\frac{\partial x}{\partial q}\right) = \frac{dK}{d\dot{x}}\frac{\partial \dot{x}}{\partial q} = \frac{\partial K}{\partial q} \tag{3.16}$$

The substitution of (3.11) and (3.16) into (3.12) yields the following equation of motion in the q-coordinate system

$$\dot{p} = F\frac{\partial x}{\partial q} + \frac{\partial K}{\partial q} \tag{3.17}$$

The first term on the right-hand side is called the *general force*

$$Q(\dot{q}, q, t) = F\frac{\partial x}{\partial q} \tag{3.18}$$

Then the equation of motion

$$\dot{p} = Q + \frac{\partial K}{\partial q} \tag{3.19}$$

is of universal form for an arbitrary choice of coordinate q. The term $\partial K/\partial q$ in this equation represents a "fictitious" force which appears whenever the coefficients $\partial x/\partial q$ or $\partial x/\partial t$ in (3.5) vary with q.

If the force F is separated into a part $-dV(x)/dx$ which is derived from a potential energy, and a part F' which cannot be expressed (or which we do not choose to express) in terms of a potential energy, the general force can be separated into corresponding parts.

$$Q = -\frac{dV(x)}{dx}\frac{\partial x}{\partial q} + F'\frac{\partial x}{\partial q} = -\frac{\partial V(q)}{\partial q} + Q' \tag{3.20}$$

What we mean by $V(q)$ is the quantity at each point that is the same as $V(x)$, that is, $V[x(q)]$. For simplicity, we use the notation $V(q)$, although it is not the same *function* as $V(x)$. Notice that the potential (conservative) part of the general force has the same form, $-\partial V/\partial q$, as the conservative cartesian force $-\partial V/\partial x$.

If (3.20) for Q is substituted into (3.19) the terms $\partial K/\partial q$ and $-\partial V/\partial q$ can be combined, giving the *Lagrange equation of motion*

$$\dot{p} = \frac{\partial L}{\partial q} + Q' \tag{3.21}$$

where

$$L(q, \dot{q}, t) \equiv K(q, \dot{q}, t) - V(q) \tag{3.22}$$

is the *Lagrangian* function. Since

$$p = \frac{\partial K}{\partial \dot{q}} = \frac{\partial L}{\partial \dot{q}} \tag{3.23}$$

follows from (3.7) and $\partial V(q)/\partial \dot{q} = 0$, the Lagrange equation of motion (3.21) can also be written

$$\frac{d}{dt}\left(\frac{\partial L}{\partial \dot{q}}\right) - \frac{\partial L}{\partial q} = Q' \tag{3.24}$$

The general force Q' must include all forces F' on the particle which are not included in the potential energy.

3.3 Lagrange's Equations in Several Dimensions

The derivation of the Lagrange equation in § 3.2 was for the motion of one particle in one dimension. The generalization to the motion of N particles in three dimensions is made by repeating, step by step, the derivation in the one-dimenensional case.

There are now $3N$ cartesian components x_k and likewise $3N$ general coordinates q_j. In analogy to (3.6) the kth cartesian momentum is $p_k = \partial K / \partial \dot{x}_k$, and from (3.8) and (3.10) the general momentum can be written as

$$p_j = \frac{\partial K}{\partial \dot{q}_j} = \frac{\partial K}{\partial \dot{x}_k} \frac{\partial \dot{x}_k}{\partial \dot{q}_j} = p_k \frac{\partial x_k}{\partial q_j} \tag{3.25}$$

where the generalization of (3.9), the identity $\partial \dot{x}_k / \partial \dot{q}_j = \partial x_k / \partial q_j$, has been used. In (3.25) and subsequent equations a summation over repeated indices (in this case k) is implied. As in (3.12) the time derivative of the general momentum is

$$\dot{p}_j = \dot{p}_k \frac{\partial x_k}{\partial q_j} + p_k \frac{\partial \dot{x}_k}{\partial q_j} \tag{3.26}$$

In parallel to the derivation in one dimension we find

$$\begin{aligned} \dot{p}_j &= \left(-\frac{\partial V}{\partial x_k} + F_k' \right) \frac{\partial x_k}{\partial q_j} + \frac{\partial K}{\partial \dot{x}_k} \frac{\partial \dot{x}_k}{\partial q_j} \\ &= -\frac{\partial V}{\partial q_j} + Q_j' + \frac{\partial K}{\partial q_j} \\ &= \frac{\partial}{\partial q_j} (K - V) + Q_j' \\ &= \frac{\partial}{\partial q_j} L + Q_j' \end{aligned} \tag{3.27}$$

where L is the Lagrangian

$$L(\{q\}, \{\dot{q}\}; t) = K(\{q\}, \{\dot{q}\}; t) - V(\{q\}) \tag{3.28}$$

The general forces derived from a potential are

$$Q_j^{\text{pot}} = F_i \frac{\partial x_i}{\partial q_j} = -\frac{\partial V}{\partial q_j} \tag{3.29}$$

and

$$Q'_j = F'_i \frac{\partial x_i}{\partial q_j} \tag{3.30}$$

are the other general forces. It follows from (3.25) and $\partial V/\partial \dot{q}_j = 0$ that the general momentum can be written as

$$p_j = \frac{\partial L}{\partial \dot{q}_j} \tag{3.31}$$

and so (3.27) becomes

$$\frac{d}{dt}\left(\frac{\partial L}{\partial \dot{q}_j}\right) - \frac{\partial L}{\partial q_j} = Q'_j \tag{3.32}$$

As an elementary application of Lagrangian techniques, we determine the r and θ equations of motion for a particle moving in a plane under the influence of a central potential energy $V(r)$. As general coordinates we take

$$q_1 = r, \qquad q_2 = \theta \tag{3.33}$$

in terms of which the cartesian coordinates are

$$x = r\cos\theta$$
$$y = r\sin\theta \tag{3.34}$$

The kinetic energy

$$K = \tfrac{1}{2}m(\dot{x}^2 + \dot{y}^2) \tag{3.35}$$

is easily expressed in polar coordinates by taking the time derivative of (3.34) to get the cartesian velocities

$$\dot{x} = \dot{r}\cos\theta - r\dot{\theta}\sin\theta$$
$$\dot{y} = \dot{r}\sin\theta + r\dot{\theta}\cos\theta \tag{3.36}$$

and therefore

$$K = \tfrac{1}{2}m(\dot{r}^2 + r^2\dot{\theta}^2) \tag{3.37}$$

This result for K also follows from (2.126) with $K = \tfrac{1}{2}m(v_r^2 + v_\theta^2)$. We note that K is a function of q_1, $\dot{q}_1$, and $\dot{q}_2$ but not of q_2. The Lagrangian is

$$L = K - V = \tfrac{1}{2}m(\dot{r}^2 + r^2\dot{\theta}^2) - V(r) \qquad (3.38)$$

In this case there are no constraint forces or non-conservative forces so that $Q'_r = Q'_\theta = 0$. Using this in the Lagrange equations of motion (3.32), with $Q'_j = 0$

$$\frac{d}{dt}\left(\frac{\partial L}{\partial \dot{r}}\right) = \frac{\partial L}{\partial r}$$

$$\frac{d}{dt}\left(\frac{\partial L}{\partial \dot{\theta}}\right) = \frac{\partial L}{\partial \theta} \qquad (3.39)$$

we find

$$m\ddot{r} - mr\dot{\theta}^2 = -\frac{\partial V}{\partial r}$$

$$\frac{d}{dt}(mr^2\dot{\theta}) = r(mr\ddot{\theta} + 2m\dot{r}\dot{\theta}) = 0 \qquad (3.40)$$

These correspond to (2.130), obtained from direct application of Newton's Laws, with $F_r = -\partial V/\partial r$ and $F_\theta = 0$.

Since L does not depend on θ, $\dot{p}_\theta = 0$ from (3.39); hence the general momentum p_θ is constant,

$$p_\theta = \frac{\partial L}{\partial \dot{\theta}} = mr^2\dot{\theta} = \text{constant} \qquad (3.41)$$

This conserved quantity is the angular momentum L_z.

This conservation law is an example of a general principle that can be deduced from the Lagrange equation (3.32): if a general coordinate q_j does not appear in the Lagrangian and $Q'_j = 0$, the corresponding general momentum $p_j = \partial L/\partial \dot{q}_j$ is constant in time — it is a constant of the motion (a conserved quantity).

3.4 Constraints

As a simple example of a constrained system we return to the simple pendulum. A mass m moves in a vertical plane as illustrated in Fig. 3-1, subject to gravitational force and to the tension force of an attached string of length ℓ which constrains the mass to always be at a distance ℓ from the other end of the string. To begin with, we shall suppose that this other end of the string is held at a fixed position; later it will be allowed to move arbitrarily. The essence of the problem is that we are to find the motion of the mass according to Newton's equations, but we are not given all the forces; instead, we are given partial information about the motion, namely the constraint(s). The unknown force, in this case the tension, is called the constraint force, and is whatever it has to be for the given motion to obey the constraint(s). Note that the number of the unknown components of constraint force must be the same as the number of constraints on the motion, otherwise the motion will be over- or under-determined.

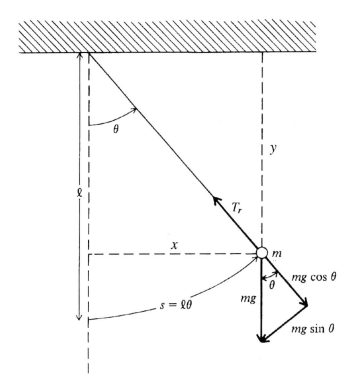

FIGURE 3-1.   Simple pendulum

How can we systematically treat a mechanical system with constraints? The first step is to find combinations of the Newton equations in which the constraint forces are absent. As we shall more-or-less demonstrate below, these are precisely Lagrange equations, for an appropriate choice of coordinates. One has enough information to solve these equations (while assuming the constraints to hold), and the solutions to these determine the motion. The remaining equations, into which one substitutes the solution to the motion, then determine the constraint forces; if only the solution for the motion were desired, this step would be unnecessary. The above procedure was carried out already for the pendulum in § 2.7. By expressing Newton's laws in polar coordinates we found (2.133) that the θ equation did not contain the string tension. We solved this equation for $\dot{\theta}$ as a function of θ and then substituted back into the radial equation (2.132) to find the tension, the constraint force.

We now proceed with the simple pendulum using the Lagrangian method. As in Fig. 3-1, let x (horizontal) and y (vertically downward) be cartesian coordinates of the mass and let the origin be at the other (fixed) end of the string, so that the constraint is $\ell \equiv r = \sqrt{x^2 + y^2}$. We first do the calculation in an awkward but informative way. Consider the following Lagrangian in polar coordinates

$$L' = \frac{1}{2}m\left(\dot{r}^2 + r^2\dot{\theta}^2\right) + mgr\cos\theta - V^{\text{constraint}}(r) \qquad (3.42)$$

where $V(r,\theta) = -mgr\cos\theta$ is the gravitational potential energy $-mgy$ and $V^{\text{constraint}}(r)$ is a potential energy that will enforce the constraint $r = \ell$ by having a deep and narrow minimum at $r = \ell$. It is understood here that only motions with low energies are being considered so that vibrations about the constraint (here $r = \ell$) are of negligible amplitude. In the real world a little friction rapidly damps these high frequency motions.

The radial Lagrange equation for the pendulum is

$$\frac{d}{dt}\left(\frac{\partial L'}{\partial \dot{r}}\right) - \frac{\partial L'}{\partial r} = 0 \qquad (3.43)$$

or

$$m\ddot{r} - mr\dot{\theta}^2 - mg\cos\theta + \frac{dV^{\text{constraint}}}{dr} = 0 \qquad (3.44)$$

Due to the deep and narrow minimum of $V^{\text{constraint}}(r)$ only $r = \ell$ is allowed. Thus the constraint force necessary to keep $r = \ell$ is

$$Q_r^{\text{constraint}} \equiv -\frac{dV^{\text{constraint}}}{dr} = -m\ell\dot{\theta}^2 - mg\cos\theta \qquad (3.45)$$

This constraint force is directed inward if the string is taut and is the negative of the string tension T_r. One sees from (3.43) and (3.44) that the condition for $r = $ constant is $\frac{\partial L'}{\partial r} = 0$, i.e. $\frac{\partial L}{\partial r} - \frac{\partial V^{\text{constraint}}}{\partial r} = 0$ or

$$Q_r^{\text{constraint}} \equiv -\left.\frac{\partial L}{\partial r}\right|_{r=\ell} \tag{3.46}$$

The angular Lagrange equation for the pendulum is

$$\frac{d}{dt}\left(\frac{\partial L}{\partial \dot\theta}\right) - \frac{\partial L}{\partial \theta} = 0$$
$$\frac{d}{dt}\left(r^2\dot\theta\right) + gr\sin\theta = 0 \tag{3.47}$$

Imposing the constraint $r = \ell$ here leads to the usual pendulum equation

$$\ddot\theta + \frac{g}{\ell}\sin\theta = 0 \tag{3.48}$$

The result of the above excercise is that:

1. We can impose constraints directly in the Lagrangian and determine the correct equations of motion without ever explicitly referring to the constraint forces.

2. If we wish to find the force required to enforce a constraint, we choose an additional general coordinate (in this case r) so that when it is held to be a particular constant ($r = \ell$ here) the constraint is maintained. The constraint force then follows as in (3.46).

We can now describe the general case: let the system, with $3N$ degrees of freedom, have C constraints, that is, the motion $x_k(t)$, $k = 1, 2, \ldots, 3N$ is to satisfy

$$f_j(\{x_k(t)\}, t) = 0, \qquad j = 1, 2, \ldots, C \tag{3.49}$$

Choose general coordinates so that C of them are

$$q_j = f_j(\{x_k(t)\}, t) \qquad j = 1, 2, \ldots, C \tag{3.50}$$

so that the constraint conditions read

$$q_j = 0 \qquad j = 1, 2, \ldots, C \tag{3.51}$$

In our pendulum case this would mean $q_1 = r - \ell$. The constraint forces can be imagined to be from a potential $V^{\text{constraint}}(q_1, q_2, \ldots, q_C; t)$ which has a deep and narrow minimum at $q_1 = q_2 = \cdots = q_C = 0$; the Lagrangian is

$$L' = K - V - V^{\text{constraint}} = L - V^{\text{constraint}} \tag{3.52}$$

with $L = K - V$. Then the combinations of Newton's equations in which no constraint forces appear, and which therefore determine the motion, are the Lagrange equations for the "non-constraint" coordinates,

$$\frac{d}{dt}\left(\frac{\partial L}{\partial \dot{q}_j}\right) = \frac{\partial L}{\partial q_j} \qquad j = C+1,\ C+2,\ldots,3N \tag{3.53}$$

where $L = K - V$ and the "constraint" coordinates q_j are taken to obey (3.51). Consequently to determine the motion K and V need only be known for constrained configurations of the system. As in the pendulum example, the constraint forces are given by $0 = \frac{d}{dt}\left(\frac{\partial L'}{\partial \dot{q}_j}\right) = \frac{\partial L'}{\partial q_j} = \frac{\partial L}{\partial q_j} - \frac{\partial V^{\text{constraint}}}{\partial q_j}$, i.e.

$$Q_j^{\text{constraint}} = -\frac{\partial L}{\partial q_j} \qquad j = 1, 2, \ldots, C \tag{3.54}$$

As in (3.53) $L = K - V$ and the q_j are all zero. (Note that $Q_j^{\text{constraint}}$ is a general force, so for example it will be a torque if q_j is an angle.)

The type of constraint $f_j(\{x_k\}, t) = 0$ is called a *holonomic* constraint. An important holonomic constraint is the rigid body constraint, in which the distances between every point in the body remain constant. The rigid body constraint can be expressed as

$$|\boldsymbol{r}_i - \boldsymbol{r}_j| = d_{ij} \tag{3.55}$$

where d_{ij} is the constant distance between particles i and j. As we will discuss in Chapter 6, the result of the rigid body constraints is that the configuration of a rigid body is described by six general coordinates — three angles and the three coordinates of the center of mass.

For completeness we should mention that some mechanical systems have constraints which *cannot* be expressed as relations among the coordinates; these are called *non-holonomic* constraints. An important class of these are expressed as constraints on the velocities

$$a_{ij}\,\dot{q}_j + b_i = 0 \qquad (3.56)$$

(the a_{ij} and b_i may depend on the q_i's) where the equations cannot be integrated. If such a set of relations could be integrated, the result would be relations between coordinates and the constraints would actually be holonomic. Consider the example of a ball rolling without slipping on a surface. There are two types of constraints here. One is that the ball touches the surface; this is holonomic and can be expressed by saying that the distance of the center of the ball above the surface is always equal to the radius of the ball. The other is the "rolling without slipping" constraint which can be expressed by saying that the ball at the point of contact must be at rest relative to the plane. This constraint is non-holonomic because it cannot be integrated to a relation among the coordinates. This is evident from the fact that the ball can be rolled to any position and orientation (note that the ball can rotate *around* the point of contact).

An example of another class of non-holonomic constraints is given by a pendulum bob on a flexible string; the distance r between the bob and the other end of the string cannot exceed the length ℓ of the string. For some initial conditions the string may not remain taut and the bob will then fall inward (*i.e.*, $r < \ell$). When this happens the string no longer exerts any force on the bob and the bob moves as a projectile until the string becomes taut once more. The precise transition from constrained to unconstrained motion and back to constrained motion requires the solution of the equations of motion at each step and cannot be cast in the usual holonomic form.

3.5 Pendulum With Oscillating Support

You may have wondered if the Lagrangian method is actually advantageous, since the examples we have solved are just as easy to do by Newton's second law. To illustrate the merits of the Lagrangian approach, we shall treat the motion of a pendulum with an oscillating support. This example also provides a simple demonstration of the forced harmonic oscillator.

The point of suspension (x_s, y_s) of a simple pendulum is moved as a specified function of time, as shown in Fig. 3-2. We take as coordinates x, y the relative coordinates of the bob to the point of suspension; thus

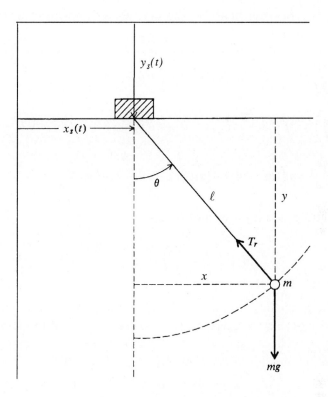

FIGURE 3-2. Simple pendulum with a moving support. The direction of positive y_s and y is downwards.

the bob has position $(x+x_s, y+y_s)$ with respect to a fixed system, where $x_s(t)$ and $y_s(t)$ are given functions of time representing the horizontal and vertically downward coordinates of the support and

$$x = \ell \sin \theta$$
$$y = \ell \cos \theta$$
(3.57)

The bob's kinetic energy is

$$K = \tfrac{1}{2}m\left[(\dot{x} + \dot{x}_s)^2 + (\dot{y} + \dot{y}_s)^2\right] = \tfrac{1}{2}m(\dot{x}^2 + \dot{y}^2 + 2\dot{x}\dot{x}_s + 2\dot{y}\dot{y}_s + \dot{x}_s^2 + \dot{y}_s^2)$$
(3.58)

and its potential energy is

$$V = -mg\ell \cos \theta - mgy_s$$
(3.59)

Thus the Lagrangian $L = K - V$ is

$$L = \tfrac{1}{2}m(\ell^2 \dot{\theta}^2 + 2\ell\dot{\theta}\dot{x}_s \cos \theta + 2\ell\dot{\theta}\dot{x}_s \sin \theta + \dot{x}_s^2 + \dot{y}_s^2) + mg\ell \cos \theta + mgy_s \quad (3.60)$$

For the general coordinate θ, the derivatives appearing in the Lagrange equation are

$$\frac{\partial L}{\partial \dot\theta} = m\ell^2 \dot\theta + m\ell \dot x_s \cos\theta + m\ell \dot y_s \sin\theta$$

$$\frac{d}{dt}\left(\frac{\partial L}{\partial \dot\theta}\right) = m\ell^2 \ddot\theta + m\ell \ddot x_s \cos\theta + m\ell \ddot y_s \sin\theta + m\ell \dot\theta \dot y_s \cos\theta - m\ell \dot\theta \dot x_s \sin\theta$$

$$\frac{\partial L}{\partial \theta} = m\ell \dot\theta \dot y_s \cos\theta - m\ell \dot\theta \dot x_s \sin\theta - mg\ell \sin\theta$$

$$(3.61)$$

The resulting equation of motion is

$$\ddot\theta + \left(\frac{g}{\ell} + \frac{\ddot y_s}{\ell}\right)\sin\theta = -\frac{\ddot x_s}{\ell}\cos\theta \tag{3.62}$$

(Note the cancellation of $\dot\theta \dot x_s$ and $\dot\theta \dot y_s$ terms.) For small angular displacements ($\theta \ll 1$) and a horizontal sinusoidal motion of the support,

$$x_s = x_0 \cos\omega t \qquad\qquad y_s = 0$$
$$\ddot x_s = -\omega^2 x_0 \cos\omega t \tag{3.63}$$

the equation of motion (3.62) becomes

$$\ddot\theta + \omega_0^2 \theta = \frac{x_0}{\ell}\omega^2 \cos\omega t \tag{3.64}$$

where $\omega_0 = \sqrt{g/\ell}$ is the natural frequency. This equation is mathematically identical with that of a forced harmonic oscillator [see (1.115)]. Because of this similarity a pendulum with a horizontally oscillating support can be used to demonstrate the properties of a driven harmonic oscillator.

We stress that the advantage of using Lagrangian methods is the methodical and straightforward procedure. Once a Lagrangian function is constructed from the kinetic and potential energies, the task of obtaining the equations of motion is simply a matter of differentiation. In complex problems there is less chance of error using this method.

3.6   Hamilton's Principle and Lagrange's Equations

An elegant method known as *Hamilton's Principle* provides an instructive demonstration that Lagrange's equations are equivalent to Newton's equations. Given the Lagrangian $L(\{q\}, \{\dot{q}\}, t)$, one defines for a given motion $q_j(t)$ of the system between the times t_1 and t_2 the quantity S called the *action* of the motion

$$S = \int_{t_1}^{t_2} L(\{q\}, \{\dot{q}\}, t)\, dt \qquad (3.65)$$

We will be interested in how S changes when the motion is changed to another motion. More specifically, we will be interested only in a slightly restricted class of motions, namely those which have a specified initial point, $q_j(t_1) = q_j^{\text{initial}}$, and likewise a specified final point, $q_j(t_2) = q_j^{\text{final}}$, as illustrated in Fig. 3-3. As the motion is varied, subject to the fixed-end-point conditions just stated, the value of S varies (in general). We can now state Hamilton's principle as the following theorem: If any small variation (satisfying the fixed-end-point conditions) of a motion produces no variation of S (to first order) then the motion satisfies Lagrange's equations and vice versa.

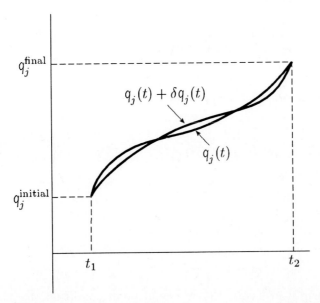

FIGURE 3-3. Two nearby trajectories having the same initial and final values, for the case of one coordinate.

In other words, the action S is *stationary* at just those motions which satisfy Lagrange's equations. [Often, a stationary value of S is a minimum value, that is, all variations of the motion raise the value of S. For this reason, Hamilton's principle is often (misleadingly) called the Least Action Principle.]

The demonstration is as follows. We consider a particular motion $q_j(t)$ and a slightly different motion $\tilde{q}_j(t) = q_j(t) + \delta q_j(t)$, where the difference of motions, $\delta q_j(t)$, is small. To satisfy the fixed-end-point conditions, δq_j must vanish at t_1 and t_2. The variation of the action when the motion is varied from $q_j(t)$ to $\tilde{q}_j(t)$ is

$$
\delta S = \int_{t_1}^{t_2} dt L\left(\{\tilde{q}\}, \{\dot{\tilde{q}}\}, t\right) - \int_{t_1}^{t_2} dt L\left(\{q\}, \{\dot{q}\}, t\right) = \int_{t_1}^{t_2} dt\, \delta L
$$

$$
= \int_{t_1}^{t_2} dt \left(\frac{\partial L}{\partial q_j}\delta q_j + \frac{\partial L}{\partial \dot{q}_j}\delta \dot{q}_j \right)
$$

(3.66)

where sums over j are implied. The final step used chain differentiation, i.e. $\delta f(x, y, \ldots) = \frac{\partial f}{\partial x}\delta x + \frac{\partial f}{\partial y}\delta y + \cdots$. Clearly, for δS to vanish it is sufficient that $\partial L/\partial q_j = 0$ and $\partial L/\partial \dot{q}_j = 0$; however, this is not a necessary condition because the functions δq_j and $\delta \dot{q}_j$ are not independent. This is dealt with as follows. Note that

$$
\delta \dot{q}_j = \frac{\delta q_j(t + dt) - \delta q_j(t)}{dt} = \frac{d}{dt}(\delta q_j)
$$

(3.67)

Thus the terms in $\delta \dot{q}_j$ can be integrated by parts

$$
\int dt \frac{\partial L}{\partial \dot{q}_j}\delta \dot{q}_j = \int dt \frac{\partial L}{\partial \dot{q}_j}\frac{d}{dt}\delta q_j = \frac{\partial L}{\partial \dot{q}_j}\delta q_j - \int dt \frac{d}{dt}\left(\frac{\partial L}{\partial \dot{q}_j}\right)\delta q_j
$$

(3.68)

and (3.66) becomes

$$
\delta S = \int_{t_1}^{t_2} dt \left[\frac{\partial L}{\partial q_j} - \frac{d}{dt}\left(\frac{\partial L}{\partial \dot{q}_j}\right) \right] \delta q_j + \left. \frac{\partial L}{\partial \dot{q}_j}\delta q_j \right|_{t_1}^{t_2}
$$

(3.69)

The last term vanishes by the fixed-end-point condition, $\delta q_j = 0$ at $t = t_2$ and $t = t_1$. Thus δS vanishes for arbitrary small $\delta q_j(t)$ if and only if the square-bracketed expression vanishes for all t between t_1 and t_2 and for all values of the coordinate index j. This vanishing is Lagrange's equations (3.32),

$$\frac{d}{dt}\left(\frac{\partial L}{\partial \dot{q}_j}\right) - \frac{\partial L}{\partial q_j} = 0 \tag{3.70}$$

and so Hamilton's Principle has been shown. We remark that Hamilton's Principle is an example of a *variational principle*, and our treatment of it is an example of the *calculus of variations*.

It is now easy to see that Lagrange's equations have the same form in any coordinate system. According to Hamilton's Principle, the statement that a motion $q(t)$ satisfies Lagrange's equations (3.32) is equivalent to the statement that the action of the motion is stationary. The latter statement is independent of the choice of coordinates, and so the former statement must be as well. To be more explicit, if we change coordinates from q to $\bar{q}$ [*cf.* Eq. (3.1)] then the reexpression of the action in terms of the new description $\bar{q}(t)$ of the orbit goes as follows:

$$
\begin{aligned}
S &= \int_{t_1}^{t_2} dt\, L\big(\{q(t)\}, \{\dot{q}(t)\}, t\big) \\
&= \int_{t_1}^{t_2} dt\, L\big(\{q(\bar{q}(t))\}, \{\dot{q}(\bar{q}(t), \dot{\bar{q}}(t), t)\}, t\big) \qquad (3.71) \\
&= \int_{t_1}^{t_2} dt\, \bar{L}\big(\{\bar{q}(t)\}, \{\dot{\bar{q}}(t)\}, t\big)
\end{aligned}
$$

where the last step defines $\bar{L}$ to be the function of the $\bar{q}_j$, $\dot{\bar{q}}_j$ and t which results from substituting into $L(\{q\}, \{\dot{q}\}, t)$ the expressions of the old coordinates and velocities q, $\dot{q}$ in terms of the new,

$$
\begin{aligned}
q_j &= q_j\left(\{\bar{q}\}, t\right) \\
\dot{q}_j &= \frac{\partial q_j}{\partial \bar{q}_i}\dot{\bar{q}}_i + \frac{\partial q_j}{\partial t}
\end{aligned} \tag{3.72}
$$

Thus Hamilton's Principle tells us that if the motion $q(t)$ satisfies Lagrange's equations with Lagrangian $L(\{q\}, \{\dot{q}\}, t)$ then its description in terms of the new coordinates, $\bar{q}(t)$, satisfies Lagrange's equations with Lagrangian $\bar{L}(\{\bar{q}\}, \{\dot{\bar{q}}\}, t)$.

Now that we have shown that Lagrange's equations have the same form in any coordinates, it follows that they are equivalent to Newton's equations if $L = K - V$ because if the $\{q\}$ and $\{\dot{q}\}$ are chosen to be cartesian coordinates, then $K = \frac{1}{2}\sum_k m_k \dot{x}_k^2$ and Lagrange's equations (3.70) become

$$
\dot{p}_k = \frac{\partial(K - V)}{\partial x_k} = -\frac{\partial V}{\partial x_k} = F_k
$$
$$
p_k = \frac{\partial L}{\partial \dot{x}_k} = \frac{\partial K}{\partial \dot{x}_k} = m\dot{x}_k
$$

(3.73)

A final remark is that if the motion is to satisfy holonomic constraints, the equations which determine the motion of the system result from using in Hamilton's Principle only motions which satisfy the constraints.

3.7 Hamilton's Equations

The Lagrange equations of motion, which are equivalent to Newton's equations, are a set of second-order differential equations. An alternative formulation of Newton's law consists of twice as many first-order differential equations known as *Hamilton's equations*.

In Lagrangian mechanics the independent variables are q_j, $\dot{q}_j$, and t, and the general momentum p_j is given in terms of these by (3.31).

$$
p_j = \frac{\partial K}{\partial \dot{q}_j} = \frac{\partial L}{\partial \dot{q}_j}
$$

(3.74)

In Hamiltonian mechanics, q_j, p_j, and t are chosen as independent variables and $\dot{q}_j$ is a dependent quantity.

$$
\dot{q}_j = \dot{q}_j(q_1, q_2, \ldots, q_n ; p_1, p_2, \ldots, p_n ; t)
$$

(3.75)

The Hamiltonian function H is defined as

$$
H(q_1, q_2, \ldots, q_n ; p_1, p_2, \ldots, p_n ; t) \equiv p_j \dot{q}_j - L
$$

(3.76)

where a summation over the repeated index j on the right-hand side is implied. In this definition the variables $\dot{q}_j$ are understood to be functions of the general coordinates, momenta, and time as in (3.75). From (3.76) the total differential of H is

$$dH = \dot{q}_j\, dp_j + p_j\, d\dot{q}_j - \frac{\partial L}{\partial q_j}\, dq_j - \frac{\partial L}{\partial \dot{q}_j}\, d\dot{q}_j - \frac{\partial L}{\partial t}\, dt$$

$$= \dot{q}_j\, dp_j - \frac{\partial L}{\partial q_j}\, dq_j - \frac{\partial L}{\partial t}\, dt \qquad (3.77)$$

where in the second line we have used $p_j = \partial L/\partial \dot{q}_j$. Since the independent differentials are now dp_j, dq_j and dt, we see that the replacement of the $\dot{q}_j$ by the p_j as the fundamental variables is achieved by the definition $H = p_j\dot{q}_j - L$. This cancellation of the coefficient of $d\dot{q}$ is an example of a *Legendre transformation*. Such variable changes are encountered frequently in the study of thermodynamics.

To derive Hamilton's equations of motion we compare (3.77) with the total differential

$$dH = \frac{\partial H}{\partial p_j}\, dp_j + \frac{\partial H}{\partial q_j}\, dq_j + \frac{\partial H}{\partial t}\, dt \qquad (3.78)$$

and use (3.32) to find

$$\frac{\partial H}{\partial p_j} = \dot{q}_j$$

$$\frac{\partial H}{\partial q_j} = -\frac{\partial L}{\partial q_j} = -\dot{p}_j + Q'_j \qquad (3.79)$$

$$\frac{\partial H}{\partial t} = -\frac{\partial L}{\partial t}$$

To establish the physical significance of the Hamiltonian we relate the quantities on the right-hand side of (3.76) to the kinetic and potential energies of the system. In cartesian coordinates the kinetic energy is

$$K = \sum_k \tfrac{1}{2} m_k \dot{x}_k^2 = \tfrac{1}{2} p_k \dot{x}_k \qquad (3.80)$$

By use of the chain rule expression for $\dot{x}_k$ the kinetic energy can be rewritten in the form

$$K = \frac{1}{2} p_k \left(\frac{\partial x_k}{\partial q_j} \dot{q}_j + \frac{\partial x_k}{\partial t} \right) = \frac{1}{2} p_j \dot{q}_j + \frac{1}{2} p_k \frac{\partial x_k}{\partial t} \tag{3.81}$$

Solving for $p_j \dot{q}_j$ from (3.81),

$$p_j \dot{q}_j = 2K - p_k \frac{\partial x_k}{\partial t} \tag{3.82}$$

and substituting the result into (3.76), we obtain

$$\begin{aligned} H &= 2K - p_k \frac{\partial x_k}{\partial t} - L \\ &= K + V - p_k \frac{\partial x_k}{\partial t} \end{aligned} \tag{3.83}$$

Hence if the transformation between the cartesian and general coordinates in (3.1) has no explicit time dependence,

$$\frac{\partial x_k}{\partial t} = 0 \tag{3.84}$$

the Hamiltonian is the total energy of the system,

$$H = K + V \tag{3.85}$$

Of course this equation holds only in cases where the potential energy exists.

To find the conditions under which the Hamiltonian is a conserved quantity, we compute the total time derivative.

$$\frac{dH}{dt} = \frac{\partial H}{\partial q_j} \dot{q}_j + \frac{\partial H}{\partial p_j} \dot{p}_j + \frac{\partial H}{\partial t} \tag{3.86}$$

Upon use of (3.79), this reduces to

$$\frac{dH}{dt} = Q'_j \dot{q}_j + \frac{\partial H}{\partial t} \tag{3.87}$$

Thus, if the forces are derivable from a potential energy ($Q'_j = 0$) and H has no explicit time dependence ($\partial H / \partial t = 0$), the Hamiltonian is a constant of the motion. This constant is the total energy of the system if (3.84) holds.

As an elementary example of the Hamiltonian method, we consider the one-dimensional harmonic oscillator, for which

$$K = \tfrac{1}{2}m\dot{x}^2$$
$$V = \tfrac{1}{2}kx^2 \tag{3.88}$$
$$L = K - V = \tfrac{1}{2}m\dot{x}^2 - \tfrac{1}{2}kx^2$$

In this case $q = x$ and $Q' = 0$. The momentum is found by differentiation according to (3.74),

$$p = \frac{\partial K}{\partial \dot{x}} = m\dot{x} \tag{3.89}$$

The Hamiltonian from (3.76) is

$$H = p\dot{x} - \left(\tfrac{1}{2}m\dot{x}^2 - \tfrac{1}{2}kx^2\right)$$
$$= \frac{p^2}{2m} + \tfrac{1}{2}kx^2 \tag{3.90}$$

where (3.89) has been used to eliminate $\dot{x}$ in favor of p. This Hamiltonian is immediately recognizable as the total energy of the oscillator.

Hamilton's equations of motion from (3.79),

$$\frac{\partial H}{\partial p} = \dot{x} \qquad \frac{\partial H}{\partial x} = -\dot{p} \tag{3.91}$$

yield

$$\frac{p}{m} = \dot{x} \qquad kx = -\dot{p} \tag{3.92}$$

When p is eliminated between these two first-order equations, we obtain the usual second-order differential form of Newton's second law:

$$m\frac{d^2 x}{dt^2} = -kx \tag{3.93}$$

Because of the similar role that coordinate and momentum play in Hamilton's equations they provide the jumping off point for the formulation of abstract mechanics, celestial mechanics, and quantum mechanics. In the latter the generalization to subatomic mechanics begins with the classical Hamiltonian. The coordinates and momenta are now operators; for example, in coordinate space $x_{\text{op}} = x$ and $p_{x\,\text{op}} = -i\hbar\frac{\partial}{\partial x}$, where $\hbar$ is the reduced Planck's constant $\hbar = h/2\pi$. In quantum mechanics all information on a physical system resides in the wavefunction $\psi(\mathbf{r}, t)$ which satisfies the Schrödinger equation

$$H\psi = i\hbar \frac{\partial \psi}{\partial t} \tag{3.94}$$

Since $\mathbf{r}$ and $\mathbf{p}$ have operator forms this amounts to a partial differential equation for the wavefunction. The introduction of quantum mechanics, when combined with special relativity, has extended the range of experimental validity of Hamiltonian mechanics down to at least 10^{-18} m.

PROBLEMS

3-1. Two equal masses $m_1 = m_2 = m$ with coordinates x_1 and x_2 in one dimension are connected by a spring of spring constant k. Use Lagrangian methods to find the equations of motion. What is the angular frequency of simple harmonic motion for the relative displacement $x_1 - x_2$ of the two masses?

3-2. Two equal masses are constrained by the spring-and-pulley system shown in the accompanying sketch. Assume a massless pulley and a frictionless surface. Let x be the extension of the spring from its relaxed length. Derive the equations of motion by Lagrangian methods. Solve for x as a function of time with the boundary conditions $x = 0$, $\dot{x} = 0$ at $t = 0$.

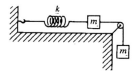

3-3. Use Lagrangian methods to find the equations of motion for Problem 2-21.

3-4. Use Lagrangian methods to find the equations of motion for Problem 2-22.

3-5. Two masses m_1 and m_2 are connected by a spring of rest length ℓ and spring constant k. The system slides without friction on a horizontal surface in the direction of the spring's length.

a) Set up the Lagrangian for the motion.

b) Find the normal modes of this system and the corresponding frequencies.

c) Give general solutions to the equations of motion. Note that an equation of motion with a zero angular frequency is not simple harmonic.

d) For the initial conditions $x_1(0) = 0$, $\dot{x}_1(0) = v_0$, $x_2(0) = 0$, $\dot{x}_2(0) = 0$, find the subsequent motion.

e) Using the solution from part d), evaluate the center of mass coordinate $x_{CM} = [m_1 x_1 + m_2(x_2 + \ell)]/(m_1 + m_2)$ and the relative coordinate $x_2 - x_1$ as a function of time.

3-6. A bead of mass m is constrained to move without friction on a hoop of radius R. The hoop rotates with constant angular velocity ω around a vertical diameter of the hoop. Use a polar angle θ and an azimuthal angle ϕ to describe the position of the bead on the hoop, with $\omega = \dot{\phi}$. Take $\theta = 0$ at the bottom of the hoop.

a) Set up the Lagrangian and obtain the equation of motion on the bead.

b) Find the critical angular velocity $\omega = \Omega$ below which the bottom of the hoop provides a stable equilibrium position for the bead.

c) Find the stable equilibrium position for $\omega > \Omega$.

3-7. A double pendulum consists of two weightless rods connected to each other and a point of support, as illustrated. The masses m_1 and m_2 are not equal, but the lengths of the rods are. The pendulums are free to swing only in one vertical plane.

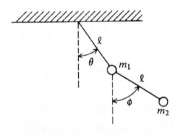

a) Set up the Lagrangian of the system for arbitrary displacements and derive the equations of motion from it.

b) Find the normal-mode frequencies of the system when both angles of oscillation are small.

c) Show that the frequencies become approximately equal if $m_1 \gg m_2$; interpret this. For $m_2 \gg m_1$ interpret the normal-mode frequencies and describe the motion of each mass.

3-8. A model of a ring molecule consists of three equal masses m which slide without friction on a fixed circular wire of radius R. The masses are connected by identical springs of spring constant $m\omega_0^2$. The angular positions of the three masses, θ_1, θ_2 and θ_3, are measured from a rest position.

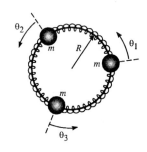

a) Write down the Lagrangian and show that the equations of motion are

$$\ddot{\theta}_1 + \omega_0^2(2\theta_1 - \theta_2 - \theta_3) = 0$$
$$\ddot{\theta}_2 + \omega_0^2(2\theta_2 - \theta_1 - \theta_3) = 0$$
$$\ddot{\theta}_3 + \omega_0^2(2\theta_3 - \theta_1 - \theta_2) = 0$$

b) Show that the mode in which $\theta_1 = \theta_2 = \theta_3$ corresponds to constant total angular momentum $L = \sum_{i=1}^{3} p_{\theta_i}$.

c) Assume the total angular momentum is zero and that $\theta_1 + \theta_2 + \theta_3 = 0$. Find two degenerate oscillatory modes and their frequency.

3-9. A triangular molecule has three identical atoms with rest separations a as shown. The molecule is represented as a mechanical system of masses and springs with the springs representing the chemical bonds. For small motions in the x, y plane about the equilibrium configuration, the kinetic and potential energies are

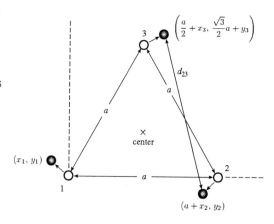

$$K = \frac{1}{2}m\sum_{i=1}^{3}\left(\dot{x}_i^2 + \dot{y}_i^2\right)$$

$$V = \frac{1}{2}k\left[(x_2 - x_1)^2 + \frac{1}{4}\left(\sqrt{3}y_3 - \sqrt{3}y_1 + x_3 - x_1\right)^2 \right.$$
$$\left. + \frac{1}{4}\left(\sqrt{3}y_3 - \sqrt{3}y_2 + x_2 - x_3\right)^2\right]$$

Use Lagrange's equations to find the equations of motion.

3.4 Constraints

3-10. A bead slides without friction on a parabolic wire of shape $y = ax^2$ with the force of gravity in the negative y direction. Write down the Lagrangian in terms of x and y coordinates. Then use the constraint equation to express the Lagrangian solely in terms of x. Find the equation of motion and then simplify it for the case of small oscillations.

3.6 Hamilton's Principle and Lagrange's Equations

3-11. A particle of mass m falls vertically in a constant gravity field g. Assume that the position as a function of time is

$$y(t) = c_0 + c_1 t + c_2 t^2 + c_3 t^3$$

where y increases upward and the $\{c_k\}$ are constant coefficients to be determined.

a) Evaluate the action S between $t = 0$, where $y(0) = 0$, and $t = T$ where $y(T) = \ell$. As a function of c_2, c_3, T and ℓ show that

$$\frac{S}{mT^3} = \frac{\ell}{2T^4}\left(\ell - gT^2\right) + \left(\frac{g}{6}\right) c_2 + \left(\frac{1}{6}\right) c_2^2$$
$$+ \left(\frac{T}{2}\right) c_2 c_3 + \left(\frac{gT}{4}\right) c_3 + \left(\frac{2T^2}{5}\right) c_3^2$$

b) For fixed T and ℓ show that the action is an extremum for $c_2 = -g/2$ and $c_3 = 0$.

c) What kind of extremum does S have at this point?

3-12. Using the methods of the calculus of variations show that the curve of shortest length connecting the two points (x_1, y_1) and (x_2, y_2) in the x, y plane is a straight line.

Hint: the length is $s = \int_{x_1}^{x_2} L\left(y, \frac{dy}{dx}, x\right) dx$ *where* $L = \sqrt{1 + \left(\frac{dy}{dx}\right)^2}$.

3-13. A bead slides without friction on a wire in the vertical x, y plane as shown. The elapsed time for the trip between the origin $(0,0)$ and the point (x_0, y_0) is $t = \int_{(0,0)}^{(x_0, y_0)} \frac{ds}{v}$, where $ds = dx\sqrt{1 + \left(\frac{dy}{dx}\right)^2}$ is the element of arc length and v is the velocity ($v = \sqrt{2gy}$ from energy conservation). Assuming that the bead is released at rest, the shape of the wire $y(x)$ is to be found for which the elapsed time is minimum. This famous *brachistochrone* problem (or curve of quickest descent), first proposed and solved by John Bernoulli in 1696, led to the development of the calculus of variations.

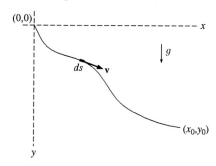

a) Show that the differential equation defining the wire shape is

$$2y\frac{d^2 y}{dx^2} + \left(\frac{dy}{dx}\right)^2 + 1 = 0$$

b) Demonstrate that the solution is a *cycloid*

$$x = a(\phi - \sin \phi)$$
$$y = a(1 - \cos \phi)$$

How are the parameter a and the values of ϕ at the endpoints determined?

3.7 Hamilton's Equations

3-14. For a particle moving in a plane under the influence of a central potential energy $V(r)$, find the Hamiltonian as a function of r, θ, p_r, and p_θ. Find the four Hamilton equations of motion. Show that the results are equivalent to Eqs. (3.40).

3-15. In a $2N$-dimensional *phase space* with coordinates (q_j, p_j) show that the "flow velocity" $(\dot{q}_j, \dot{p}_j)$ in this space satisfies $\frac{\partial \dot{q}_j}{\partial q_j} + \frac{\partial \dot{p}_j}{\partial p_j} = 0$, assuming that the general forces appear only in H. This indicates that the "flow" in this space is incompressible. This result is fundamental to statistical mechanics.

Chapter 4

MOMENTUM CONSERVATION

The conservation of linear momentum is a universal law for all of physics. In classical mechanics this conservation law is a direct consequence of Newton's laws. In the absence of external forces, the equation of motion

$$\frac{d\mathbf{p}}{dt} = \mathbf{F} = 0 \tag{4.1}$$

implies that $\mathbf{p}$ is independent of time. In other words, a particle with definite mass moves with constant velocity $\mathbf{v}$ in a force-free region. The most interesting ramifications of momentum conservation concern systems of more than one particle.

For a two-particle system, internal forces $\mathbf{F}^{\text{int}}$ between the particles and external forces $\mathbf{F}^{\text{ext}}$ on the particles can be present. The laws of motion for particles 1 and 2 are

$$\frac{d\mathbf{p}_1}{dt} = \mathbf{F}_1^{\text{int}} + \mathbf{F}_1^{\text{ext}}$$
$$\frac{d\mathbf{p}_2}{dt} = \mathbf{F}_2^{\text{int}} + \mathbf{F}_2^{\text{ext}} \tag{4.2}$$

The total momentum of the system

$$\mathbf{P} = \mathbf{p}_1 + \mathbf{p}_2 \tag{4.3}$$

obeys an equation given by the sum of (4.2)

$$\frac{d\mathbf{P}}{dt} = (\mathbf{F}_1^{\text{int}} + \mathbf{F}_2^{\text{int}}) + (\mathbf{F}_1^{\text{ext}} + \mathbf{F}_2^{\text{ext}}) \tag{4.4}$$

If the total external force is zero,

$$\mathbf{F}^{\text{ext}} = \mathbf{F}_1^{\text{ext}} + \mathbf{F}_2^{\text{ext}} = 0 \tag{4.5}$$

and the internal forces cancel,

$$\mathbf{F}_1^{\text{int}} = -\mathbf{F}_2^{\text{int}} \tag{4.6}$$

as implied by Newton's third law, then the momentum is conserved

$$\frac{d\mathbf{P}}{dt} = 0 \tag{4.7}$$

The preceding argument can be generalized in a straightforward way to demonstrate the conservation of total momentum for a system consisting of an arbitrary number of particles.

From the above we see that Newton's third law was "cooked up" to keep isolated systems from spontaneously accelerating. There is a deeper and more general way of expressing this idea. For simplicity consider two particles moving in one dimension which interact by a potential energy depending only on the difference of their coordinates. The Lagrangian is then

$$L = \frac{1}{2}m_1\dot{x}_1^2 + \frac{1}{2}m_2\dot{x}_2^2 - V(x_1 - x_1) \tag{4.8}$$

The resulting equations of motion are

$$\dot{p}_1 = -\frac{\partial V(x_2 - x_1)}{\partial x_1}$$
$$\dot{p}_2 = -\frac{\partial V(x_2 - x_1)}{\partial x_2} \tag{4.9}$$

Adding the above equations gives

$$\frac{d}{dt}(p_1 + p_2) = 0 \tag{4.10}$$
$$P = p_1 + p_2 = \text{constant} \tag{4.11}$$

The generalization to three dimensions and many particles proceeds along similar lines.

The big step here is the assumption that the potential energy depends only on differences in coordinates. The Lagrangian is then independent of the choice of the origin of the coordinate system. To formally express this: adding a constant vector **a** to every particle coordinate does not change the Lagrangian

$$L(\{\mathbf{r}_i + \mathbf{a}\}) = L(\{\mathbf{r}_i\}) \tag{4.12}$$

This is called *translational invariance*. This invariance of the Lagrangian is associated with a conservation law, conservation of momentum,

$$\mathbf{P} = \sum_{i=1}^{N} \mathbf{p}_i = \text{constant vector} \qquad (4.13)$$

where the sum is over the momenta of all particles in the system.

4.1 Rocket Motion

As an illustration of momentum-conservation methods, we apply (4.13) to the following problem with a time-varying mass. An open gondola freight car of mass m_0 coasts along a level straight frictionless track with initial velocity v_0. Rain starts falling straight down at time $t_0 = 0$, and water accumulates in the gondola at the rate (mass per unit time) σ. The problem is to find the velocity of the gondola.

Since the falling rain has no horizontal component of velocity, it has no component of momentum along the track. Hence the momentum along the track of the gondola is unchanged by the accumulating rain

$$P = mv = \text{constant} = m_0 v_0 \qquad (4.14)$$

Here m is the mass of the total system, gondola and accumulated rain. Since the time rate of change of m is given by

$$\frac{dm}{dt} = \sigma \qquad (4.15)$$

we have

$$m = m_0 + \sigma t \qquad (4.16)$$

From (4.14) and (4.16), we find the velocity of the gondola at time t to be

$$v = v_0 \frac{m_0}{m_0 + \sigma t} \qquad (4.17)$$

More commonly, situations involving a time-varying mass are encountered in conjunction with an externally applied force. Rocket motion is such an example. The time variation of the mass in rocket motion is due to the expulsion of the exhaust. The external forces are primarily due to gravity and air resistance. To derive the fundamental equation for linear rocket motion, we treat the rocket as a system of a large number of particles. The generalized form of (4.4) and (4.6) is

$$\frac{d\mathbf{P}}{dt} = \mathbf{F}^{\text{ext}} \tag{4.18}$$

where $\mathbf{P}$ is the total momentum and $\mathbf{F}^{\text{ext}}$ is the net external force.

At time t, the rocket of mass m is moving with a velocity v relative to a fixed coordinate system, as illustrated in Fig. 4-1. The exhaust is ejected with a constant velocity u opposite to the rocket velocity. In an infinitesimal time interval dt a mass $-dm$ is ejected as exhaust. Note that the change in the rocket mass dm will be negative for the normal burn of a rocket. The residual rocket mass $m + dm$ has velocity $v + dv$ and the exhaust mass $-dm$ has velocity $v - u$ in the fixed reference frame. The momentum $P + dP$ of the system at time $t + dt$ is thus

$$P + dP = (m + dm)(v + dv) + (-dm)(v - u) \tag{4.19}$$

To first order in the differentials

$$P + dP = mv + m(dv) + u(dm) \tag{4.20}$$

The momentum of the system at the time $t = 0$ was $P = mv$. Thus the change in momentum of the rocket-plus-exhaust system is

$$dP = m(dv) + u(dm) \tag{4.21}$$

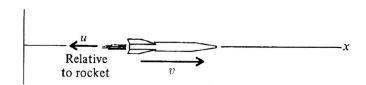

FIGURE 4-1. Motion of a rocket with velocity v relative to a fixed coordinate system. The velocity of the exhaust relative to the rocket is $-u$.

The time rate of change of momentum is

$$\frac{dP}{dt} = m\frac{dv}{dt} + u\frac{dm}{dt} = F^{ext} \tag{4.22}$$

where we have used (4.18) and (4.21). This fundamental equation of rocket motion is usually written

$$m\frac{dv}{dt} = F^{ext} - u\frac{dm}{dt} \tag{4.23}$$

The term $u(-dm/dt)$ increases the velocity v of the rocket and is called the *thrust* of the rocket. In the case of a jet plane, the exiting air and burned fuel contribute positively to the thrust while the air entering the engines contributes negatively (since $dm/dt > 0$ and $u > 0$ for the entering air).

In the absence of external forces (4.23) can be readily solved. Multiplying through by dt and rearranging factors we have

$$dv = -u\frac{dm}{m} \tag{4.24}$$

The result of integration is

$$v_f - v_i = u\ln\frac{m_i}{m_f} \tag{4.25}$$

where the i and f subscripts label initial and final values. The exhaust velocity u depends on the type of rocket fuel that is burned. Fuels with low molecular weights generally have higher exhaust velocities, and thus yield high rocket velocities. Present rocket technology gives exhaust velocities close to the thermodynamic limit for chemical fuels. The result in (4.25) places a limit on the velocities which can be reached with a single-stage rocket. Velocities several orders of magnitude greater than u cannot be achieved. The mass m_f of the payload plus empty rocket is an important factor in determining the final velocity that the rocket reaches.

4.2 Frames of Reference

Newton's law stipulates that the equation of motion applies only in an inertial frame. Since a frame moving with a constant velocity relative to an inertial frame is also inertial, considerable latitude exists in choosing a reference coordinate frame. Frequently, the solution to a problem can be simplified by a suitable choice of coordinate frame.

The Galilean transformation of classical mechanics relates positions of a point as measured from two coordinate systems in relative translational motion. We take S' to denote a coordinate frame whose origin moves with constant velocity $\mathbf{V}_0$ relative to the origin of a fixed coordinate frame S, as illustrated in Fig. 4-2. If $\mathbf{r}'$ is the location of a point in space as measured in S', and $\mathbf{r}$ is the location of the same point as measured in S, the Galilean transformation is

$$\mathbf{r} = \mathbf{r}' + \mathbf{V}_0 t \qquad (4.26)$$

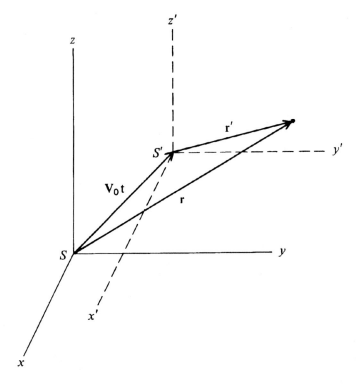

FIGURE 4-2. Galilean transformation between two coordinate frames S and S' in relative translational motion.

When we differentiate (4.26) with respect to t, we get the velocity-addition law

$$\mathbf{v} = \mathbf{v}' + \mathbf{V}_0 \qquad (4.27)$$

A second differentiation gives

$$\frac{d^2\mathbf{r}}{dt^2} = \frac{d^2\mathbf{r}'}{dt^2} \qquad (4.28)$$

due to the constancy of $\mathbf{V}_0$, both in magnitude and direction. Since the mass of a particle is a constant (in particular, it is independent of the velocity), m is unchanged under a Galilean transformation. Combining the invariance of the mass with the invariance of acceleration [(4.28)], we see that the ma term of Newton's second law is invariant under Galilean transformations. The force $\mathbf{F}'(\mathbf{r}', \mathbf{v}', t)$ on the particle in the S' frame is the same as the force $\mathbf{F}(\mathbf{r}, \mathbf{v}, t)$ in the S frame if the force is velocity-independent:

$$\mathbf{F}'(\mathbf{r}', t) = \mathbf{F}(\mathbf{r}, t) \tag{4.29}$$

This follows from the fact that $\mathbf{r}$ and $\mathbf{r}'$ denote the same physical location in space. Thus, if the force is independent of velocity, the equation of motion is Galilean-invariant. All consequences of Newton's law, such as conservation of energy and momentum, must hold in all inertial frames.

Even though the Galilean transformation of (4.26) and (4.27) seems self-evident, it must be regarded as a postulate. In fact, for a velocity $\mathbf{v}$ comparable with the speed of light, the equations of motion are not invariant to a Galilean transformation. The Lorentz transformation of Einstein's special relativity theory, to which the equations of motion are always invariant, reduces to the Galilean transformation in the limit of small velocity.

To show the advantage of a judicious choice of coordinate frame, we discuss the lift experienced by seagulls rising in a uniform wind. On a windy day along the beach, seagulls on the ground are often observed to extend their wings and to be carried aloft without the need of flapping. Since the seagull does no work in the ascent, energy-conservation methods should apply. However, in a coordinate frame which is at rest with respect to the ground, the gull gains energy, both kinetic and potential!

The energy gained by the gull is given up by the wind but it is difficult to achieve a quantitative estimate. However, if we make a Galilean transformation to a frame which moves with the air, the situation becomes clearer. In this frame the air has no kinetic energy initially but the seagull's initial mechanical energy is

$$K_i(\text{gull}) = \frac{1}{2} m v_w^2 \tag{4.30}$$

where v_w is the wind speed. If the wind is sufficiently strong when the gull extends its wings the aerodynamic lift force on the gull will exceed its weight and the gull will rise. Energy conservation in the air rest frame then yields

$$\frac{1}{2}mv_w^2 = mgx + K_f(\text{gull}) + K_f(\text{air}) \tag{4.31}$$

where at height x the gull's initial kinetic energy now appears partly as the gull's potential energy plus its remaining kinetic energy plus the kinetic energy of the air which has been deflected by its wings. When the gull has nearly come to rest relative to the wind $K_f(\text{gull}) \approx 0$ and assuming it has made only a small disturbance to the motion of the air, $K_f(\text{air}) \approx 0$, the initial kinetic energy is converted largely into potential energy and the gull reaches a limiting height $x = h$ given by

$$mgh = \tfrac{1}{2}mv_w^2 \tag{4.32}$$

or

$$h = \frac{v_w^2}{2g} \tag{4.33}$$

In a brisk wind of 40 km/h an ideal seagull can glide up to a height of

$$h = \frac{[40(1,000/3,600)]^2}{2(9.8)} = 6.3\,\text{m}\,. \tag{4.34}$$

4.3 Elastic Collisions: Lab and CM Systems

Collisions provide some especially interesting examples of momentum-conservation methods. In collisions where no external forces are involved and the internal forces satisfy the third law, the total momentum of the colliding objects is conserved, as in (4.7). The following discussion, which is based on momentum conservation, applies to all collisions, no matter what the detailed interactions are, so long as the interactions are sufficiently short-ranged so that the bodies can be treated as free before and after the collision. Of course if there are non-contact forces acting it is not necessary that the surfaces of the colliding particles actually touch for a collision to take place. Even in cases where external forces are present the collision approximation may still be valid. If the collision takes place over a short enough time the collision force will be much larger than the external force during the collision and the motion can be separated into three time segments: before and after the collision the particles move independently under the action of the external forces and during the collision the particles collide as if they were acted upon only by their interparticle forces.

Two colliding particles can approach each other from any direction. We have our choice however (at least theoretically) of viewing the collision from any convenient Galilean reference frame. All such frames are equivalent in the sense that momentum is conserved in each.

In the description of two-particle collisions, the most common choices of coordinate frames of reference are the laboratory (lab) frame and the center-of-mass (CM) frame. Hereafter, we label the lab coordinate system by S and the CM coordinate system by S'. In the lab system the target particle m_2 is initially at rest, $\mathbf{v}_{2i} = 0$, and the incident particle m_1 has velocity $\mathbf{v}_{1i}$. This system is so named because most experiments in the laboratory are performed with these initial conditions. After the collision, when the forces are no longer acting, the final lab velocities are $\mathbf{v}_{1f}$ and $\mathbf{v}_{2f}$. Conservation of total momentum in the lab frame implies

$$\mathbf{P} = m_1\mathbf{v}_{1i} = m_1\mathbf{v}_{1f} + m_2\mathbf{v}_{2f} \tag{4.35}$$

where we assume that the masses are unchanged by the collision.

The center-of-mass frame is the system of coordinates for which the total momentum of the two particles is zero.

$$\mathbf{P}' = \mathbf{p}'_{1i} + \mathbf{p}'_{2i} = 0 = \mathbf{p}'_{1f} + \mathbf{p}'_{2f} \tag{4.36}$$

In this frame the CM is at rest; the CM frame is also sometimes called the *center-of-momentum frame*. To determine the Galilean transformation velocity $\mathbf{V}_0$ between the lab and CM frames, we use (4.27) to relate the initial velocities.

$$\begin{aligned} \mathbf{v}'_{1i} &= \mathbf{v}_{1i} - \mathbf{V}_0 \\ \mathbf{v}'_{2i} &= \mathbf{v}_{2i} - \mathbf{V}_0 = -\mathbf{V}_0 \end{aligned} \tag{4.37}$$

When we impose the $\mathbf{P}' = 0$ requirement of (4.36) for the initial velocities in (4.37),

$$m_1(\mathbf{v}_{1i} - \mathbf{V}_0) + m_2(-\mathbf{V}_0) = 0 \tag{4.38}$$

the transformation velocity is obtained as

$$\mathbf{V}_0 = \frac{m_1}{m_1 + m_2}\mathbf{v}_{1i} \tag{4.39}$$

Of course $\mathbf{V}_0$ is the velocity of the CM in the lab frame. From (4.37) and

(4.39), we find

$$\mathbf{v}'_{1i} = \frac{m_2}{m_1 + m_2}\mathbf{v}_{1i}$$

$$\mathbf{v}'_{2i} = -\frac{m_1}{m_1 + m_2}\mathbf{v}_{1i}$$

(4.40)

for the initial velocities of the particles in the CM frame. A kinematical diagram of the initial and final velocities in the lab and CM frames is given in Figs. 4-3.

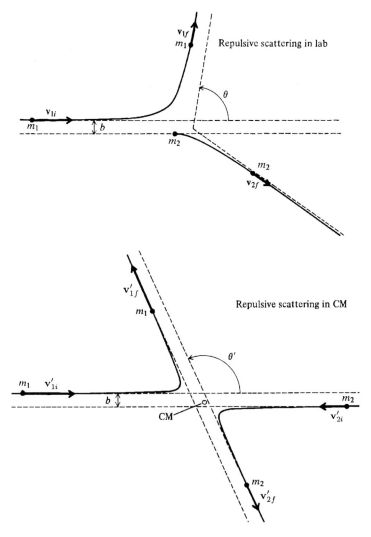

FIGURE 4-3(a). Repulsive scattering of two particles as viewed from the laboratory and center-of-mass coordinate frames.

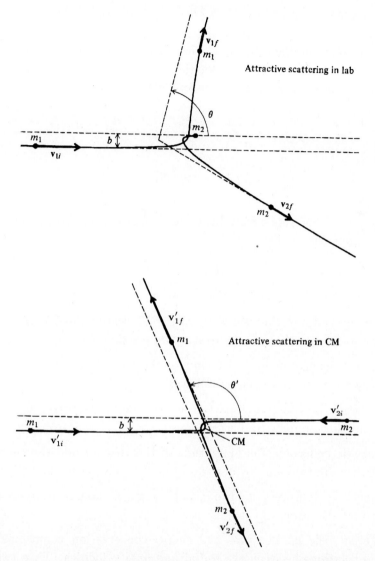

FIGURE 4-3(b). Attractive scattering of two particles in lab and CM frames.

Collisions in which kinetic energy as well as momentum is conserved are called *elastic* collisions. In an elastic collision of two particles, the momentum-conservation condition in (4.35) or (4.36) is supplemented by the energy conservation condition,

$$\tfrac{1}{2}m_1(\mathbf{v}_{1i})^2 = \tfrac{1}{2}m_1(\mathbf{v}_{1f})^2 + \tfrac{1}{2}m_2(\mathbf{v}_{2f})^2 \qquad (4.41)$$

in the lab, or equivalently,

$$\tfrac{1}{2}m_1(\mathbf{v}'_{1i})^2 + \tfrac{1}{2}m_2(\mathbf{v}'_{2i})^2 = \tfrac{1}{2}m_1(\mathbf{v}'_{1f})^2 + \tfrac{1}{2}m_2(\mathbf{v}'_{2f})^2 \qquad (4.42)$$

in the CM. The momentum- and energy-conservation conditions lead to simple relations between the velocities in the CM system. From the momentum-conservation condition of (4.36), we have

$$\mathbf{v}'_{2i} = -\frac{m_1}{m_2}\mathbf{v}'_{1i}$$

$$\mathbf{v}'_{2f} = -\frac{m_1}{m_2}\mathbf{v}'_{1f} \qquad (4.43)$$

By substitution of this result into the energy-conservation relation of (4.42), we obtain

$$v'_{1i} = v'_{1f}$$

$$v'_{2i} = v'_{2f} \qquad (4.44)$$

for the magnitudes of the velocity vectors. Viewed in the CM frame, the energy of each particle is unchanged by an elastic collision.

The significance of the CM frame for theoretical analyses of collision problems can be summarized as follows:

1. The motion both before and after the collision is *collinear* in the CM frame, and thus simpler to discuss.

2. For elastic collisions, the magnitude of the velocity and the energy of each particle are the same before and after the collision in the CM frame.

3. For inelastic collisions, *all* the initial energy in the CM frame is available for inelastic processes, as discussed in § 4.5.

Together, the momentum- and energy-conservation conditions provide four relations between the initial and final velocities. Thus two out of the six final velocity components are unspecified by the initial velocities. This indeterminacy is easily visualized in the CM system. By (4.43) the two final velocities must be opposite in direction, with magnitudes in ratio $v'_{2f}/v'_{1f} = m_1/m_2$. Thus momentum conservation leaves only one independent final-velocity vector, say $\mathbf{v}'_{1f}$. The magnitude of this velocity vector is specified by the energy-conservation condition. The angles (θ', ϕ') of this velocity are undetermined by energy and momentum conservation but are determined by the solutions to the equation of motion for specified initial conditions.

The angle between the final and initial velocities of particle 1 is defined to be the *scattering angle*, as illustrated in Fig. 4-3. The scattering angle θ' in the CM system can be related to the lab scattering angle θ through the velocity-transformation equation

$$\mathbf{v}_{1f} = \mathbf{v}'_{1f} + \mathbf{V}_0 \tag{4.45}$$

of (4.27). When we take components of this equation along directions perpendicular and parallel to the initial velocities as shown in Fig. 4-4, we obtain

$$
\begin{aligned}
v_{1f} \sin \theta &= v'_{1f} \sin \theta' \\
v_{1f} \cos \theta &= v'_{1f} \cos \theta' + V_0
\end{aligned}
\tag{4.46}
$$

The ratio of these equations is

$$\tan \theta = \frac{\sin \theta'}{\cos \theta' + V_0/v'_{1f}} \tag{4.47}$$

From (4.37), (4.43), and (4.44) the velocity ratio V_0/v'_{1f} is given by

$$\frac{V_0}{v'_{1f}} = \frac{v'_{2i}}{v'_{1f}} = \frac{m_1}{m_2} \frac{v'_{1i}}{v'_{1f}} = \frac{m_1}{m_2} \tag{4.48}$$

With this substitution in (4.47) we arrive at the desired relation between the lab and CM scattering angles.

$$\tan \theta = \frac{\sin \theta'}{\cos \theta' + m_1/m_2} \tag{4.49}$$

For a fixed-target particle (that is, $m_2 = \infty$), the lab and CM scattering angles are equal. For equal masses the application of standard trigonometric identities [$\sin \theta' = 2 \sin \frac{1}{2}\theta' \cos \frac{1}{2}\theta'$, $\cos \theta' = 2 \cos^2 \frac{1}{2}\theta' - 1$] to (4.49) shows that the lab angle is half the CM angle, $\theta = \theta'/2$. For most mass ratios (4.49) must be solved numerically.

As the CM scattering θ' varies from 0 to π, the vector $\mathbf{v}'_{1f}$ traces out the dashed circles in the velocity diagrams of Fig. 4-4. If $m_1 < m_2$, the lab angle also varies from 0 to π, as can be deduced from Fig. 4-4(a) or from (4.49). For $m_1 > m_2$, illustrated in Fig. 4-4(b), θ increases from 0 to a maximum value θ_{max} and then decreases back to 0 as θ' goes from 0 to π. The maximum lab angle occurs when $\mathbf{v}_{1f}$ and $\mathbf{v}'_{1f}$ are perpendicular

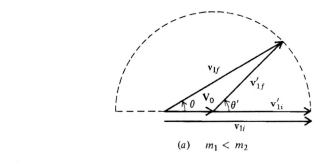

(a) $m_1 < m_2$

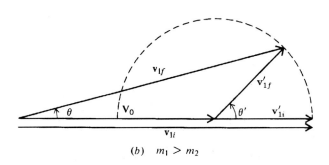

(b) $m_1 > m_2$

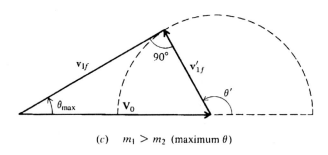

(c) $m_1 > m_2$ (maximum θ)

FIGURE 4-4. Velocity diagram illustrating the lab and CM quantities for elastic scattering.

$(\mathbf{v}_{1f} \cdot \mathbf{v}'_{1f} = 0)$, as indicated in Fig. 4-4(c). This orthogonality condition leads to the following expression for θ_{max}:

$$\sin \theta_{max} = \frac{m_2}{m_1}, \qquad m_1 > m_2 \qquad (4.50)$$

For example, the maximum laboratory scattering angle of a proton incident on an electron is 0.03°.

In the lab system the kinetic energy transferred to the target particle is given by

$$K_{2f} = \tfrac{1}{2}m_2 v_{2f}^2 = \tfrac{1}{2}m_2(\mathbf{v}_{2f}' - \mathbf{v}_{2i}')^2$$

$$= m_2 v_{2i}^{2\prime}(1 - \cos\theta') = 2m_2 v_{2i}^{2\prime}\sin^2\frac{\theta'}{2} \tag{4.51}$$

Here we have made use of $\mathbf{v}_{2f} = \mathbf{v}_{2f}' + \mathbf{V}_0$ in (4.27) and $\mathbf{V}_0 = -\mathbf{v}_{2i}'$ in (4.37). This energy transfer is a maximum for $\theta' = \pi$, corresponding to backward scattering in the CM system. The ratio of the final energy of the target particle K_{2f} to the incident energy K_{1i} of the projectile is

$$\frac{K_{2f}}{K_{1i}} = \frac{2m_2 v_{2i}^{2\prime}\sin^2(\theta'/2)}{\tfrac{1}{2}m_1 v_{1i}^2} \tag{4.52}$$

Using (4.40), this ratio simplifies to

$$\frac{K_{2f}}{K_{1i}} = \frac{4m_1 m_2}{(m_1 + m_2)^2}\sin^2\frac{\theta'}{2} \tag{4.53}$$

Thus the greatest energy transfer in the lab frame occurs when target and projectile have equal mass. This fact is of great importance in nuclear physics where hydrogen rich materials such as paraffin are used in slowing neutrons.

4.4 Collisions of Billiard Balls

As an illustration of scattering, we consider the collisions of billiard balls on a smooth table. The initial conditions simulate the lab frame since the target ball is at rest and the cue ball in motion with velocity $\mathbf{v}_{1i}$. For a head-on collision the velocity of the cue ball is directed at the center of the target ball. To analyze the subsequent collision we transform to the CM frame. In a frame moving with velocity $\mathbf{V}_0$ relative to the lab system, the velocities of the balls are

$$\mathbf{v}_{1i}' = \mathbf{v}_{1i} - \mathbf{V}_0$$
$$\mathbf{v}_{2i}' = -\mathbf{V}_0 \tag{4.54}$$

For the CM frame the vanishing of the total momentum requires

$$\mathbf{v}_{1i}' = -\mathbf{v}_{2i}' \tag{4.55}$$

because of the equality of the masses. By comparison of (4.54) and (4.55) we obtain

$$\mathbf{V}_0 = \frac{\mathbf{v}_{1i}}{2} = \mathbf{v}'_{1i} = -\mathbf{v}'_{2i} \tag{4.56}$$

After the collision, the balls are moving away from each other with equal velocities in the CM frame

$$\mathbf{v}'_{1f} = -\mathbf{v}'_{2f} \tag{4.57}$$

Billiard balls have both linear and rotational kinetic energy

$$K = \tfrac{1}{2}mv^2 + \tfrac{1}{2}I\omega^2 \tag{4.58}$$

where I is the moment of inertia and ω is the angular velocity of rotation (*i.e.*, the spin). Since the balls are assumed to be smooth, no spin transfer occurs in the collision, and the rotational energies of the individual balls remain unchanged by the collision. Because the ratio of translational velocity to spin has been altered by the collisions the balls will slip instead of roll on the table and friction forces will come into play. Some of these effects will be discussed in Chapter 6, but we will ignore these complications for now. Assuming that the collision is elastic, the translational kinetic energies must satisfy the conservation condition

$$\tfrac{1}{2}m\left[(v'_{1i})^2 + (v'_{2i})^2\right] = \tfrac{1}{2}m\left[(v'_{1f})^2 + (v'_{2f})^2\right] \tag{4.59}$$

From (4.55), (4.57), and (4.59), the magnitudes of the initial and final velocities must be equal; thus the balls must reverse direction in the CM frame and the velocity vectors are

$$\begin{aligned}
\mathbf{v}'_{1f} = \mathbf{v}'_{1i} = -\frac{\mathbf{v}_{1i}}{2} \\
\mathbf{v}'_{2f} = -\mathbf{v}'_{2i} = \frac{\mathbf{v}_{1i}}{2}
\end{aligned} \tag{4.60}$$

To express this result in terms of what is observed on the billiard table, we transform back to the lab frame, using (4.27), (4.56) and (4.60)

$$\begin{aligned}
\mathbf{v}_{1f} = \mathbf{v}'_{1f} + \mathbf{V}_0 = 0 \\
\mathbf{v}_{2f} = \mathbf{v}'_{2f} + \mathbf{V}_0 = \mathbf{v}_{1i}
\end{aligned} \tag{4.61}$$

The cue ball stops, and the target ball moves forward with the initial velocity of the cue ball as a result of this head-on collision; see Fig. 4-5.

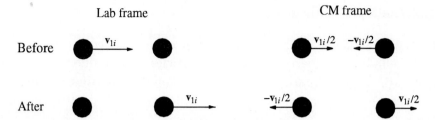

FIGURE 4-5. Head-on collisions of billiard balls in the lab and CM frames.

Of course, we could have solved this collision problem directly in the lab frame, but the physical interpretation is more transparent in the CM frame.

In noncentral collisions, as any good billiard player knows, the balls leave the point of impact at right angles. We can prove this fact directly by momentum and energy conservation in the lab system. From momentum conservation,

$$(\mathbf{p}_{1i})^2 = (\mathbf{p}_{1f} + \mathbf{p}_{2f})^2 = (\mathbf{p}_{1f})^2 + (\mathbf{p}_{2f})^2 + 2\mathbf{p}_{if} \cdot \mathbf{p}_{2f} \qquad (4.62)$$

and from energy conservation,

$$(\mathbf{p}_{1i})^2 = (\mathbf{p}_{if})^2 + (\mathbf{p}_{2f})^2 \qquad (4.63)$$

We obtain

$$\mathbf{p}_{1f} \cdot \mathbf{p}_{2f} = 0 \qquad (4.64)$$

from the difference of (4.62) and (4.63). This establishes that the final velocities are at right angles if neither velocity is zero. A velocity diagram for the collision is given in Fig. 4-6.

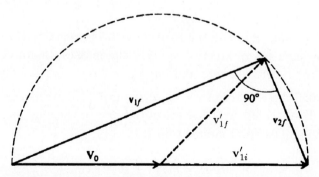

FIGURE 4-6. Velocity diagram for elastic scattering of two equal-mass particles. Since v_{1f} and v_{2f} form the sides of a triangle inscribed in a semicircle, these two vectors are perpendicular by geometry.

4.5 Inelastic Collisions

In many instances mechanical energy is not conserved in the collision and the individual masses of the final particles may be different from the masses of the initial particles. In this more complex situation we can write an energy-balance equation for the colliding system and surroundings of the form

$$K_1 = K_3 + K_4 + Q \tag{4.65}$$

in the lab system, or

$$K_1' + K_2' = K_3' + K_4' + Q \tag{4.66}$$

in the CM system. In these equations K designates particle kinetic energy and Q is the energy released (if $Q > 0$) or absorbed (if $Q < 0$) by the collision in the form of heat (or changes in rest energy of the particles in nuclear collisions). The value of Q is independent of reference frame. Collisions in which energy is transferred ($Q \neq 0$), or the final masses are different from the initial masses, are called *inelastic*. In elastic processes $Q = 0$, and the masses are not changed by the collision.

A common method of describing an inelastic collision is with the *coefficient of restitution e* defined by

$$e \equiv \frac{u_f \text{ (relative velocity after)}}{u_i \text{ (relative velocity before)}} = \left| \frac{\mathbf{v}_{2f} - \mathbf{v}_{1f}}{\mathbf{v}_{2i} - \mathbf{v}_{1i}} \right| \tag{4.67}$$

This coefficient has the same value in any coordinate system. If $e = 0$ there is no final relative velocity and the final particles are at rest in the CM system (*i.e.*, they stick together). This corresponds to a completely inelastic collision. If $e = 1$ the relative velocity remains the same which implies conservation of kinetic energy if the masses do not change in the collision. For e between zero and one the collision is partially inelastic.

To make these statements more quantitative we consider two free particles. These could be either the initial pair or the final pair of particles in a collision. The total momentum is

$$\mathbf{P} = m_1 \mathbf{v}_1 + m_2 \mathbf{v}_2 \tag{4.68}$$

Their relative velocity is

$$\mathbf{u} = \mathbf{v}_2 - \mathbf{v}_1 \tag{4.69}$$

and the kinetic energy is given by

$$2K = m_1 v_1^2 + m_2 v_2^2 \tag{4.70}$$

By direct substitution it is easy to see that (4.68)–(4.70) imply the identity

$$P^2 + m_1 m_2 u^2 = 2(m_1 + m_2)K \tag{4.71}$$

In a collision P^2 is always conserved. Hence if the particle masses are unchanged in the collision the energy released by the collision is

$$Q \equiv K_i - K_f = \frac{1}{2}\left(\frac{m_1 m_2}{m_1 + m_2}\right)(u_i^2 - u_f^2)$$

$$= \frac{1}{2}\left(\frac{m_1 m_2}{m_1 + m_2}\right)u_i^2(1 - e^2) \tag{4.72}$$

Thus kinetic energy is conserved if $e = 1$. Note that the Q value is independent of coordinate frame since the relative velocities are frame independent. From (4.71) if both P^2 and K are conserved in a collision the relative velocity $\mathbf{u}$ just reverses sign in any coordinate frame.

As an example we consider an inelastic collision between two equal mass putty balls. In the lab system the projectile putty ball has velocity $\mathbf{v}_{1i}$ and the target putty ball is at rest. In the center-of-mass system the initial velocities are

$$\mathbf{v}'_{1i} = \frac{\mathbf{v}_{1i}}{2}$$

$$\mathbf{v}'_{2i} = -\frac{\mathbf{v}_{1i}}{2} \tag{4.73}$$

Upon impact the two putty balls stick together to form a mass $2m$, as shown in Fig. 4-7. The final velocity of the aggregate putty ball is zero in the CM frame. The energy released by the collision is deduced to be

$$Q = K'_1 + K'_2 - K'_3$$
$$= \tfrac{1}{2}m[(\mathbf{v}'_{1i})^2 + (\mathbf{v}'_{2i})^2] - 0 \tag{4.74}$$
$$= \tfrac{1}{4}m(\mathbf{v}_{1i})^2$$

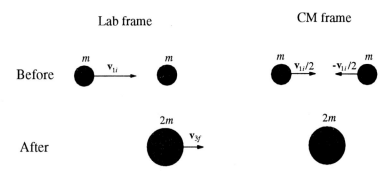

FIGURE 4-7. Putty-ball inelastic collision in lab and CM frames.

Since there is no final kinetic energy in the CM system, all the initial kinetic energy in the CM frame has been converted to heat, and the collision is called *completely inelastic*. This result also follows from (4.72) with $e = 0$. The final velocity of the aggregate putty ball in the lab system is

$$\mathbf{v}_{3f} = 0 + \frac{\mathbf{v}_{1i}}{2} \tag{4.75}$$

To achieve the maximal possible available energy in collisions of particles, accelerators are constructed to use colliding beams with nearly equal and opposite momenta. In such colliders the laboratory frame is in fact the center of mass frame, since the sum of the momenta of the colliding particles is nearly zero.

PROBLEMS

4.1 Rocket Motion

4-1. Material drops from a hopper at a constant rate $dm/dt = \sigma$ onto a conveyor belt moving with constant velocity v parallel to the ground. What power motor would be needed to drive the belt? *Note: Power is the time rate of doing work:* $P = \frac{dW}{dt} = F\frac{dx}{dt}$. Show that half of the required power equals the time rate of change of mechanical energy (the rest of the input power is converted to heat).

4-2. A vertical drain in the floor of the gondola car example in § 4.1 is open to keep rainwater from accumulating in the car. Find the velocity of the gondola at time t in terms of the initial velocity v_0, the mass of the car m_0, and the rate σ at which rain mass enters and leaves the gondola.

4-3. A forest-fire-fighting airplane gliding horizontally at 200 km/hr lowers a scoop to load water from a lake. It continuously picks up $\frac{1}{10}$ of the airplane's initial mass in water every 10 s. Neglecting friction find an expression for the airplane's speed as a function of time. What is its speed after 10 s? With a frictional force $F = -bv$ and a constant time rate of loading, find and solve the equation of motion of the airplane during the process.

4-4. During a rocket burn in free space, at what residual mass is the momentum a maximum if the rocket starts from rest? At what residual mass is the rocket kinetic energy maximum? Determine the rocket velocity in each case.

4-5. At cruising velocity of 1,000 km/h, each of the four fan jet engines on a Boeing 747 plane burns fuel at a rate of 0.3 kg/s, which is ejected through the engine turbines. The airflow through each engine turbine of 50 kg/s has an exhaust velocity of 2,100 km/h. The airflow of 250 kg/s through the fans around each turbine is exhausted with a velocity of 1,300 km/h. Calculate the net forward thrust on the plane.

4-6. It is difficult to construct a rocket which, even if it carries no payload, has a mass ratio $r = m_{\text{initial}}/m_{\text{final}}$ as large as 10. The final velocity in free space for a single-stage rocket is then $v_f \leq u \ln 10 = 2.3\,u$. Since the exhaust velocity u for chemical rockets is less than about 5 km/s, it is difficult to send a payload to the moon using a single rocket. Fortunately, the device of "staging" circumvents this difficulty. If a large rocket carries as its payload a smaller rocket which fires after the first burns out, a considerable increase in final velocity can be achieved; explain. Show that if n rockets are staged so that each has the same exhaust velocity u and the same mass ratio r (where the masses include all the upper stages), the final velocity of the last stage is $nu \ln r$.

4-7. For a rocket fired upward from the surface of the earth, the rocket equation (4.23) is

$$m\frac{dv}{dt} = -mg - u\frac{dm}{dt}$$

provided that air resistance is neglected and the earth is regarded as an inertial frame. To achive lift-off the thrust must exceed the weight of the rocket. For a constant rate of burn (*i.e.*, $dm/dt = -\alpha m_i$ with α a constant) show that the velocity at the instant of burnout is

$$\frac{v_f}{u} = -\frac{1}{T}(m_i - m_f)g + \ln \frac{m_i}{m_f}$$

where the thrust $T = u\alpha m_i$.

4-8. Using the results from the preceding problem show that the maximum altitude which can be attained by a rocket shot vertically from the surface of the earth is

$$h = \frac{u^2}{2g} \ln^2 \left(\frac{m_i}{m_f} \right) - \frac{u}{\alpha} \left[\ln \left(\frac{m_i}{m_f} \right) - \left(1 - \frac{m_f}{m_i} \right) \right]$$

in the approximation that the gravitational acceleration is constant. Take into account the height reached at fuel burnout and the additional distance coasting above that height. Calculate the numerical value of h for the first stage of the Apollo moon rocket. Compare the result with that which could be reached if the entire fuel burn occurred instantaneously (that is, $\alpha = \infty$). The initial mass of the Apollo rocket is $m_i = 2.94 \times 10^6$ kg, the final mass is 0.79×10^6 kg, the exhaust velocity is $u = 2.8$ km/s, and the thrust is $T = 37.2 \times 10^6$ Newtons.

4.3 Elastic Collisions: Lab and CM Systems

4-9. A proton of energy 4 MeV scatters off a second proton at rest. One proton comes off at an angle of 30° in the lab system. What is its energy? What is the energy and scattering angle of the second proton?

4-10. In a collision with a nucleus of unknown mass, an α particle scatters directly backward and loses 75 percent of its energy. What is the mass of the nucleus, assuming that the scattering is elastic?

4-11. Two balls of unequal mass moving with equal velocities in opposite directions collide. One ball is stationary after the collision. If the collision is elastic, what is the ratio of the masses?

4-12. In a head-on elastic collision of two masses m_1 and m_2 with initial velocities v_{1i} and v_{2i}, show that the relative velocity after collision is opposite in sign to the relative velocity before collision: $v_{1f} - v_{2f} = -(v_{1i} - v_{2i})$.

4-13. A light sphere of mass m and a heavy
sphere of mass M fall vertically at the
same velocity with the light sphere a
small distance above the heavy sphere
as shown. Just before they strike the
floor, their common velocity is v_0.
Consider what happens just after
they strike the floor. What is the velocity of the lighter mass m just after
the bounce? Assume that all collisions are perfectly elastic and that the
two spheres remain aligned vertically.
*Hint: Analyze as a sequence of two
closely-spaced impulsive collisions.*

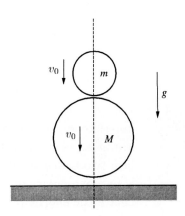

4-14. A group of n_1 identical smooth billiard balls moving along a line
in contact collide elastically with a group of n_2 stationary balls
also lined up in contact. How many masses will come out? Assume that the collision forces propagate with finite speed and treat
the collisions successively. Analyze as a sequence of closely-spaced
impulsive collisions.

4.5 Inelastic Collisions

4-15. A mass m moving horizontally with velocity v_0 strikes a pendulum
of mass m as shown.

a) If two masses stick together, find
the maximum height reached by
the pendulum.

b) If the masses scatter elastically
along the line of the initial motion, find the resulting maximum
height.

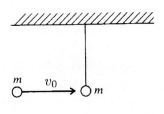

4-16. A steel ball-bearing is dropped from a height h onto a flat steel
plate. The coefficient of restitution is e. Find the total distance
traveled by the ball-bearing until it comes to rest and how long it
bounces. *Hint: first analyze the case of one bounce to show that
the new height is $h_1 = e^2 h$ and the time up is $t_1 = e\sqrt{2h/g}$. Then
apply this result successively.*

4-17. A mass $3m$ moves with velocity $2v_0\hat{z}$ and overtakes a mass m moving with velocity $v_0\hat{z}$. The masses collide with a coefficient of restitution $e = 1/2$ and in the CM frame both leave the collision in the $\hat{x}$ direction. Make a Galilean transformation to determine the velocities in the CM frame before and after the collision. Find the final velocity of $3m$ in the original coordinate system.

4-18. A beam of hydrogen molecules moves along the $\hat{z}$ direction with a kinetic energy of 1 electron volt (eV) per molecule. The molecules are in an excited state, from which they can decay and dissociate into two hydrogen atoms. When the velocity of a dissociation atom is perpendicular to the $\hat{z}$ direction, its energy is always 0.8 eV. Calculate the energy per molecule released in the dissociative reaction.

4-19. A mass m moving horizontally with velocity v strikes and sticks to a horizontal spring system of length l and spring constant k with masses m at each end, as shown. During the subsequent motion, what is the maximum compression of the spring?

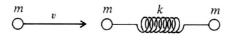

4-20. Show that the drag force on a satellite moving with velocity v in the earth's upper atmosphere is approximately $f_D = \rho A v^2$ where ρ is the atmospheric density and A is the cross-sectional area perpendicular to the direction of motion. Assume that the air molecules are moving slowly compared with v and that their collisions with the satellite are completely inelastic.

4-21. Two pucks, each of mass m, are connected by a massless string of length $2l$. The pucks lie on a horizontal frictionless sheet of ice. The string is initially straight (*i.e.*, $\theta = 90°$). A constant horizontal force F is applied in a direction perpendicular to the string. When the pucks collide they stick together. How much mechanical energy is lost in the collision?

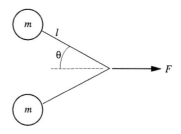

Chapter 5

ANGULAR-MOMENTUM CONSERVATION

A conservation law pertaining to angular motion can also be derived from Newton's law. As is the case for the momentum- and energy-conservation laws, the validity of angular-momentum conservation extends beyond the domain of classical mechanics.

5.1 Central Forces

For a single particle of mass m at a distance $\mathbf{r}$ from the origin of an inertial coordinate system, the law of angular-momentum conservation can be derived by taking the cross product of $\mathbf{r}$ with the equation of motion

$$\mathbf{r} \times \mathbf{F} = \mathbf{r} \times \dot{\mathbf{p}} \tag{5.1}$$

The left-hand side of this equation is known as the torque,

$$\mathbf{N} = \mathbf{r} \times \mathbf{F} \tag{5.2}$$

For illustration, see Fig. 5-1(a). The right-hand side of (5.1) can be written

$$
\begin{aligned}
\mathbf{r} \times \dot{\mathbf{p}} &= \frac{d}{dt}(\mathbf{r} \times \mathbf{p}) - \dot{\mathbf{r}} \times \mathbf{p} \\
&= \frac{d}{dt}(\mathbf{r} \times \mathbf{p}) - m\mathbf{v} \times \mathbf{v} \\
&= \frac{d}{dt}(\mathbf{r} \times \mathbf{p})
\end{aligned}
\tag{5.3}
$$

The quantity

$$\mathbf{L} = \mathbf{r} \times \mathbf{p} \tag{5.4}$$

is called the *angular momentum* about the origin, as shown in Fig. 5-1(b). Equations (5.1) to (5.4) relate the angular momentum and the torque.

$$\mathbf{N} = \frac{d\mathbf{L}}{dt} \tag{5.5}$$

If $\mathbf{N} = 0$, then $\mathbf{L}$ is constant in time. From (5.2), the angular momentum is conserved if either $\mathbf{r} = 0$ or $\mathbf{F} = 0$ or $\mathbf{F}$ is proportional to $\mathbf{r}$. The latter case is of particular interest since it corresponds to a central force, and

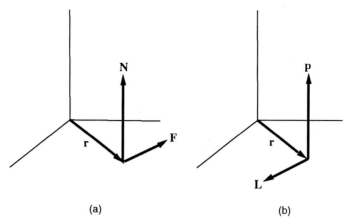

(a)   (b)

FIGURE 5-1. Diagrams illustrating two relations of (a) **r**, **F**, **N** and (b) **r**, **p**, **L** for a particle at position **r**.

many fundamental forces of classical mechanics are central. A central force is one with direction parallel to **r** with a functional dependence on the magnitude of **r** only.

The angular momentum for motion in a plane is

$$\mathbf{L} = mrv_\theta\hat{\mathbf{z}} = mr^2\dot{\theta}\hat{\mathbf{z}} \tag{5.6}$$

where $\hat{\mathbf{z}}$ is perpendicular to the (r, θ) directions in a right-hand sense. The time rate of change of L_z is determined by the torque N_z about the origin.

$$\frac{dL_z}{dt} = \frac{d}{dt}(mr^2\dot{\theta}) = N_z = rF_\theta \tag{5.7}$$

This equation is identical with (2.130). For central forces **F** is along **r**, so $F_\theta = 0$ and

$$\frac{d}{dt}(mr^2\dot{\theta}) = 0 \tag{5.8}$$

Thus the angular momentum L is conserved

$$L \equiv |\mathbf{L}| = mr^2\dot{\theta} = \text{constant} \tag{5.9}$$

From the conservation of angular momentum, we can prove that the motion of a particle under a central force occurs in a plane. From the properties of the mixed vector product in (2.63), the dot product of **L**

$= m\mathbf{r} \times \mathbf{v}$ with either $\mathbf{r}$ or $\mathbf{v}$ vanishes.

$$\mathbf{r} \cdot \mathbf{L} = 0$$
$$\mathbf{v} \cdot \mathbf{L} = 0 \tag{5.10}$$

Inasmuch as $\mathbf{L}$ is constant, its direction is fixed. The position and velocity of the particle thus remain perpendicular to the direction of $\mathbf{L}$ in the course of the motion. The trajectory is therefore confined to a plane passing through the origin and perpendicular to $\mathbf{L}$.

Expressing $\dot{\theta}$ in terms of L with (5.9) and then substituting into the radial equation of motion (2.130) gives

$$m\ddot{r} - mr \left(\frac{L}{mr^2} \right)^2 = F_r \tag{5.11}$$

The solution to this equation give $r(t)$. Then the angular dependence $\theta(t)$ can be obtained from integration of (5.9).

$$\int d\theta = \int \frac{L}{mr^2} dt \tag{5.12}$$

The angular momentum L is specified by the initial conditions on $\mathbf{r}$ and $\mathbf{v}$, and thus can be used in place of one of the initial conditions.

The radial equation is sometimes written in analogy to a one dimensional equation of motion as

$$m\ddot{r} = F_r + \frac{L^2}{mr^3} \tag{5.13}$$

Since (5.13) can be written in the form of Newton's law in one dimension, the techniques introduced in Chaps. 1 and 2 can be employed to find the radial solution.

In a one-dimensional interpretation of the radial motion, the quantity

$$F_{cf} = \frac{L^2}{mr^3} = mr\dot{\theta}^2 = \frac{mv_\theta^2}{r} \tag{5.14}$$

is a so-called *fictitious* centrifugal force that must be added to F_r. Just as the real force can be derived from a potential energy, $F_r = -\frac{dV(r)}{dr}$, the fictitious force can be derived from the fictitious potential energy, $F_{cf} = -\frac{dV_{cf}}{dr}$, with

$$V_{cf}(r) = \frac{L^2}{2mr^2} \tag{5.15}$$

Since the centrifugal force is repulsive and large at small radial distances, it repels the particle from the vicinity of $r = 0$. Consequently the term *centrifugal barrier* is often used in reference to this potential energy. The equivalent one-dimensional representation of the radial motion can be expressed in terms of an effective potential energy

$$E = \tfrac{1}{2}m\dot{r}^2 + V_{\text{eff}}(\mathbf{r}) \tag{5.16}$$

where

$$V_{\text{eff}}(\mathbf{r}) = V(\mathbf{r}) + \frac{L^2}{2mr^2} \tag{5.17}$$

In Fig. 5-2 we illustrate the effective potential with $V(r) = -\alpha/r$.

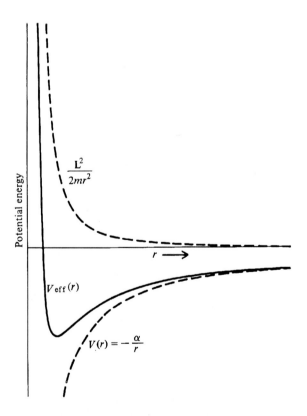

FIGURE 5-2. Effective potential energy for $V(r) = -\alpha/r$.

As an example of a system in which angular-momentum conservation plays an important role, we consider two equal point masses connected by a string which passes through a small hole on a frictionless mass on the table for vertical up-and-down motion of the suspended mass, as shown in Fig. 5-3. The equations are most simply formulated in a cylindrical coordinate system (r, θ, z) with the origin at the hole in the table and the positive z axis upward. Because the length ℓ of the string is fixed, the variables r and z are related

$$r - z = \ell \tag{5.18}$$

We shall choose r and θ as the independent variables. The gravitational potential energy associated with the mass on the table is constant; that of the suspended mass is

$$V = mgz = mg(r - \ell) \tag{5.19}$$

Since the potential energy depends only on r, the force is central and the angular momentum is conserved. The kinetic energy of the suspended mass is $K = \frac{1}{2}m\dot{z}^2 = \frac{1}{2}m\dot{r}^2$. From (3.38) and (5.19), the Lagrangian for the two-mass system is

$$L = \tfrac{1}{2}m\dot{r}^2 + \tfrac{1}{2}m(\dot{r}^2 + r^2\dot{\theta}^2) - mg(r - \ell) \tag{5.20}$$

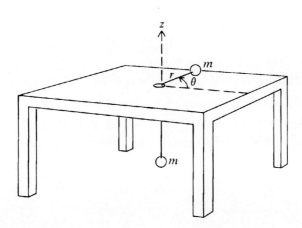

FIGURE 5-3. Central-force example.

With this expression for L, Lagrange's equations (3.39) give the radial equation

$$2m\ddot{r} - mr\dot{\theta}^2 + mg = 0 \tag{5.21}$$

and the angular equation

$$\frac{d}{dt}(mr^2\dot{\theta}) = 0 \tag{5.22}$$

The angular equation expresses the conservation of angular momentum:

$$\mathbf{L} = mr^2\dot{\theta}\hat{\mathbf{z}} \tag{5.23}$$

In terms of L the radial equation is

$$2m\ddot{r} - \frac{L^2}{mr^3} + mg = 0 \tag{5.24}$$

This corresponds to (5.10) with $F_r = -mg$ and a radially accelerated mass $2m$. Equation (5.24) is valid only for $r \leq \ell$.

The radial equation admits a solution with a circular orbit at $r = r_0$:

$$\ddot{r} = 0 \qquad L^2 = gm^2r_0^3 \tag{5.25}$$

A circular orbit of radius r_0 will be realized if the initial velocities satisfy the conditions

$$\dot{r} = 0 \qquad \dot{\theta} = \frac{L}{mr_0^2} = \sqrt{\frac{g}{r_0}} \equiv \Omega \tag{5.26}$$

In a circular orbit the centrifugal force $mr_0\dot{\theta}^2$ exactly balances the tension mg in the string.

By considering orbits having the same L but which deviate slightly from a circular orbit, we can determine whether or not the motion is stable. We substitute

$$r(t) = r_0 + \delta(t) \tag{5.27}$$

where $\delta \ll r_0$ in (5.24), and make a power series expansion in δ/r_0. In the approximation that we retain only constant and linear terms in δ, we find

$$2m\ddot{\delta} - \frac{L^2}{mr_0^3}\left(1 - \frac{3\delta}{r_0}\right) + mg = 0 \tag{5.28}$$

After we insert the circular-orbit condition from (5.25), the equation of motion for small radial deviations reduces to

$$\ddot{\delta} + \frac{3g}{2r_0}\delta = 0 \tag{5.29}$$

This describes simple harmonic motion in δ with angular frequency

$$\omega = \sqrt{\frac{3g}{2r_0}} = \sqrt{\frac{3}{2}}\,\Omega \tag{5.30}$$

Thus the particle undergoes small radial oscillations, indicating a stable configuration. Motion of this sort could be initiated from a circular orbit by a small radial impulsive blow. The general solution to (5.29) for the radial-displacement parameter δ is

$$\delta(t) = A\cos\omega t + B\sin\omega t \tag{5.31}$$

so that

$$r(t) = r_0 + A\cos\omega t + B\sin\omega t \tag{5.32}$$

When we impose the initial conditions

$$\begin{aligned} r(0) &= r_0 \\ \dot{r}(0) &= v_0 \end{aligned} \tag{5.33}$$

at $t = 0$, the solution becomes

$$\begin{aligned} \delta(t) &= \frac{v_0}{\omega}\sin\omega t \\ r(t) &= r_0 + \frac{v_0}{\omega}\sin\omega t \end{aligned} \tag{5.34}$$

where we assume that $v_0/\omega r_0 \ll 1$.

The angular motion for circular orbits with small radial oscillations can be calculated from the angular-momentum-conservation relation in

(5.12). Expanding the $1/r^2$ factor in powers of $v_0/\omega r_0$, we have

$$\int_{\theta_0}^{\theta} d\theta = \int_0^t \frac{L}{mr(t)^2} dt \simeq \int_0^t \frac{L}{mr_0^2} \left[1 - 2\frac{\delta(t)}{r_0} \right] dt \qquad (5.35)$$

When we substitute $\delta(t)$ from (5.34), the integration can be carried out. We find

$$\theta(t) = \theta_0 + \frac{L}{mr_0^2} \left[t - \frac{2v_0}{r_0\omega^2} (1 - \cos\omega t) \right] \qquad (5.36)$$

As initial conditions on the angular coordinate, we choose $\theta = 0$ at $t = 0$ and $L = mr_0^2\Omega$. With this choice, L is the angular momentum in a circular orbit with radius r_0 and angular frequency Ω. The solution for the angular position of the particle becomes

$$\theta(t) = \Omega \left[t - \frac{2v_0}{r_0\omega^2} (1 - \cos\omega t) \right] \qquad (5.37)$$

We have thus far considered only a limited class of possible solutions to (5.24). The complete solution to the radial equation can be discussed qualitatively in terms of the effective one-dimensional potential energy, using the methods of § 2.1. From (5.16), (5.17) and (5.19), the effective one-dimensional potential energy for this system is

$$V_{\text{eff}}(r) = mg(r - \ell) + \frac{L^2}{2mr^2} \qquad (5.38)$$

A sketch of $V_{\text{eff}}(r)$ and its two components in the physical range $0 \leq r \leq \ell$ is given in Fig. 5-4. A typical energy E is denoted by the dashed line in the figure. Since the system is conservative, the total energy E is given by (5.19), (5.20), and (5.23) as

$$E = K + V = m\dot{r}^2 + \tfrac{1}{2}mr^2\dot{\theta}^2 + V(r) = m\dot{r}^2 + \frac{L^2}{2mr^2} + V(r) \qquad (5.39)$$

or

$$E = m\dot{r}^2 + V_{\text{eff}}(r) \qquad (5.40)$$

The radial velocity $\dot{r}$ is

$$\dot{r} = \pm\sqrt{\frac{1}{m}[E - V_{\text{eff}}(r)]} \qquad (5.41)$$

The allowed physical region for motion is determined by

$$V_{\text{eff}}(r) \leq E \tag{5.42}$$

For the energy given by the dashed line in Fig. 5-4, the radial motion of the particle is bounded by maximum and minimum radii at the turning points, where $E = V_{\text{eff}}$. For an energy $E \geq L^2/2m\ell^2$, the maximum radius is unbounded and both masses end up on the table when r equals ℓ. The original Lagrangian must be modified for $r \geq \ell$.

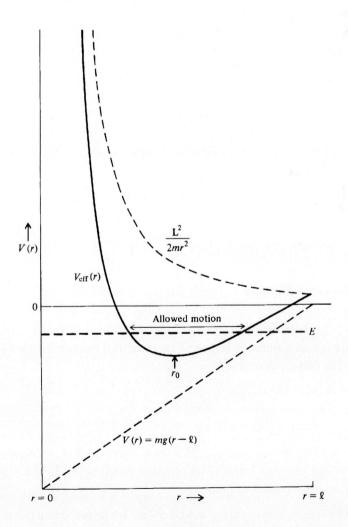

FIGURE 5-4. Effective potential energy for the two-mass system in Fig. 5-3.

The effective potential energy in Fig. 5-4 has a minimum at

$$\left(\frac{dV_{\text{eff}}}{dr}\right)_{r=r_0} = 0 = mg - \frac{L^2}{mr_0^3} \tag{5.43}$$

or

$$L^2 = gm^2r_0^3 \tag{5.44}$$

For $E = V_{\text{eff}}(r_0)$, $\dot{r} = 0$ and the rotating mass moves in a circle, as deduced previously from (5.24). For E slightly larger than $V_{\text{eff}}(r_0)$ the mass on the table undergoes small radial oscillations in the $V_{\text{eff}}(r)$ potential well about $r = r_0$. The frequency of these oscillations can be found from a series expansion as in (5.20). The spring constant k is

$$k = \left(\frac{d^2 V_{\text{eff}}}{dr^2}\right)_{r=r_0} = \frac{3mg}{r_0} \tag{5.45}$$

Since the total mass in the radial-kinetic-energy term is $2m$, the frequency of radial oscillation is

$$\omega = \sqrt{\frac{k}{2m}} = \sqrt{\frac{3g}{2r_0}} \tag{5.46}$$

in agreement with the result given in (5.30).

5.2 Planetary Motion

The most fundamental applications of classical mechanics involve the gravitational and Coulomb forces. The potential energy for these conservative forces can be written

$$V(r) = -\frac{\alpha}{r} \tag{5.47}$$

where $\alpha = Gm_1m_2$ for the gravitational force and $\alpha = -e_1e_2/(4\pi\epsilon_0)$ for the static Coulomb force. The angular momentum $\mathbf{L} = \mathbf{r} \times \mathbf{p}$ is conserved, since the force is central.

The orbit equation relating r and θ can be found from the conservation equations for angular momentum and energy. The energy equation

(5.16) gives

$$\frac{1}{2}m\left(\frac{dr}{dt}\right)^2 = -\frac{L^2}{2mr^2} + \frac{\alpha}{r} + E \tag{5.48}$$

and the angular momentum equation (5.9) gives

$$\left(mr^2\frac{d\theta}{dt}\right)^2 = L^2 \tag{5.49}$$

Dividing (5.48) by (5.49) to elimiate dt, we obtain the differential equation for the orbit

$$\left(\frac{1}{r^2}\frac{dr}{d\theta}\right)^2 = -\frac{1}{r^2} + \frac{2m\alpha}{L^2 r} + \frac{2mE}{L^2} \tag{5.50}$$

The solution for $r(\theta)$ can be found straightforwardly by separation of variables and integration. However a simpler method is to substitute $dr = -r^2 d\left(\frac{1}{r}\right)$ and complete the square in $1/r$ on the right-hand side of (5.50) to obtain

$$\left[\frac{d\left(\frac{1}{r}\right)}{d\theta}\right]^2 = -\left(\frac{1}{r} - \frac{m\alpha}{L^2}\right)^2 + \left(\frac{m\alpha}{L^2}\right)^2\left(1 + \frac{2EL^2}{m\alpha^2}\right) \tag{5.51}$$

The solution to this equation is of the form

$$\frac{1}{r} = \frac{m\alpha}{L^2} + \frac{m\alpha}{L^2}\epsilon\cos(\theta - \theta_0) \tag{5.52}$$

where θ_0 is arbitrary. By substitution we determine ϵ to be

$$\epsilon = \sqrt{1 + \frac{2EL^2}{m\alpha^2}} \tag{5.53}$$

The orbit equation (5.52) is written in the standard form as

$$r(\theta) = \frac{\lambda(1 + \epsilon)}{1 + \epsilon\cos(\theta - \theta_0)} \tag{5.54}$$

where λ is defined by

$$\lambda = \frac{L^2}{m\alpha}\frac{1}{1 + \epsilon} \tag{5.55}$$

The orbit is symmetric around the angle θ_0. For convenience the coordinate axes are often chosen in such a way that $\theta_0 = 0$.

With some effort the orbit equation (5.54) with $\theta_0 = 0$ can be cast in terms of cartesian coordinates $(x = r\cos\theta, y = r\sin\theta)$ in the forms

$$\frac{\left(x + \frac{\epsilon}{1-\epsilon}\lambda\right)^2}{\left(\frac{\lambda}{1-\epsilon}\right)^2} + \frac{y^2}{\left(\lambda\sqrt{\frac{1+\epsilon}{1-\epsilon}}\right)^2} = 1 \qquad 0 \le \epsilon < 1 \qquad (5.56)$$

$$y^2 + 4\lambda x = 4\lambda^2 \qquad \epsilon = 1 \qquad (5.57)$$

$$\frac{\left(x - \frac{\epsilon}{\epsilon-1}\lambda\right)^2}{\left(\frac{\lambda}{\epsilon-1}\right)^2} - \frac{y^2}{\left(\lambda\sqrt{\frac{\epsilon+1}{\epsilon-1}}\right)^2} = 1 \qquad \epsilon > 1 \qquad (5.58)$$

Equation (5.54) represents conic sections with a focus at $r = 0$. The type of conic section depends on the values of the parameters ϵ and λ as follows:

$$
\begin{array}{lll}
0 \le \epsilon < 1 & \lambda > 0 & \text{ellipse } (\epsilon = 0 \text{ is a circle}) \\
\epsilon = 1 & \lambda > 0 & \text{parabola} \\
\epsilon > 1 & \lambda < 0 \text{ or } \lambda > 0 & \text{hyperbola}
\end{array}
\qquad (5.59)
$$

Sketches of these orbits with $\theta_0 = 0$ are shown in Fig. 5-5. The $\lambda > 0$ requirement for the elliptical and parabolic orbits follows from the positivity requirement on r in (5.54). For $\lambda > 0$, the angle $\theta = \theta_0$ corresponds to the turning point of minimum r, with $r_{min} = \lambda$. For $\lambda < 0$, this turning point occurs at $\theta = \theta_0 + \pi$, as shown in the hyperbolic orbits of Fig. 5-5, with $r_{min} = \lambda[(1+\epsilon)/(1-\epsilon)]$.

From (5.55) we observe that λ is positive for an attractive potential energy $(\alpha > 0)$ and negative for a repulsive potential energy $(\alpha < 0)$. By reference to (5.53) and (5.59), we conclude that the α and energy ranges for the three types of orbits are

$$
\begin{array}{lll}
\text{Ellipse} & \alpha > 0 & E < 0 \\
\text{Parabola} & \alpha > 0 & E = 0 \\
\text{Hyperbola} & \alpha > 0 \text{ or } \alpha < 0 & E > 0
\end{array}
\qquad (5.60)
$$

with $V = 0$ at $r = \infty$.

The motion about the sun of the planets in our solar system is governed by the gravitational potential energy, which has $\alpha > 0$. Of the conic sections only the ellipse is an orbit of finite extent. Thus, from (5.60), all planetary orbits have $E < 0$.

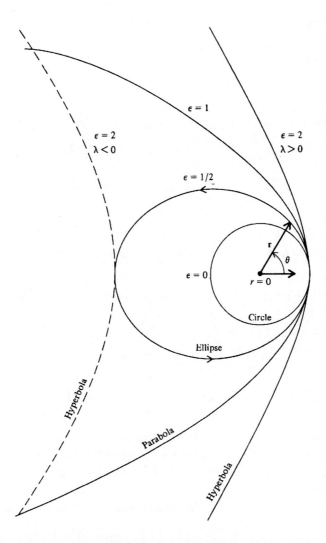

FIGURE 5-5. Sketches of representative conic-section orbits.

For an elliptic orbit, the semimajor axis a and semiminor axis b are commonly specified in place of λ and ϵ. From (5.56) the lengths of these axes are related to λ and ϵ by

$$a = \frac{\lambda}{1-\epsilon}$$

$$b = \lambda\sqrt{\frac{1+\epsilon}{1-\epsilon}}$$

(5.61)

The ratio of b to a is determined by the eccentricity ϵ

$$\frac{b}{a} = \sqrt{1 - \epsilon^2} \tag{5.62}$$

Using (5.61) and (5.62), along with (5.53) and (5.46), the semimajor and semiminor axes of the ellipse are given in terms of the energy and angular momentum,

$$a = \frac{\lambda}{1 - \epsilon} = \frac{\lambda(1 + \epsilon)}{1 - \epsilon^2} = \frac{(L^2/m\alpha)}{(-2EL^2/m\alpha^2)} = -\frac{\alpha}{2E} = \frac{\alpha}{2|E|}$$

$$b = a\sqrt{1 - \epsilon^2} = \left(\frac{-\alpha}{2E}\right)\left(\frac{-2EL^2}{m\alpha^2}\right)^{1/2} = \frac{L}{\sqrt{2m|E|}} \tag{5.63}$$

Thus the energy of the orbit is fixed by the length of the semimajor axis, independently of the value of L

$$E = -\frac{\alpha}{2a} \tag{5.64}$$

In the special case $\epsilon = 0$, an elliptical orbit reduces to a circle. The eccentricity of the moon's orbit about the earth is $\epsilon = 0.055$. The eccentricity of the earth's orbit about the sun is $\epsilon = 0.017$. Both orbits are therefore very nearly circular. In a circular orbit the semimajor axis is just the radius of the circle. From (5.64) and (5.47) we then find

$$E = \tfrac{1}{2}V(a) = -K \tag{5.65}$$

where the second equality follows from $K = E - V$.

If material orbiting the sun were in noncircular orbits there would be a much higher probability of collisions of material in different orbits. Over the five or so billion years since the solar system formed there may have been many such collisions resulting in nearly circular orbits.

We can use the circular-orbit relation of (5.65) to resolve the so-called *satellite paradox*: The effect of the slight atmospheric drag on a satellite in a circular orbit at a height of several hundred kilometers above the earth is to increase the speed of the satellite, contrary to intuition. The atmospheric drag converts mechanical energy into heat. Hence the energy E of the satellite decreases and by (5.64) the radius a of the orbit decreases. Since the atmosphere is quite thin at this altitude, the satellite makes many orbits before its orbital height is appreciably changed, and the orbit remains nearly circular. The decrease in E must be accompanied by an increase in K by (5.65). Since the kinetic energy increases, the satellite speeds up.

In Coulomb scattering of charged particles from a fixed scattering center, $E > 0$, and so the particle trajectories are hyperbolic orbits. For hyperbolic orbits the $r \to \infty$ asymptotes of (5.54) with $\theta_0 = 0$ are given by

$$\theta_{\text{asy}}^{\pm} = \pm \arccos\left(-\frac{1}{\epsilon}\right) \tag{5.66}$$

5.3   Kepler's Laws

The observed data on planetary motion were reduced by Kepler in the early seventeenth century to three empirical laws. These laws played an important role in Newton's discovery of the gravitational-force law. The first law states that the orbit of a planet is an ellipse with the sun at one focus. We have established this law in § 5.2 from the inverse-square nature of the gravitational force. The result neglects perturbations due to the presence of the other planets.

The second law of Kepler states that the time rate of change of area swept out by the radius from the sun to a planet is a constant, as illustrated in Fig. 5-6. For this to happen the planet must have higher tangential velocities at smaller radial distances from the sun. The second law is nothing but angular-momentum conservation as we now demonstrate. From Fig. 5-7 the element of area swept out in dt is

$$dA = \tfrac{1}{2}r^2 d\theta = \tfrac{1}{2}r^2 \dot{\theta} dt \tag{5.67}$$

so that by (5.10)

$$\frac{dA}{dt} = \frac{L}{2m} = \text{constant} \tag{5.68}$$

Kepler's third law states that the square of the period of revolution about the sun is proportional to the cube of the semimajor axis of the elliptical orbit. To derive the third law we use the constancy of the

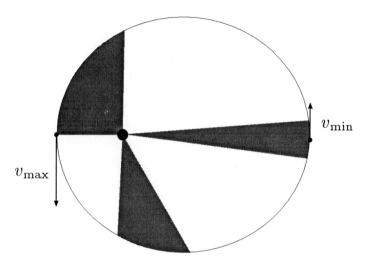

FIGURE 5-6. Areas swept by radius from the sun to a planet in equal time intervals.

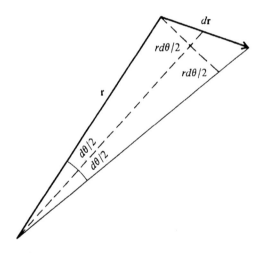

FIGURE 5-7. Element of area swept out by a radius vector in an infinitesimal time interval dt.

angular momentum and integrate (5.68) over a complete revolution.

$$\int_0^\tau dt = \frac{2m}{L} \int dA$$

$$\tau = \frac{2mA}{L} \tag{5.69}$$

where A is the area of the elliptical orbit. This area is given in terms of the semimajor and semiminor axes by

$$A = \pi ab = \pi a^2 (1 - \epsilon^2) = \frac{\pi^2 a^3 L^2}{m\alpha} \tag{5.70}$$

where (5.62), (5.53) and (5.64) have been used. Substitution into (5.69) completes the derivation of Kepler's third law.

$$\tau^2 = \left(\frac{4\pi^2 m}{\alpha} \right) a^3 \tag{5.71}$$

Since $\alpha = GmM_{\odot}$, where $M_{\odot}$ is the mass of the sun, the ratio

$$\frac{\tau^2}{a^3} = \frac{4\pi^2}{GM_{\odot}} \tag{5.72}$$

is independent of the mass of the planet, in the approximation that the position of the sun is unaffected by the planet.

In Table 5-1 we use the observed semimajor axes and periods of the planets and of the moons of Jupiter to test Kepler's third law. For each orbiting body we compute a^3/τ^2, which should be proportional to the mass of the central body. We observe from the table that the ratio of the sun's mass to Jupiter's is $M_{\odot}/M_J = 2.51 \times 10^{19} / 2.40 \times 10^{16} = 1045$

Kepler's laws neglect the motion of the sun. A correction term to (5.72) due to the sun's motion will be derived in §6.1. The correction term is proportional to $m/M_{\odot}$, and is thus very small.

TABLE 5-1

Planet	τ (d)	a (km)	a^3/τ^2 (km^3/d^2)
Mercury	87.97	0.5791×10^8	2.510×10^{19}
Venus	224.7	1.082	2.509
Earth	365.3	1.496	2.509
Mars	687.0	2.279	2.508
Jupiter	4333	7.783	2.511
Saturn	10760	14.27	2.510
Uranus	30685	28.69	2.508
Nuptune	60190	44.98	2.512
Pluto	90700	59.0	2.497

Moons of Jupiter	τ (d)	a (km)	a^3/τ^2 (km^3/d^2)
Metis	0.294	0.1280×10^6	2.43×10^{16}
Adrastea	0.297	0.1290	2.43
Amalthea	0.498	0.180	2.35
Thebe	0.674	0.222	2.41
*Io	1.769	0.422	2.40
*Europa	3.551	0.671	2.40
*Ganymede	7.155	1.070	2.39
*Callisto	16.69	1.885	2.40
Leda	240	11.11	2.38
Himalia	251	11.47	2.40
Lysithea	260	11.71	2.38
Elara	260	11.74	2.39
Ananke	631	21.20	2.39
Carrme	692	22.85	2.33
Pasiphac	735	23.33	2.35
Sinope	758	23.37	2.22

*Galilean moons

5.4 Satellites and Spacecraft

The orbits of satellites and spacecraft are interesting problems in celestial mechanics. For a satellite in a circular orbit at a distance h above the earth, the period of revolution is found from adaptation of (5.72) to be

$$\tau = \frac{2\pi}{\sqrt{GM_E}}(R_E + h)^{3/2} \qquad (5.73)$$

where M_E is the mass of the earth and R_E is its radius. From (1.7) we can write the quantity GM_E in terms of the gravitational acceleration at the surface of the earth

$$GM_E = gR_E^2 \qquad (5.74)$$

Earth satellite periods for circular orbits at various heights are given in

Table 5-2. If $h \ll R_E$, τ is given to a good approximation by

$$\tau \simeq 2\pi \sqrt{\frac{R_E}{g}} \simeq 2\pi \sqrt{\frac{6,371 \times 10^3}{9.8}} \tag{5.75}$$

$$\simeq 5,100 \text{ s} \simeq 1.4 \text{ h}$$

The velocity of the satellite in a low-altitude circular orbit about the earth is

$$v_c = (R_E + h)\omega \simeq R_E \left(\frac{2\pi}{\tau}\right) \simeq (6,371)\frac{2\pi}{5,100} \simeq 7.9 \text{ km/s} \tag{5.76}$$

This velocity is necessarily less than the escape velocity of 11.2 km/s, discussed in § 2.2.

TABLE 5-2   EARTH SATELLITE PERIODS

Altitude h (km above surface of earth)	Period τ (h)
0	1.41
200	1.47
500	1.58
1,680	2.00
35,850	24.00

To study storm systems, the U.S. National Aeronautics and Space Administration (NASA) launched a weather satellite into a polar orbit which goes out to 6545 km from the earth's surface and swings in to 200 km. Thus the major axis of the orbit is $2a = 2R_E + 200 \text{ km} + 6545 \text{ km} = 19,487 \text{ km}$. With a circular satellite orbit at an altitude of $h \approx 200 \text{ km}$ as a reference, we can calculate the period of this weather satellite from Kepler's third law in (5.71).

$$\tau' = \tau \left(\frac{a'}{a}\right)^{3/2} = (1.47 \text{ hr}) \left(\frac{9744}{6,571}\right)^{3/2} = 2.65 \text{ h} \simeq \frac{24}{9} \text{ h} \tag{5.77}$$

The satellite makes nine orbits in 24 h, and therefore as the earth rotates perigee (point of closest approach) occurs over the same nine points on

the earth each day. The velocity at any point on the elliptical orbit of the
weather satellite (or for that matter any other satellite) can be calculated
by comparing two expressions for satellite energy (5.64)

$$E = K + V = \frac{1}{2}mv^2 - \frac{\alpha}{r} = -\frac{\alpha}{2a} \tag{5.78}$$

The escape velocity from the surface of the earth is obtained with $E = 0$,

$$\frac{1}{2}mv_{esc}^2 = \frac{\alpha}{R_E} \tag{5.79}$$

Eliminating α/m in the above two equations we can express the velocity
of the satellite at any point of its orbit as

$$v = v_{esc}\sqrt{\left(\frac{1}{r} - \frac{1}{2a}\right)R_E} \tag{5.80}$$

Recall from (2-15) that the escape velocity from the earth is

$$v_{esc} = \sqrt{2gR_E} = 11.2\,\text{km/s} \tag{5.81}$$

At perigee the velocity reaches its maximum value, and at apogee (farthest distance) the velocity reaches its minimum value. For the weather
satellite discussed above the velocity at perigee ($r_p = 6571$ km) is $v_p =
9.12$ km/s, and the velocity at apogee ($r_a = 12{,}920$ km) is $v_a = 4.57$ km/s.

A spacecraft in a circular orbit of radius R_c around the earth can
be most economically inserted into an elliptical orbit at a distance of
closest approach R_c by firing rockets at perigee. A rocket burn at perigee
increases the velocity perpendicular to the radius vector, without change
in the $\mathbf{v} \cdot \mathbf{r} = 0$ condition as a turning point on the orbit. The increase in
velocity is accompanied by an increase in energy and angular momentum.
From (5.53) and (5.59) the orbit is thereby changed from circular to
elliptical. The procedure can be used in reverse to convert an elliptical
orbit to a circular one by firing retrorockets at the distance of closest
approach. This technique was followed in the lunar orbit insertion for
the Apollo lunar landings of 1969 to 1971. The Apollo spacecraft was
first inserted in an elliptical lunar orbit, as illustrated in Fig. 5-8. After
two orbits a retrograde burn was used to circularize the orbit prior to the
landing on the lunar surface.

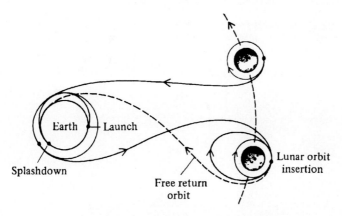

FIGURE 5-8. Schematic trajectory of the Apollo moon missions from earth launch to splashdown.

5.5 Grand Tours of the Outer Planets

Several direct spacecraft missions have been made to the two nearest planetary neighbors of Earth—Venus and Mars. A great difficulty in outer-planet exploration is the long time duration of direct flights. Fortunately, the flight times for outer-planet missions can be considerably shortened by means of *gravitational assists* as the spacecraft swings by the planets en route. In the late 1970s the outer planets lined up in a favorable configuration that permitted a single spacecraft to make a *Grand Tour* of the planets Jupiter, Saturn, Uranus, and Neptune; see Fig. 5-9. The possibility of this four-planet mission occurs only at 175-year intervals. By utilizing the gravitational energy boost obtained from a Jupiter swingby, the Grand Tour of these four planets was made in 12 years. In comparison, the flight time for a direct mission to Neptune with equivalent launch energy would take 30 years. The essential aspects of the gravity-assistance trajectory for the Grand Tour can be developed from the planetary-orbit equations derived in preceding sections. By a similar mechanism a planetary swingby can act as a brake. The Mercury Mariner voyage of 1974 used the planet Venus to reduce energy allowing a subsequent close approach to the planet Mercury.

The earth's orbital velocity represents a substantial fraction of the minimum launch velocity needed to send a spacecraft to the outer parts of our solar system. Thus in sending a spacecraft to the outer planets, the launch should be made in the direction of the earth's orbital velocity about the sun, as illustrated in Fig. 5-10. This velocity of the earth, in a

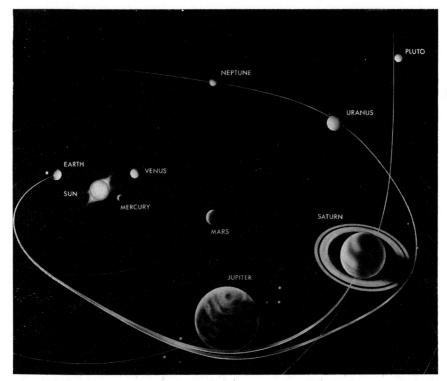

FIGURE 5-9. Trajectories of two gravitationally-assisted Grand Tours of the outer planets. *(Photo provided courtesy of the Jet Propulsion Laboratory, California Institute of Technology.)*

nearly circular orbit of radius a_E and period τ_E about the sun, is

$$V_E = \omega_E a_E = \frac{2\pi}{\tau_E} a_E = \frac{(2\pi)(1.5 \times 10^8 \text{km})}{(365 \times 24 \times 3,600\text{s})} = 30 \,\text{km/s} \qquad (5.82)$$

For a spacecraft of mass m at an initial distance a_E from the sun to completely escape the gravitational pull of the sun, the minimum initial velocity necessary is determined by

$$E = 0 = \tfrac{1}{2}m(v_{\text{esc}}^\odot)^2 - \frac{GmM_\odot}{a_E} \qquad (5.83)$$

On the other hand, for the circular orbit of the earth about the sun,

$$\frac{m_E(V_E)^2}{a_E} = \frac{Gm_E M_\odot}{a_E} \qquad (5.84)$$

From (5.82) to (5.84) we find

$$v_{\text{esc}}^\odot = \sqrt{\frac{2GM_\odot}{a_E}} = \sqrt{2}\, V_E = 42 \,\text{km/s} \qquad (5.85)$$

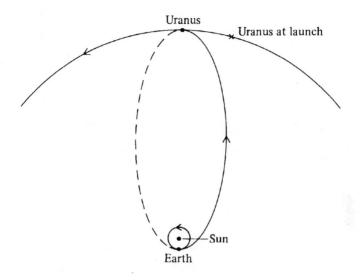

FIGURE 5-10. Elliptical orbit from Earth to Uranus.

This relation between the escape velocity at a distance r from a gravitational source and the velocity in a circular orbit at radius r is always true; that is, $v_{\text{esc}}(r) = \sqrt{2}\, V_E(r)$. By making the launch from the moving earth, the initial velocity required for escape from the gravitational pull of the sun can be reduced to

$$(\sqrt{2} - 1)V_E = 12\,\text{km/s} \tag{5.86}$$

The spacecraft must have additional initial velocity to escape from the gravitational attraction of the earth.

Direct Uranus Mission

To send a direct mission to the planet Uranus with a minimum amount of propulsion energy, the spacecraft should be launched in the direction of the earth's orbital motion into an elliptical orbit about the sun with perihelion at the earth's orbit and aphelion at the orbit of Uranus, as shown in Fig. 5-10. The launch must be made at the proper time in order that Uranus and the spacecraft arrive together at the aphelion of the spacecraft's orbit. The minimum and maximum values of the distance r from the sun on this spacecraft orbit are

$$r_{\text{min}} = 1\,\text{AU} \qquad \text{(at Earth)}$$
$$r_{\text{max}} = 19.2\,\text{AU} \qquad \text{(at Uranus)}$$

where AU stands for the *astronomical unit* of length, namely, the sun to earth distance of $a_E \simeq 1.5 \times 10^8$ km. The parameters λ and ϵ of the orbit equation (5.54) can be determined from the minimum and maximum values of r,

$$r_{\min} = \lambda$$

$$r_{\max} = \lambda \frac{1+\epsilon}{1-\epsilon} \tag{5.87}$$

giving

$$\lambda = 1\,\text{AU}$$

$$\epsilon = 0.9 \tag{5.88}$$

Thus the spacecraft orbit from Earth to Uranus is

$$r = \frac{1.9}{1 + 0.9\cos\theta} \tag{5.89}$$

The semimajor axis of the orbit is

$$a = \tfrac{1}{2}(r_{\min} + r_{\max}) = 10.1\,\text{AU} \tag{5.90}$$

The velocity of the spacecraft at any point on the orbit can be found by adapting (5.80) to this case. The perihelion velocity necessary for insertion of the spacecraft at $r = a_E$ into the elliptical orbit to Uranus as calculated from (5.80), taking $r = R_E$ and substituting (5.85) and (5.90), is

$$v_p = v_{\text{esc}}^{\odot} \sqrt{\frac{19.2}{20.2}} = 41\,\text{km/s} \tag{5.91}$$

The time after earth launch at which the spacecraft reaches Uranus is just the half period $\tau/2$ of the elliptical orbit. We can use Kepler's third law in (5.72) to calculate the time duration of this mission from the radius a_E and period τ_E of the earth's orbit about the sun.

$$\frac{\tau}{2} = \frac{\tau_E}{2} \left(\frac{a}{a_E} \right)^{3/2} = \tfrac{1}{2}(10.1)^{3/2} \approx 16 \text{ years} \tag{5.92}$$

Gravity Boost Mission to Uranus

For the same launch energy as needed for the elliptical orbit, the duration of flight to Uranus can be cut from 16 years to about 5 years on a gravity-assistance orbit which swings by Jupiter, as we will now demonstrate. The spacecraft is initially launched from Earth into an elliptical orbit about the sun. For our Grand Tour comparison the first portion of the gravity boost orbit is taken to be the same as the direct mission to Uranus. The launch time is chosen such that the spacecraft will make a close encounter with Jupiter, as illustrated in Fig. 5-11. As a result of the encounter the heliocentric velocity of the spacecraft is changed. In our discussion, we can neglect the slight change in the direction of Jupiter's velocity during the encounter, since the time duration of the encounter is short compared with Jupiter's period of revolution around the sun. Until the spacecraft reaches the immediate vicinity of Jupiter, the spacecraft's orbit is governed by the strong gravititational field of the sun. In the vicinity of Jupiter the sun's gravitational force on the spacecraft changes slowly compared to Jupiter's gravitational force so the spacecraft's orbit relative to Jupiter is essentially determined by Jupiter's gravitational field. [The equation of motion for the relative coordinate will be discussed in §6.1.]

If we let $\mathbf{v}_i$ and $\mathbf{v}_f$ denote the spacecraft momenta in the heliocentric (sun-centered) inertial frame just before and just after the Jovian encounter, we can write the Galilean transformation

$$\mathbf{v}_i = \mathbf{u}_i + \mathbf{V}_J$$
$$\mathbf{v}_f = \mathbf{u}_f + \mathbf{V}_J$$

$$(5.93)$$

where $\mathbf{V}_J$ is the velocity of Jupiter about the sun and $\mathbf{u}_i$ and $\mathbf{u}_f$ are the spacecraft velocities relative to Jupiter (*i.e.*, in the reference frame in which Jupiter is at rest). From (5.93) the change in velocity during the encounter,

$$\Delta \mathbf{v} \equiv \mathbf{v}_f - \mathbf{v}_i \tag{5.94}$$

is the same in both frames

$$\Delta \mathbf{v} = \Delta \mathbf{u} \tag{5.95}$$

and for an elastic collision in the CM (Jupiter fixed) frame

$$u_f = u_i \tag{5.96}$$

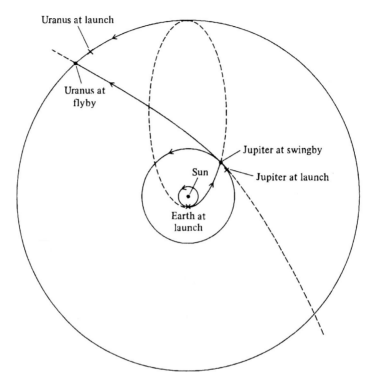

FIGURE 5-11. Orbit to Uranus on a gravity-assistance trajectory which swings past Jupiter.

The change in squared heliocentric velocity measures the "gravity boost" in kinetic energy $\Delta K = \frac{1}{2}m(v_f^2 - v_i^2)$. By squaring the two equations (5.93)

$$
\begin{aligned}
v_f^2 &= u_f^2 + 2\mathbf{u}_f \cdot \mathbf{V}_J + V_J^2 \\
v_i^2 &= u_i^2 + 2\mathbf{u}_i \cdot \mathbf{V}_J + V_J^2
\end{aligned}
\tag{5.97}
$$

and then subtracting we obtain

$$
v_f^2 - v_i^2 = 2\Delta\mathbf{u} \cdot \mathbf{V}_J = 2\Delta\mathbf{v} \cdot \mathbf{V}_J
\tag{5.98}
$$

The magnitude of the heliocentric velocity increases (or decreases) depending on whether the projection of $\Delta\mathbf{u}$ on $\mathbf{V}$ is positive (or negative). It is easiest to appreciate the implications of (5.98) in the planet rest frame. If the spacecraft crosses the planet's orbit behind the planet, then $\Delta\mathbf{u} \cdot \mathbf{V}$ will be positive and a graviational boost will result. If, on the other hand, the spacecraft crosses the planet's path in front, $\Delta\mathbf{u} \cdot \mathbf{V}$ will be negative and the effect will be to brake the spacecraft. The two situations are illustrated in Fig. 5-12 by the actual cases of the Voyager mission with a gravitational boost from Jupiter, and the Mercury Mariner mission, which used Venus as a brake.

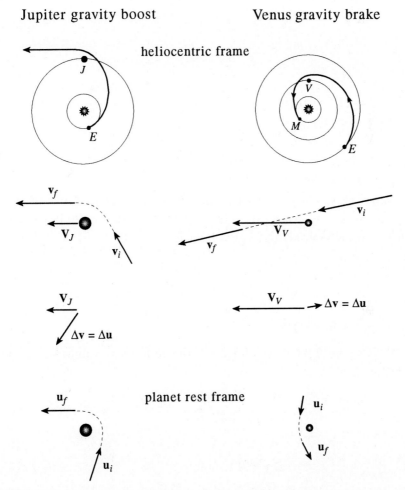

FIGURE 5-12. Velocity diagrams illustrating spacecraft velocities for a gravity boost trajectory around Jupiter and a gravity brake trajectory around Venus.

For a given $\mathbf{v}_i$ the maximum gravity boost is achieved with $\mathbf{v}_f$ parallel to $\mathbf{V}_J$, as can be seen from the velocity diagram in Fig. 5-12. Since $\mathbf{v}_f = \mathbf{u}_f + \mathbf{V}_J$, $\mathbf{u}_f$ is also parallel to $\mathbf{V}_J$ and

$$v_f = V_J + u \qquad\qquad (5.99)$$

From $\mathbf{u}_i = \mathbf{v}_i - \mathbf{V}_J$ the value of u is

$$u = \sqrt{V_J^2 + v_i^2 - 2\mathbf{v}_i \cdot \mathbf{V}_J} \qquad (5.100)$$

The magnitude of Jupiter's nearly circular orbit is

$$V_J = \frac{2\pi}{\tau_J} a_J \qquad (5.101)$$

where $\tau_J = 11.9$ years is Jupiter's period. The velocity v_i of the spacecraft as it nears Jupiter is

$$v_i = v_{esc}^\odot \sqrt{\frac{1}{a_J} - \frac{1}{2a}} \qquad (5.102)$$

The value of $\mathbf{v}_i \cdot \hat{\mathbf{V}}_J = v_{i\theta}$ can be found from the conserved angular momentum of the spacecraft orbit, equating L at the Jupiter encounter with its perihelion value at earth launch

$$\frac{L}{m} = v_{i\theta} a_J = v_p a_E \qquad (5.103)$$

with v_p given by (5.91).

With $a_E = 1\,\mathrm{AU}$, $a_J = 5.2\,\mathrm{AU}$ the numerical values of the above quantities are

$$\begin{aligned}
V_J &= 13\,\mathrm{km/s} \\
v_i &= 16\,\mathrm{km/s} \\
v_{i\theta} &= \mathbf{v}_i \cdot \hat{\mathbf{V}}_J = 8\,\mathrm{km/s} \qquad (5.104) \\
u &= 14.7\,\mathrm{km/s} \\
v_f &= V_J + u = 27.7\,\mathrm{km/s}
\end{aligned}$$

The spacecraft leaves the region of Jupiter's influence with its exit velocity $\mathbf{v}_f$ parallel to Jupiter's velocity. The outgoing orbit of the spacecraft around the sun is another conic section with a turning point at the location of the encounter with Jupiter (that is, $\dot{r} = 0$ at $r = a_J$ since $\mathbf{v}_f$ is parallel to the circular orbit of Jupiter). The type of new heliocentric conic section of the spacecraft orbit depends on the amount of velocity boost in the encounter. Since the escape velocity $V_{esc}^\odot$ from the solar system at Jupiter's orbit is

$$V_{\text{esc}}^{\odot} = \sqrt{2}V_J \tag{5.105}$$

the new orbit of the spacecraft is related to the velocity v_f as follows:

$$
\begin{array}{ll}
v_f < \sqrt{2}V_J & \text{ellipse} \\
v_f = \sqrt{2}V_J & \text{parabola} \\
v_f > \sqrt{2}V_J & \text{hyperbola}
\end{array} \tag{5.106}
$$

For the encounter considered above,

$$\frac{v_f}{\sqrt{2}V_J} = 1.5 \tag{5.107}$$

and the new orbit is hyperbolic, as illustrated in Fig. 5-11.

The preceding analysis shows how gravity assisted dynamics works. It remains to be shown that the distance of closest approach to Jupiter exceeds its radius. In fact $r_{\min} = 1.85 R_J$ for this slingshot orbit. Once we have worked out the orbit parameters, the time of flight in any portion of the orbit can be computed. In our example these times are

$$
\begin{aligned}
t \text{ (earth to Jupiter)} &\simeq 1.3 \, \text{years} \\
t \text{ (Jupiter to Uranus)} &\simeq 3.7 \, \text{years}
\end{aligned} \tag{5.108}
$$

yielding a total trip time to Uranus of about 5 years compared with the 16 years required for a direct mission. Of course, these numbers are approximate, since the gravitational influence of Jupiter and the sun on the spacecraft were treated independently. Numerical methods can be used to make precise calculations of the orbit without such an approximation.

In the late 1970s two Grand Tour missions by NASA were launched, Voyagers 1 and 2 in 1979, which have provided vast amounts of new information on Jupiter, Saturn, Uranus, Neptune and their associated moons and rings. Voyager 1 left the earth September 5, 1977; visited Jupiter March 5, 1979; Saturn November 12, 1980; then left the solar system. Voyager 2 left the earth August 20, 1977; visited Jupiter July 9, 1979; Saturn August 2,5 1981; Uranus January 24, 1986; then Neptune on August 24, 1989, before leaving the solar system. Other spacecraft to use gravitational boosts from Jupiter were the Pioneer 10 and 11, launched in 1973 and 1974, which flew by Jupiter and Saturn. The trajectories for these Grand Tour missions are illustrated in Fig. 5-13. After the last planetary encounter, the spacecraft on these missions continued to travel away from the sun, escaping the solar gravitational field and entering interstellar space.

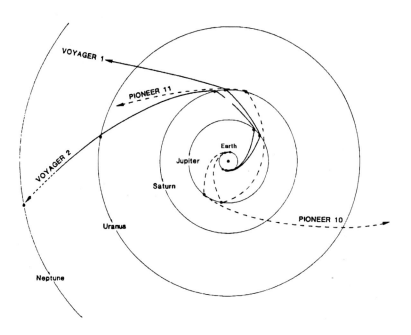

FIGURE 5-13. Flight paths of the *Pioneer* and *Voyager* spacecraft among the outer planets.

5.6 Rutherford Scattering

An important physical example in which hyperbolic orbits are realized is the Coulomb scattering of charged particles. For the scattering of a light particle with charge e_1 and mass m by a heavy particle of charge e_2 and mass M, the location of the scattering center can be regarded as essentially fixed at at the position of M at $r = 0$. The potential energy of the interaction is then given by (5.47), with $\alpha = -e_1 e_2/4\pi\epsilon_0$. For e_1 and e_2 of the same (opposite) sign, the interaction is repulsive (attractive). The initial conditions can be specified in various ways, as for example by the energy E and the angular momentum L, or by the initial velocity v_0 and impact parameter b.

$$E = \tfrac{1}{2}mv_0^2$$
$$L = mv_0b \tag{5.109}$$

We need to find the relation between the impact parameter and the laboratory scattering angle in order to calculate the number of particles scattered into a given angular range. From (5.66) the asymptotes ($r \to \infty$) of the trajectory occur at angles $\theta_{asy}^{\pm} = \pm \arccos(-1/\epsilon)$. From the geometry of Fig. 5-14, the angle of scattering θ_s of the particle from the incident direction is

$$\theta_s = 2 \text{arc cos} \left(-\frac{1}{\epsilon} \right) - \pi \tag{5.110}$$

for both attractive and repulsive potentials. The result can be written

$$\cos \left(\frac{\theta_s}{2} + \frac{\pi}{2} \right) = -\frac{1}{\epsilon} \tag{5.111}$$

or

$$\sin \frac{\theta_s}{2} = \frac{1}{\epsilon} \tag{5.112}$$

Since $\epsilon = \sqrt{1 + \frac{2EL^2}{m\alpha^2}}$ we can express θ_s in terms of b and v_0 as

$$\sin \frac{\theta_s}{2} = \frac{1}{\sqrt{1 + (mv_0^2 b/\alpha)^2}} \tag{5.113}$$

Solving this equation, the impact parameter b and scattering angle θ_s of of a single incident particle are related by

$$b = \frac{|\alpha|}{mv_0^2} \cot \frac{\theta_s}{2} \tag{5.114}$$

Smaller values of b correspond to larger values of θ_s.

All incident particles with impact parameters less than or equal to some particular impact parameter b will have scattering angles greater than or equal to θ_s given by (5.113) with a maximum $\theta_s = \pi$ when $b = 0$. The incident intensity (particles per second per target area) is

$$I_0 = \frac{N_0}{A} \tag{5.115}$$

where N_0 is the number of incident particles per second and A is the area of the beam; see Fig. 5-15. The number per second that are scattered

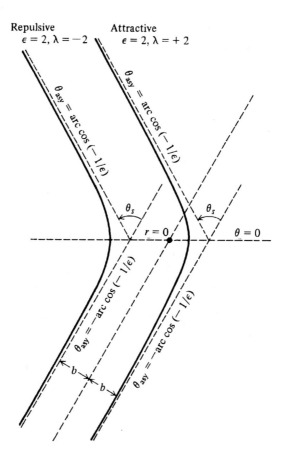

FIGURE 5-14. Hyperbolic orbits for scattering from an attractive- or repulsive-force center at $r = 0$.

into angles greater than θ_s is

$$N = I_0 \pi b^2 \tag{5.116}$$

The quantity πb^2, with units of area, is called the *cross section*

$$\sigma = \pi b^2 = N/I_0 \tag{5.117}$$

For more than one scattering center, the cross section is defined as the number scattered per target particle per second divided by the incident intensity. The incident intensity or flux is a property of the experimental conditions and not of the force laws between the particles. For this reason, comparisons of theory and experiment are made for the cross section

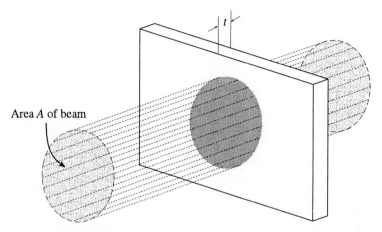

FIGURE 5-15. Particle flux in a scattering experiment. The beam has area A and N_0 particles per second enter the target.

rather than the number of scattered particles. We have assumed that the incident particles are uniformly distributed over the target area A.

Under normal experimental conditions there are multiple scattering centers of atomic dimensions. In these circumstances it is not possible to make measurements of the impact parameters of individual particles and the cross section concept is essential. For a foil of thickness t and n nuclei per unit volume, the total number of nuclei is nAt. Then if the foil is sufficiently thin that the nuclei do not overlap, the number of scatters through angles greater than θ_s is

$$N = I_0 \pi b^2 nt A \qquad (5.118)$$

The number of incident particles is $N_0 = I_0 A$, so the fraction of particles scattered at angles greater than θ_s is

$$f = \pi b^2 nt \qquad (5.119)$$

The particles that are incident in a range of impact parameters between b and $b - db$ are scattering into the angular range θ_s to $\theta_s + d\theta_s$, as shown in Fig. 5-16. The differential area of this annulus is

$$d\sigma = 2\pi b |db| \qquad (5.120)$$

Here the absolute value of the db differential is used to insure that the area $d\sigma$ is positive. From the relation between b and θ_s in (5.114), the expression for $d\sigma$ in terms of θ_s is

$$d\sigma = \pi \left(\frac{\alpha}{2mv_0^2}\right)^2 \frac{\sin\theta_s}{\sin^4(\theta_s/2)} d\theta_s \tag{5.121}$$

The quantity $d\sigma$ is called the *differential cross section.*

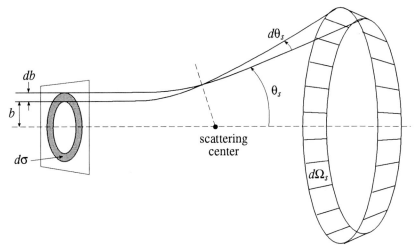

FIGURE 5-16. Repulsive scattering of particles with impact parameters between b and $b - db$.

The number scattered into angles between θ_s and $\theta_s + d\theta_s$ per target nucleus is

$$dN = I_0 d\sigma \tag{5.122}$$

The area subtended by a detector at a distance r from the nucleus that can measure the scattered particles is

$$dA_s = (2\pi r \sin\theta_s)(r d\theta_s) = r^2 d\Omega_s \tag{5.123}$$

where $d\Omega_s$ is the solid angle. Thus the number of scattered per unit area of the detector is

$$\frac{dN}{dA_s} = \frac{I_0}{r^2}\frac{d\sigma}{d\Omega_s} \tag{5.124}$$

where the differential cross section is given by

$$\frac{d\sigma}{d\Omega_s} = \left(\frac{\alpha}{2mv_0^2}\right)^2 \frac{1}{\sin^4(\theta_s/2)} \tag{5.125}$$

The dependence of $d\sigma/d\Omega_s$ on the scattering angle is illustrated in Fig. 5-17. This result was derived in 1911 by Rutherford to explain the experimental results of Geiger and Marsden on the scattering of α particles by heavy nuclei. In the derivation of the Rutherford formula, the assumption was made that no incident particle interacted with more than one target nucleus, which is valid if the target is thin and the scattering angle is not too small.

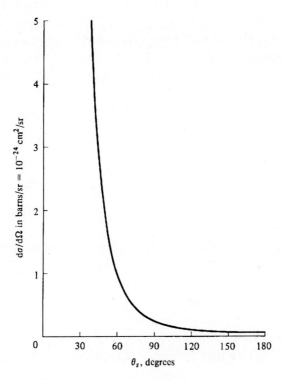

FIGURE 5-17. Rutherford scattering differential cross section for $|\alpha|/E = 10^{-14}$ m.

In the repulsive scattering of α particles ($e_1 = 2e$) by atomic nuclei ($e_2 = Ze$), the distance of closest approach is found from (5.54)to be

$$r_{\min} = \lambda \left(\frac{1+\epsilon}{1-\epsilon} \right) = \frac{\lambda(1+\epsilon)^2}{1-\epsilon^2} \tag{5.126}$$

From (5.53) and (5.55) we note that

$$1 - \epsilon^2 = -\frac{2EL^2}{m\alpha^2} = -\frac{2E\lambda(1+\epsilon)}{\alpha} \tag{5.127}$$

so that

$$r_{min} = \left(-\frac{\alpha}{2E} \right)(1+\epsilon) \tag{5.128}$$

The distance of closest approach can be expressed in terms of the scattering angle θ_s by use of (5.112)

$$r_{min} = \frac{|\alpha|}{2E}\left(1 + \frac{1}{\sin(\theta_s/2)}\right) \tag{5.129}$$

The scatterer probes closest to the nucleus in the large-angle events. At $\theta_s = \pi$, the region down to $r_{min} = |\alpha|/E$ is probed; this minimum distance can also be deduced from conservation of energy. By study of backward scattering events, Rutherford found that the Coulomb potential result in (5.125) held only for energies with $|\alpha|/E > 10^{-14}$ m. This established the size of the typical atomic nucleus to be 10^{-14} m, instead of 10^{-10} m (the size of the atom) as was previously believed.

The integrated Rutherford scattering cross section

$$\sigma = \int_{\theta_{min}}^{\theta_{max}} \frac{d\sigma}{d\Omega_s} d\Omega_s = 2\pi \left(\frac{\alpha}{2mv_0^2}\right)^2 \int_{\theta_{min}}^{\theta_{max}} \frac{\sin\theta_s d\theta_s}{\sin^4(\theta_s/2)} \tag{5.130}$$

is infinite for $\theta_{min} = 0$. This is a consequence of the infinite range of the Coulomb force. Nuclei in ordinary matter are surrounded by an electronic cloud within a radius of 10^{-10} m, forming an electrically neutral atom. Outside the atom the charge of the nucleus is *screened* by the electrons and the Coulomb force no longer holds. Thus θ_{min} is set by the atomic size. The value of θ_{max} is set by the nucleon size since the derivation of the Rutherford formulas fails once the incident particle penetrates the nucleus.

In the scattering off a nucleus in the target, the momentum

$$\mathbf{q} = \mathbf{p}_f - \mathbf{p}_0 \tag{5.131}$$

is transferred to the α particle. The magnitude of $\mathbf{q}$ is related to the scattering angle by

$$q^2 = (\mathbf{p}_f - \mathbf{p}_0)^2 = p_f^2 + p_0^2 - 2\mathbf{p}_f \cdot \mathbf{p}_0 = p_f^2 + p_0^2 - 2p_f p_0 \cos\theta \tag{5.132}$$

For our idealization of an infinitely heavy nucleus ($M_N \to \infty$), the energy-conservation condition

$$\frac{p_0^2}{2m} = \frac{p_f^2}{2m} + \frac{q^2}{2M_N} \tag{5.133}$$

gives

$$p_f = p_0 = mv_0 \tag{5.134}$$

The square of the momentum transfer in (5.132) then reduces to

$$q^2 = 2p_0^2(1 - \cos\theta_s) = 4p_0^2 \sin^2 \frac{\theta_s}{2} \tag{5.135}$$

In terms of this variable the expression for the Rutherford differential cross section in (5.125) simplifies to

$$\frac{d\sigma}{d\Omega_s} = \left(\frac{2\alpha m}{q^2}\right)^2 \tag{5.136}$$

The calculation of Rutherford scattering in quantum mechanics coincidentally gives the same result, though the physical principles are radically different.

PROBLEMS

5.1 Central Forces

5-1. A particle of mass m is subject to two forces, a central force $\mathbf{f}_1$ and a frictional force $\mathbf{f}_2$, with

$$\mathbf{f}_1 = F(r)\hat{\mathbf{r}}$$
$$\mathbf{f}_2 = -b\mathbf{v}, \quad b > 0$$

If the particle initially has angular momentum $\mathbf{L}_0$ about $\mathbf{r} = 0$, find the angular momentum for all subsequent times. *Hint: use* $\mathbf{N} = \dot{\mathbf{L}}$.

5-2. Find the condition for stable circular orbits for a potential energy of the form

$$V(r) = -\frac{c}{r^\lambda}$$

where $\lambda < 2$, $\lambda \neq 0$ and the constant c is positive (negative) if λ is positive (negative). Show that the angular frequency for small radial oscillations ω_r is related to the orbit angular frequency ω_θ by

$$\omega_r = \omega_\theta\sqrt{2 - \lambda}$$

This result implies that the orbit is closed and the motion is periodic only if $\sqrt{2 - \lambda}$ is a rational number. Sketch the orbits for $\lambda = 1$ (Coulomb potential energy), $\lambda = -2$ (harmonic oscillator), $\lambda = -7$, and $\lambda = \frac{7}{4}$.

5-3. A particle of mass m moves under the influence of the force

$$\mathbf{F} = -c^2 \frac{\mathbf{r}}{r^{5/2}}$$

a) Calculate the potential energy.

b) By means of the effective potential energy discuss the motion.

c) Find the radius of any circular orbit in terms of the angular momentum and calculate the period for the orbit.

d) Derive the frequency for small radial oscillations about the circular orbit of part c).

5-4. Find the force law for a central force which allows a particle to move in a spiral orbit given by $r = C\theta^2$, where C is a constant. *Hint: use (5.9), (5.16) and (5.17) to find $V(r)$ in terms of C and the angular momentum L.*

5.2 Planetary Motion

5-5. A planet moves in a circular orbit about a massive star with force law given by

$$\mathbf{F}(\mathbf{r}) = -\frac{\alpha}{r^2}\hat{\mathbf{r}}$$

The star evolves into a supernova and blows off half its mass in a time short compared to the planet's orbit period. (Assume that the supernova explosion is spherically symmetric.) Show that the planet's orbit becomes parabolic.

5-6.a) Calculate the orbital speed and period of revolution of the moon assuming the earth is fixed and the orbit is circular. The earth-moon distance is approximately 384,000 km.

b) Compare the orbital velocity of a satellite in a circular orbit 200 km above the surface of the earth with the orbital velocity in a circular orbit at a similar distance from the surface of the moon. The ratio of lunar to earth mass is $M_L/M_E \approx 1/81.6$. The radii are $R_L = 1741$ km and $R_E = 6,371$ km.

5-7. By jumping, an astronaut can rise vertically 50 cm on earth. Is he in danger of not returning if he jumps while exploring a spherical asteroid of radius 4 km and uniform density $2\,\text{gm}/\text{cm}^3$? Can he by his own exertions launch himself into any orbit? Assume that the initial velocity of the the jump on the asteroid is the same as on the earth.

5-8. The point of a planet's closest approach to the sun is known as the perihelion. The aphelion, which exists only for the bound orbits, is the point furthest away from the sun. Note that in general the perihelion distance is *not* the same as the semiminor axis *b* of an ellipse because the semiminor and semimajor axes are measured from the *center* of the ellipse and the center of force is located at the *focus* of the ellipse. For circular and parabolic orbits about the sun having the same angular momentum show that the perihelion distace of the parabola is one half the radius of the circular orbit, $r_p = \frac{1}{2} r_c$.

5.3 Kepler's Laws

5-9. Suppose the earth's orbital motion about the sun suddenly stops. How much time would elapse before the earth falls into the sun? *Hint: consider a very elongated elliptical orbit with the sun at the focus near one end and the earth on the ellipse at the other end.*

5-10. Consider the motion of a particle in the central force.

$$\mathbf{F} = -k\mathbf{r}$$

Show that:

a) The orbit is an ellipse with the force center at the center of the ellipse.

b) The period is independent of the orbit parameters. *Hint: use cartesian coordinates to solve for the orbit equation.*

5-11. A communications satellite always remains vertically above an observer on the earth's surface. Where on the earth must the observer be located? What direction does the satellite move? What is the radius of the orbit?

5-12. Halley's comet has a period of revolution $\tau = 76$ years around the sun.

a) Determine the semimajor axis a of its orbit in A.U.

b) The observed minimum distance r_{min} of the comet from the center of the sun is 0.6 A.U. What is the maximum distance r_{max}?

c) From r_{min} and r_{max} determine the eccentricity of the orbit.

5-13. The eccentricity of the moon's orbit about the earth is $\epsilon = 0.055$. An undergraduate astronomy student takes telescopic pictures of the full moon six months apart. If the first image, taken when the moon is closest to the earth, is two centimeters in diameter what is the diameter of the second picture? Would the difference be obvious?

5.4 Satellites and Spacecraft

5-14. For a short rocket blast show that the most efficient way to change the energy of an orbit is to fire the rocket parallel (opposite) to the motion at perigee if the energy is to be increased (decreased). *Hint: the change in kinetic energy is the change in total energy.*

5-15. Show that the close-circular-orbit period of a pebble around a boulder is roughly the same as the period of a low-altitude earth satellite (*i.e.*, show that the close-orbit period depends only on the density of the large body).

5-16. The first artificial earth satellite, Sputnik I, was launched in 1957 into an orbit with a perigee (closest approach to earth) of 227 km above the earth's surface. At perigee its speed was 8 km/s. Find its apogee (maximum distance from the earth's center) and its period of revolution.

5-17. For a spacecraft launched from the surface of the earth, find the minimum velocity needed for escape from the solar system. Take into account the gravitational attraction of the earth. *Hint: first for an initial velocity v find the velocity after escape from the earth relative to the earth, then go to a sun-fixed frame and calculate the escape from the sun.*

5-18. For an inverse square elliptic orbit show that the time averages of the kinetic and potential energies satsify

$$2 \langle K \rangle_t = - \langle V \rangle_t$$

This is known as the *Clausius Virial theorem*. The time average is defined by $\langle F \rangle_t \equiv \frac{1}{\tau} \int_0^\tau F(t)\,dt$ where τ is the period. *Hint: rewrite the time average of V as an integration over angle θ. Use the integral $\int_0^{2\pi} \frac{d\theta}{a+b\cos\theta} = \frac{2\pi}{\sqrt{a^2-b^2}}$ for $b < a$. Then $\langle K \rangle_t = \langle E \rangle - \langle V \rangle_t$ with $E = -\alpha/2a$ for the elliptical orbit.*

5.5 Grand Tours of the Outer Planets

5-19. The Mercury Mariner Voyage of 1974 used the planet Venus as a "gravitational brake" to save fuel. The orbit of Mercury is elliptical so it can most easily be approached at its aphelion of 0.47 A.U.

a) A direct mission from Earth to Mercury (initiated beyond the Earth's gravitational influence) might be done by firing a retro rocket to produce an elliptical transfer orbit from Earth to Mercury as shown. Find the orbit parameters and the velocity change needed in the rocket burn.

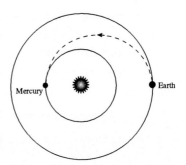

b) In the Mariner mission a retrofire from an Earth orbit around the sun produces an elliptical orbit crossing Venus's orbit at an angle $\theta_V = 110°$ as shown. The spacecraft passes in front of Venus as in Fig. 5-12 giving a braking effect and enters a new orbit passing by Mercury. Find the parameters of the Earth-Venus orbit and compare the retrofire velocity change to that of part a.

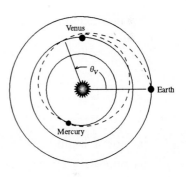

5.6 Rutherford Scattering

5-20. A parallel beam of small projectiles is fired from space toward the moon with initial velocity v_0. What is the collision cross section σ for the projectiles to hit the moon? Express σ in terms of the moon's radius R_L, the escape velocity from the moon v_{esc}^L, and v_0. Neglect the motion of the moon.

5-21. The interaction between an atom and an ion at distances greater than contact is given by the potential energy $V(r) = -C/r^4$. [$C = (e^2/2)P_a^2$, where e is the ion charge and P_a is the polarizability of the atom.] Make a sketch of the effective potential energy versus the radial coordinate. Note that if the total energy of the ion exceeds the maximum value of the effective potential energy, the ion spirals inward to the atom. Find the cross section for an ion of velocity v_0 to strike an atom. Assume that the ion is much lighter than the atom.

5-22. A typical heavy atom consists of a nucleus of radius 10^{-14} m surrounded by an electronic cloud of radius 10^{-10} m which neutralizes the nuclear electric charge for radii larger than 10^{-10} m. For an alpha particle $(Z = 2)$ of 3.6 MeV kinetic energy striking a Magnesium atom $(Z = 12)$ in what range of scattering angles will the Rutherford formula be valid?

5-23. a) From the conic section solution (5.54) for hyperbolic orbits $(\epsilon > 1)$ show that the distance of closest approach is

$$r_{min} = \begin{cases} \lambda & \text{attractive force} \\ (-\lambda)\left(\frac{\epsilon+1}{\epsilon-1}\right) & \text{repulsive force} \end{cases}$$

Using the expressions for λ and $\epsilon^2 = 1 + \frac{2EL^2}{m\alpha^2}$ establish that

$$r_{min} = \begin{cases} \dfrac{\alpha}{2E}(\epsilon - 1) & \text{attractive force} \\ \dfrac{(-\alpha)}{2E}(\epsilon + 1) & \text{repulsive force} \end{cases}$$

b) For a hyperbolic orbit with asymptotic energy E and impact parameter b show that $L^2 = 2mb^2E$ and $\epsilon = 1 + \left(\frac{2Eb}{\alpha}\right)^2$. Then using energy and angular momentum conservation show that the distance of closest approach (where $\dot{r} = 0$) is given by the result in part a.

5-24. In the Jupiter flyby discussed in § 5.5, the spacecraft approaches Jupiter moving at 16 km/s with a component of 8 km/s in Jupiter's direction.

a) Jupiter moves with velocity 13 km/s. Find the velocity and direction of the spacecraft relative to Jupiter.

b) Assuming that the spacecraft emerges in the orbital direction of Jupiter, find the eccentricity of the orbit during the Jupiter flyby. *Hint: use the Rutherford relation between θ_s and ϵ.*

c) Using the fact that $v_{esc}^J = 60$ km/s, find the closest approach to Jupiter in terms of Jupiter's radius. Use the result of the previous problem.

Chapter 6

PARTICLE SYSTEMS AND RIGID BODIES

The physical systems of interest in the real world are collections of many particles. Nevertheless, we do not need to cast away the knowledge accumulated up to this point. The center of mass (often abbreviated CM) of a system of particles moves as if the total external force were applied at the CM point. When the system consists of only two particles the relative motion of the two particles often satisfies an equation of motion analogous to that of a single particle. With rigid bodies the possible motions are translation of the CM and rotations about the CM.

6.1 Center of Mass and the Two-Body Problem

We begin with a concept fundamental to a multiparticle system—that there exists a point, the CM point, which moves exactly like a point particle under the action of the total external force. A physical attribute of the system, such as its energy or angular momentum, can be thought of as this attribute of the CM point plus this same attribute relative to the CM point. For the case of a two-particle system a complete analytic solution of the system is possible.

For the ith particle m_i, Newton's law is

$$\mathbf{F}_i^{\text{ext}} + \mathbf{F}_i^{\text{int}} = m_i \ddot{\mathbf{r}}_i \tag{6.1}$$

where the external force $\mathbf{F}_i^{\text{ext}}$ is the force on m_i from outside the system. The internal force $\mathbf{F}_i^{\text{int}}$ is the force on m_i due to other particles in the system, and is given by

$$\mathbf{F}_i^{\text{int}} = \sum_{j \neq i} \mathbf{F}_i^{[j]} \tag{6.2}$$

where $\mathbf{F}_i^{[j]}$ is the force on m_i due to particle j. Summing the equations of motion (6.1) for all particles yields

$$\sum_i \mathbf{F}_i^{\text{ext}} + \sum_i \mathbf{F}_i^{\text{int}} = \frac{d^2}{dt^2}\left(\sum_i m_i \mathbf{r}_i\right) \tag{6.3}$$

The first term is the total external force acting on the system

$$\mathbf{F}^{\text{ext}} = \sum_i \mathbf{F}_i^{\text{ext}} \tag{6.4}$$

The second term vanishes by use of Newton's third law

$$\sum_i \mathbf{F}_i^{\text{int}} = \sum_{\text{pairs}} \left(\mathbf{F}_i^{[j]} + \mathbf{F}_j^{[i]} \right) = 0 \tag{6.5}$$

Alternatively, the total internal force must vanish as a consequence of translational invariance, as pointed out in § 4.1. Then (6.3) can be rewritten as

$$\mathbf{F}^{\text{ext}} = M\ddot{\mathbf{R}} \tag{6.6}$$

$$\mathbf{R} = \frac{\sum_i m_i \mathbf{r}_i}{\sum_i m_i} = \frac{\sum_i m_i \mathbf{r}_i}{M} \tag{6.7}$$

where the total system mass is $M = \sum_i m_i$. Consequently, the second law of motion holds, not just for a particle, but for an arbitrary body, if the position of the body is interpreted to mean the position of its center of mass. In continuous systems we replace the mass elements m_i by $dm = \rho(r)dV$, where $\rho(r)$ is the mass density and dV is the differential volume element. Then the CM vector is given by

$$\mathbf{R} = \frac{\int \mathbf{r}\rho(\mathbf{r})dV}{\int \rho(\mathbf{r})dV} \tag{6.8}$$

We can measure each particle's position relative to the CM position as

$$\mathbf{r}_i = \mathbf{R} + \mathbf{r}_i' \tag{6.9}$$

If we multiply (6.9) by m_i and sum over all particles we obtain

$$\sum_i m_i \mathbf{r}_i = M\mathbf{R} + \sum_i m_i \mathbf{r}_i' \tag{6.10}$$

which by comparison to the CM point definition (6.7) yields the condition

$$\sum_i m_i \mathbf{r}_i' = 0 \tag{6.11}$$

Time differentiation of $\sum m_i \mathbf{r}_i = M\mathbf{R}$ gives the system momentum

$$\mathbf{P} = \sum m_i \dot{\mathbf{r}}_i = M\dot{\mathbf{R}} = M\mathbf{V} \tag{6.12}$$

From time differentiation of (6.11) the total momentum relative to the

CM point is

$$\sum_i \mathbf{p}'_i = 0 \tag{6.13}$$

Thus the CM is equally the center of momentum. From the time derivative of (6.9) the velocity relation $\mathbf{v_i} = \mathbf{V} + \mathbf{v}'_i$ is obtained. The system kinetic energy can then be expressed in CM coordinates as follows

$$K = \frac{1}{2}\sum_i m_i v_i^2 = \frac{1}{2}\sum_i m_i \mathbf{v_i} \cdot \mathbf{v_i} = \frac{1}{2}\sum_i m_i \left(V^2 + 2\mathbf{V} \cdot \mathbf{v}'_i + v_i'^2\right) \tag{6.14}$$

The middle term in the right-hand side of (6.14) vanishes because of (6.13). Thus the system kinetic energy is the kinetic energy of the CM plus the kinetic energy relative to the CM

$$K = \frac{1}{2}MV^2 + K'$$

$$K' = \frac{1}{2}\sum_i m_i v_i'^2 \tag{6.15}$$

A similar calculation can be done for the system angular momentum,

$$\mathbf{L} = \sum_i m_i \mathbf{r_i} \times \mathbf{v_i} = \sum_i m_i (\mathbf{R} + \mathbf{r}'_i) \times (\mathbf{V} \times \mathbf{v}'_i) \tag{6.16}$$

giving

$$\mathbf{L} = M\,(\mathbf{R} \times \mathbf{V}) + \mathbf{L}'$$

$$\mathbf{L}' = \sum_i \mathbf{r}'_i \times \mathbf{v}'_i \tag{6.17}$$

The cross terms in (6.16) vanish due to the CM condition (6.11). The angular momentum about the CM plus the angular momentum of the CM point about the origin gives the total system angular momentum about the origin of $\mathbf{R}$.

Now we return to the two-particle system and treat the part of the motion of the system which is not described by (6.6). For two particles the equations of motion are

$$\mathbf{F}_1^{\text{ext}} + \mathbf{F}_1^{\text{int}} = m_1 \ddot{\mathbf{r}}_1 \tag{6.18}$$

$$\mathbf{F}_2^{\text{ext}} + \mathbf{F}_2^{\text{int}} = m_2 \ddot{\mathbf{r}}_2 \tag{6.19}$$

In the most important physical applications, the force of one particle on the other depends only on the relative-position coordinate

$$\mathbf{r} \equiv \mathbf{r}_1 - \mathbf{r}_2 \tag{6.20}$$

of the two particles, as illustrated in Fig. 6-1. Hence we want to form a combination of (6.18) and (6.19) to obtain an equation of motion for $\mathbf{r}$. This is achieved by dividing (6.18) by m_1 and (6.19) by m_2 and then subtracting the two equations. Using Newtons's third law $\mathbf{F}_2^{int} = -\mathbf{F}_1^{int}$, we find

$$\ddot{\mathbf{r}} = \ddot{\mathbf{r}}_1 - \ddot{\mathbf{r}}_2 = \left(\frac{1}{m_1} + \frac{1}{m_2}\right)\mathbf{F}_1^{int} + \left(\frac{\mathbf{F}_1^{ext}}{m_1} - \frac{\mathbf{F}_2^{ext}}{m_2}\right) \tag{6.21}$$

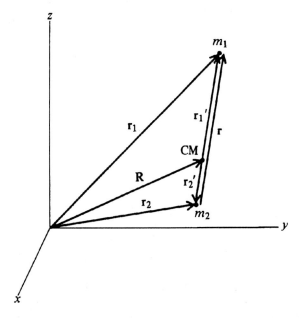

FIGURE 6-1. Center-of-mass vector $\mathbf{R}$ and relative coordinate vector $\mathbf{r}$ for a two-particle system.

This equation can be rewritten in the form

$$\mu\ddot{\mathbf{r}} = \mathbf{F}_1^{int} + \mu\left(\frac{\mathbf{F}_1^{ext}}{m_1} - \frac{\mathbf{F}_2^{ext}}{m_2}\right) \tag{6.22}$$

where μ is the *reduced mass*.

$$\frac{1}{\mu} = \frac{1}{m_1} + \frac{1}{m_2}$$

$$\mu = \frac{m_1 m_2}{m_1 + m_2} \tag{6.23}$$

Due to the presence of the external-force term on the right-hand side of (6.22), the motion of $\mathbf{r}$ is not independent of $\mathbf{R}$ in general, but that term vanishes or is insignificant in two important cases:

1. There is no external force on either particle

$$\mathbf{F}_1^{\text{ext}} = \mathbf{F}_2^{\text{ext}} = 0 \tag{6.24}$$

2. The external forces are gravitational and due to distant sources, so that

$$\frac{\mathbf{F}_1^{\text{ext}}}{m_1} \approx \frac{\mathbf{F}_2^{\text{ext}}}{m_2} \tag{6.25}$$

If $\mathbf{F}^{\text{int}}$ depends only on $\mathbf{r} = \mathbf{r}_1 - \mathbf{r}_2$, the equation of motion for the relative coordinate $\mathbf{r}$ simplifies in these special cases to

$$\mu\ddot{\mathbf{r}} = \mathbf{F}_1^{\text{int}}(\mathbf{r}) \tag{6.26}$$

which has the same form as that for a single particle of mass μ in the influence of a force center at $\mathbf{r} = 0$. For a central force of interaction,

$$\mathbf{F}^{\text{int}} = F(r)\hat{\mathbf{r}} \tag{6.27}$$

and we can directly apply the methods of Chapter 5 to the solution of the two-body problem. For example, the exact form of Kepler's third law for the period of revolution of two bodies of masses m_1, m_2 going around one another is immediately found from (5.71) by replacing m by $\mu = m_1 m_2/(m_1 + m_2)$, giving

$$\tau^2 = \left(\frac{4\pi^2 \mu}{\alpha}\right) a^3 = \frac{4\pi^2}{G(m_1 + m_2)} a^3 \tag{6.28}$$

Where m appears in the gravitational force (*i.e.*, in α), it is left alone since the Newtonian gravity force does not depend on the state of motion.

The center of mass location of a two-particle system is given by

$$MR = m_1\mathbf{r}_1 + m_2\mathbf{r}_2 \tag{6.29}$$

The total momentum of the system

$$\mathbf{P} = m_1\dot{\mathbf{r}}_1 + m_1\dot{\mathbf{r}}_2 = M\dot{\mathbf{R}} \tag{6.30}$$

is just the momentum of the center of mass. In the center of mass coordinate system the particle locations are measured from the CM point.

$$\begin{aligned}\mathbf{r}_1' &= \mathbf{r}_1 - \mathbf{R}\\r_2' &= \mathbf{r}_2 - \mathbf{R}\end{aligned} \tag{6.31}$$

Using (6.29) the CM coordinates can be expressed in terms of the relative coordinate $\mathbf{r} = \mathbf{r}_1 - \mathbf{r}_2$ as

$$\mathbf{r}_1' = \frac{m_2}{M}\mathbf{r} \tag{6.32}$$

$$\mathbf{r}_2' = \frac{m_1}{M}\mathbf{r} \tag{6.33}$$

and hence the vectors $\mathbf{r}_1'$ and $\mathbf{r}_2'$ are colinear, as indicated in Fig. 6-1. In (6.15) and (6.17) we showed that the kinetic energy and angular momentum of a general system can be separated into parts associated with the center-of-mass motion and the relative motion. For a two-body sytem this result is

$$\begin{aligned}K &= \frac{1}{2}m_1\dot{\mathbf{r}}_1^2 + \frac{1}{2}m_2\dot{\mathbf{r}}_2^2\\&= \frac{1}{2}M\dot{\mathbf{R}}^2 + \frac{1}{2}\mu\dot{\mathbf{r}}^2\end{aligned} \tag{6.34}$$

$$\begin{aligned}\mathbf{L} &= m_1(\mathbf{r}_1 \times \dot{\mathbf{r}}_1) + m_2(\mathbf{r}_2 \times \dot{\mathbf{r}}_2)\\&= M(\mathbf{R} \times \dot{\mathbf{R}}) + \mu(\mathbf{r} \times \dot{\mathbf{r}})\end{aligned} \tag{6.35}$$

In (6.34) we use the notation $\dot{\mathbf{r}}_1^2 = \dot{\mathbf{r}}_1 \cdot \dot{\mathbf{r}}_1 = v_1^2$.

Although in the many-particle situation it is always possible to separate out the CM motion, as in (6.6) and (6.7), the remaining coordinates cannot in general be further separated. A complex system of coupled equations usually remains which frequently is not soluble by analytic techniques, and one must make approximations or resort to a numerical treatment. Only in the two-body problem is the relative motion simple after the CM motion is separated out. An important exception to the above is the case of a rigid body. If the effect of the internal forces is to keep all interparticle distances fixed, the dynamics can be described in terms of moments of inertia, as will be addressed later in this chapter.

In the CM system the scattering angle (angle between initial and final directions of particle 1) is given by

$$\cos \theta' = \hat{\mathbf{r}}'_{1i} \cdot \hat{\mathbf{r}}'_{1f} \qquad (6.36)$$

By (6.32) $\mathbf{r}'_1$ always points in the same direction as the relative coordinate $\mathbf{r} = \mathbf{r}_1 - \mathbf{r}_2$. Thus the CM scattering angle is the same as the scattering angle of the relative motion. Consequently, the CM scattering angle can be directly deduced from the equivalent particle scattering angle found in the solution to (6.26). The scattering angle in the laboratory system can in turn be determined from its relation to the CM angle given in (4.49).

We can apply these techniques for solving the two-body problem to Rutherford scattering on a target particle of finite mass. The projectile mass m in (5.125) gets replaced by the reduced mass μ of the equivalent two-body problem. The scattering angle in the laboratory system in (5.125) is replaced by the scattering angle θ' of the relative motion (which is also the CM scattering angle). The initial relative velocity v_0 is the same as the initial lab velocity. With these changes the CM differential cross section is given by

$$\frac{d\sigma}{d\Omega'} = \left(\frac{\alpha}{2\mu v_0^2}\right)^2 \frac{1}{\sin^4(\theta'/2)} \qquad (6.37)$$

We can use (4.49) to write $d\sigma$ in terms of the laboratory angle. For equal masses for projectile and target, the conversion is particularly simple.

$$\theta' = 2\theta$$
$$d\Omega' = 4\cos\theta \, d\Omega \qquad (6.38)$$

The resulting differential cross-section expression in the lab system is

$$\frac{d\sigma}{d\Omega} = \left(\frac{2\alpha}{mv_0^2}\right)^2 \frac{\cos\theta}{\sin^4\theta} \qquad (6.39)$$

6.2 Rotational Equation of Motion

We now return to the general case of a system composed of many particles and derive an equation to describe the rotational motion analogous to (5.5) for a single particle. In analogy to the total momentum

$$\mathbf{P} = \sum \mathbf{p}_i = \sum_i m_i \dot{\mathbf{r}}_i \tag{6.40}$$

the total angular momentum about a point p (with coordinate $\mathbf{r}_p$) is the sum of the angular momenta about p of the particles in the system.

$$\mathbf{L} = \sum_i (\mathbf{r}_i - \mathbf{r}_p) \times m_i(\dot{\mathbf{r}}_i - \dot{\mathbf{r}}_p) \tag{6.41}$$

From this, we compute the time derivative of $\mathbf{L}$ to be

$$\dot{\mathbf{L}} = \sum_i (\mathbf{r}_i - \mathbf{r}_p) \times m_i(\ddot{\mathbf{r}}_i - \ddot{\mathbf{r}}_p) \tag{6.42}$$

We have allowed for the possibility that $\mathbf{r}_p$ is not a fixed point. If we now use the equations of motion

$$m_i \ddot{\mathbf{r}}_i = \mathbf{F}_i^{\text{ext}} + \mathbf{F}_i^{\text{int}} \tag{6.43}$$

in (6.42) we get

$$\dot{\mathbf{L}} = \mathbf{N}^{\text{ext}} + \mathbf{N}^{\text{int}} + M(\mathbf{r}_p - \mathbf{R}) \times \ddot{\mathbf{r}}_p \tag{6.44}$$

where

$$\mathbf{N}^{\text{ext}} = \sum_i (\mathbf{r}_i - \mathbf{r}_p) \times \mathbf{F}_i^{\text{ext}}$$
$$\mathbf{N}^{\text{int}} = \sum_i (\mathbf{r}_i - \mathbf{r}_p) \times \mathbf{F}_i^{\text{int}} \tag{6.45}$$

are the total external and internal torques, respectively, about p. The internal torque will vanish under a wide range of conditions. If it did not, an isolated body might spontaneously begin to rotate.

A simple way to see that $\mathbf{N}^{\text{int}}$ vanishes is to consider an object which is free to rotate about a fixed axis. Choosing the rotation angle ϕ as the general coordinate, the general force (a torque in this case) is

$$N_\phi^{\text{int}} = -\frac{\partial V^{\text{int}}(\phi)}{\partial \phi} \tag{6.46}$$

The internal potential energy is the net result of all the internal forces. Since the value of V^{int} cannot depend on the orientation of the body, V^{int} must be a constant and the torque due to the internal forces vanishes. This absence of a preferred or absolute angle of orientation is known as *rotational invariance*.

Setting the total internal torque to zero, (6.44) simplifies to

$$\dot{\mathbf{L}} = \mathbf{N}^{\text{ext}} + \mathbf{M}(\mathbf{r}_p - \mathbf{R}) \times \ddot{\mathbf{r}}_p \tag{6.47}$$

The last term in (6.47) vanishes if

1. $\ddot{\mathbf{r}}_p = 0$: the reference point is fixed or moving with constant velocity.
2. $\mathbf{r}_p = \mathbf{R}$: the reference point is the center of mass.
3. $\mathbf{r}_p - \mathbf{R}$ is parallel to $\ddot{\mathbf{r}}_p$.

The first two instances are of most practical importance. In any of these cases the rotational equation of motion is

$$\dot{\mathbf{L}} = \mathbf{N} \qquad \ddot{\mathbf{r}}_p = 0 \ \text{ or } \ \mathbf{r}_p = \mathbf{R} \tag{6.48}$$

From now on we will dispense with the label "external" on forces or torques since the net internal forces and torques always vanish and do not appear in the equations of interest.

The equation $\dot{\mathbf{L}} = \mathbf{N}$ bears a close resemblance to $\dot{\mathbf{P}} = \mathbf{F}$. For an isolated system both $\mathbf{P}$ and $\mathbf{L}$ are constants, since $\mathbf{F} = 0$ and $\mathbf{N} = 0$. Even though the two equilibrium conditions $\mathbf{F} = 0$ and $\mathbf{N} = 0$ appear similar, they exhibit some interesting differences for systems in which internal motion is possible. The vanishing of the net force assures that the center of mass once fixed will remain so, no matter what the internal forces or internal motion. If the net external torque is zero, the total angular momentum is constant, and if initially zero, will remain zero. However, $\mathbf{L} = 0$ does not preclude changes in orientation of the system by exclusive use of internal forces and motions. This can be demonstrated by a person sitting on a piano stool that is free to rotate, with a dumb-bell in each hand. The dumbbells are initially held close to the body. When the dumbbells are extended radially outward at arm's length, the stool remains stationary. Then, when the dumbbells are moved by the person in a circular arc parallel to the floor, the person is rotated on the stool in the opposite direction to the motion of the dumbbells. The angular momentum from the rotation of the person cancels the angular momentum due to the dumbbell motion, so that the total momentum remains zero. If the dumbbells are then drawn radially back to the body, a net rotation has been achieved, with no change in the system or angular momentum. Repetition of the process enables the person to face in any arbitrary direction.

A dramatic illustration of the possibility of a change in orientation by exclusive use of internal forces is the ability of a cat to turn itself in midair and land upright on its feet, even when dropped vertically from an upside-down position. By contrast, the cat can do nothing whatever to alter its fall, that is, to change the motion of its CM. Gymnasts and divers also use internal motions to change their orientation in mid-air.

For an isolated system the condition that

$$\mathbf{P} = M\dot{\mathbf{R}} = \text{ constant} \tag{6.49}$$

leads to uniform motion of the CM.

$$\mathbf{R} = \mathbf{R}_0 + \frac{\mathbf{P}}{M}t \tag{6.50}$$

This can be stated as a conservation law:

$$M\mathbf{R} - \mathbf{P}t = \text{ constant} \tag{6.51}$$

No corresponding result can be derived from $\mathbf{L} = $ constant, in general, because there is no *rotational coordinate* analogous to $\mathbf{R}$.

6.3 Rigid Bodies: Static Equilibrium

Applications of mechanics to many particle systems commonly deal with the motion of rigid bodies. A rigid body is a system of particles whose distances from one another are fixed. The position of every particle in a rigid body is determined by the position of any one point of the body (such as the CM) plus the orientation of the body about that point. Six coordinates are needed to specify the motion of a rigid body. The position of one particle in the body requires the specification of three coordinates. The position of a second particle can be specified by two angular coordinates since it lies at a fixed distance from the first. The position of a third particle is determined by only one coordinate because its distances from the first and second particles are fixed. The positions of any other particles in the rigid body are completely fixed by their distances from the first three particles. Thus $3 + 2 + 1 = 6$ coordinates determine the positions of all particles in a rigid body. Consequently, the motion of a rigid body is controlled by only six equations of motion. The translational motion of the CM is determined by

$$\dot{\mathbf{P}} = \mathbf{F} \tag{6.52}$$

and the rotational motion about the CM, or a fixed point is determined by

$$\dot{\mathbf{L}} = \mathbf{N} \tag{6.53}$$

These six equations, which hold for any system of particles, completely describe the motion of a rigid body.

The conditions under which a rigid body remains in equilibrium under the action of a set of forces are of great practical importance in the design of permanent structures. From (6.52) and (6.53) the six conditions for complete equilibrium of a rigid body are

$$\mathbf{F} = \sum_i \mathbf{F}_i = 0 \tag{6.54}$$

$$\mathbf{N}_{\text{CM}} = \sum_i (\mathbf{r_i} - \mathbf{R}) \times \mathbf{F}_i = 0 \tag{6.55}$$

The net external force must vanish in order that the CM move with constant velocity. The torque about the CM point $\mathbf{R}$ must vanish in order that the angular momentum about the CM (and thereby the rotational motion) does not change. For static equilibrium, the CM must be initially at rest and the total angular momentum about the CM must initially be zero.

The torque about an arbitrary point p can be easily related to the torque about the CM. If the vector distance between $\mathbf{r}_p$ and $\mathbf{R}$ is

$$\mathbf{d} = \mathbf{r}_p - \mathbf{R} \tag{6.56}$$

then the torque about p is

$$\mathbf{N}_p = \sum_i (\mathbf{r}_i - \mathbf{r}_p) \times \mathbf{F}_i = \sum_i (\mathbf{r}_i - \mathbf{R}) \times \mathbf{F}_i - \mathbf{d} \times \sum_i \mathbf{F}_i \tag{6.57}$$

or

$$\mathbf{N}_p = \mathbf{N}_{\text{CM}} - \mathbf{d} \times \mathbf{F} \tag{6.58}$$

For a body in complete equilibrium, $\mathbf{F} = 0$ and $\mathbf{N}_{\text{CM}} = 0$ from (6.54) and (6.55). In this case we find from (6.58) that the external torque about an arbitrary point p also vanishes.

$$\mathbf{N}_p = 0 \tag{6.59}$$

Often, however, equilibrium is desired only for a subset of the six independent directions of motion. The drag-strip racer of § 1.3 is such an example. The external force in the direction of the racer's motion was nonzero, but equilibrium was to be maintained in all other directions. For rotational equilibrium, the torque about the CM of the racer vanishes according to (6.55). However, by (6.58) the torque about another point p may not be zero.

A set of two antiparallel forces with equal magnitudes $\mathbf{F}$ and $-\mathbf{F}$, separated by a vector $\mathbf{c}$ as in Fig. 6-2, is called a *couple*. Since the net external force of a couple is zero, by (6.58) the torque produced is independent of the point of reference. The torque of the couple is

$$\mathbf{N}_{\text{couple}} = \mathbf{c} \times \mathbf{F} \tag{6.60}$$

It is sometimes convenient to represent a system of forces by a single total-force vector on a given point plus a couple with the equivalent torque of the original forces about that point.

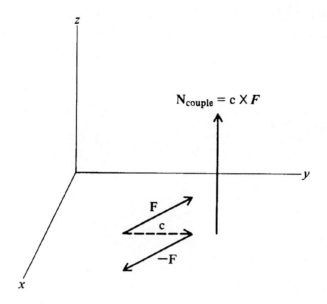

FIGURE 6-2. A vector force couple.

As a simple application, we discuss the motion which results when a light bar magnet is rigidly attached near one end of a rectangular board floating in a pan of water. A couple acts on the magnet, tending to align it in a north-south direction. Since there is no net external force on the board, the CM of the board must remain stationary. The torque produced by the couple therefore acts to rotate the board about its CM point (not about the end of the board where the magnet is placed) until the magnet reaches a north-south alignment.

6.4 Rotations of Rigid Bodies

A *rotation* can be defined as the motion of a point p about a line such that the distance from p to each point on the line is constant. In an infinitesimal rotation, the rotational displacement $d\mathbf{r}$ must thus be perpendicular to a vector $\mathbf{r}$ from a point O on the line to the point p. Furthermore, $d\mathbf{r}$ must be perpendicular to a unit vector $\hat{\mathbf{n}}$ directed along the line, as illustrated in Fig. 6-3. These two conditions can be mathematically expressed as

$$d\mathbf{r} \equiv \hat{\mathbf{n}} \times \mathbf{r}d\phi \qquad (6.61)$$

Since $|\hat{\mathbf{n}} \times \mathbf{r}| = r \sin \theta = \rho$ is the perpendicular radius of rotation about the line, and $|d\mathbf{r}| = \rho d\phi$, the quantity $d\phi$ is the infinitesimal angle of rotation about the axis $\hat{\mathbf{n}}$. The velocity of the point p relative to the point O due to rotation is

$$\mathbf{v} = \frac{d\mathbf{r}}{dt} = \hat{\mathbf{n}} \times \mathbf{r}\dot{\phi} \tag{6.62}$$

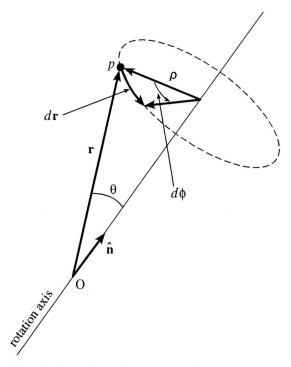

FIGURE 6-3. Infinitesimal rotation about an axis $\hat{\mathbf{n}}$.

If we introduce an angular velocity vector

$$\boldsymbol{\omega} = \hat{\mathbf{n}}\omega \equiv \hat{\mathbf{n}}(t)\dot{\phi} \tag{6.63}$$

for the rotation about $\hat{\mathbf{n}}$, the rotational velocity of p can be expressed as

$$\mathbf{v}_{\text{rot}} = \boldsymbol{\omega} \times \mathbf{r} \tag{6.64}$$

If the point O is moving with a translational velocity $\mathbf{v}_0$ relative to a fixed reference frame, the velocity of the point p in the fixed frame is the vector sum of the translational velocity $\mathbf{v}_0$ of the point O and the rotational velocity about O.

$$\mathbf{v} = \mathbf{v}_0 + \boldsymbol{\omega} \times \mathbf{r} \qquad (6.65)$$

Over a period of time both the direction and magnitude of $\boldsymbol{\omega}$ may change as the body rotates.

In a rigid body the distance between any two points $\mathbf{r}_A$, $\mathbf{r}_B$ stays constant.

$$\frac{d}{dt}|\mathbf{r}_A - \mathbf{r}_B|^2 = 0 \qquad (6.66)$$

Since

$$\frac{d}{dt}|\mathbf{r}_A - \mathbf{r}_B|^2 = 2(\mathbf{r}_A - \mathbf{r}_B) \cdot (\mathbf{v}_A - \mathbf{v}_B) \qquad (6.67)$$

the rigid-body constraint in (6.66) is satisfied by the velocity field in the (6.65), provided that $\boldsymbol{\omega}$ is the same for all points in the body.

$$(\mathbf{r}_A - \mathbf{r}_B) \cdot (\mathbf{v}_A - \mathbf{v}_B) = (\mathbf{r}_A - \mathbf{r}_B) \cdot \boldsymbol{\omega} \times (\mathbf{r}_A - \mathbf{r}_B) = 0 \qquad (6.68)$$

Hence $\mathbf{v}(\mathbf{r})$ as given by (6.65) is the velocity field of a rigid body. The velocities of all particles in a rigid body can be specified by six independent numbers, the components of $\mathbf{v}_0$ and $\boldsymbol{\omega}$.

As an example, we consider a wheel of radius R which rolls without slipping on a level surface, as illustrated in Fig. 6-4. For no slipping we have

$$dx = R d\theta \qquad (6.69)$$

where dx and $d\theta$ are infinitesimal horizontal and angular displacements, respectively. Dividing both sides by the time interval dt, we get

$$\mathbf{v}_0 = R\omega \hat{\mathbf{x}} \qquad (6.70)$$

where $\mathbf{v}_0$ is the velocity of the center of mass which coincides in this case with the rotation axis. The velocity of a point on the wheel relative to the CM is

$$\mathbf{v}_{\text{rot}} = \boldsymbol{\omega} \times \mathbf{r} = \frac{v_0}{R} \hat{\boldsymbol{\omega}} \times \mathbf{r} \qquad (6.71)$$

as illustrated in Fig. 6-5. The direction of $\boldsymbol{\omega}$ points into the page. With respect to a reference frame at rest on the level surface, the velocity of the point on the wheel is given by

$$\mathbf{v} = \mathbf{v}_0 + \mathbf{v}_{\text{rot}} = \mathbf{v}_0 + \frac{v_0}{R}\hat{\boldsymbol{\omega}} \times \mathbf{r} \tag{6.72}$$

We note that $\mathbf{v} = 0$ at the point of contact. No slipping means that there is no relative motion of the wheel and surface at the contact point.

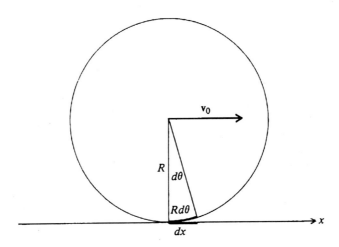

FIGURE 6-4. Wheel rolling without slipping on a level surface.

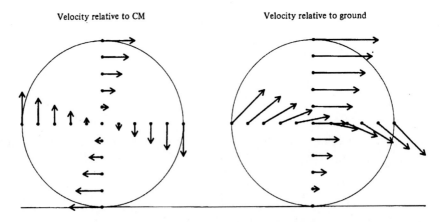

FIGURE 6-5. Velocity of points on a wheel relative to the center of mass and relative to the ground.

In general, the angular velocity $\boldsymbol{\omega}$ is not the time derivative of any angular coordinate ϕ. As a consequence, angular displacements are quite different in nature from translational displacements. Only in the special case of fixed axis of rotation is it possible to express $\boldsymbol{\omega}$ as the time derivative of a coordinate. We can show that it is not possible to write $\boldsymbol{\omega} = \dot{\boldsymbol{\phi}}$ in general by the following simple example. If such a representation were possible, we could compute the coordinates ϕ_x, ϕ_y, ϕ_z describing the orientation of the rigid body by integrating over the components of $\boldsymbol{\omega}$. For example,

$$\phi_x(t) - \phi_x(0) = \int_0^t \omega_x(t')dt' \tag{6.73}$$

This would say that the change in orientation $\boldsymbol{\phi}(t) - \boldsymbol{\phi}(0)$ resulting from a motion $\boldsymbol{\omega}$ depends only on the three numbers $\int_0^t \omega_i(t')dt'$. That this is not true can be seen from the following demonstration. Take a book and choose fixed axes $\hat{\mathbf{x}}$, $\hat{\mathbf{y}}$, $\hat{\mathbf{z}}$. First rotate the book by 90° around the $\hat{\mathbf{x}}$ axis and then by 90° around the $\hat{\mathbf{y}}$ axis. Then start again from the original orientation and make the same rotations in the opposite order. In the two cases the resulting orientations of the book are different, but the integral $\int \boldsymbol{\omega} dt$ is the same, namely,

$$\int \omega_x dt = \frac{\pi}{2}, \qquad \int \omega_y dt = \frac{\pi}{2}, \qquad \int \omega_z dt = 0 \tag{6.74}$$

6.5 Gyroscope Effect

As an interesting application of the rotational equation of motion (6.48) we will discuss the gyroscope effect experienced by a wheel spinning in a vertical plane, as illustrated in Fig. 6-6. With a counterclockwise spin, the angular-momentum vector points along the positive x axis. When a torque which tends to turn the wheel in a counterclockwise sense about the positive y axis is applied, the wheel is observed to precess about the z axis. We can predict this precession from (6.48) and derive an expression for the precession frequency. According to (6.48), the change in angular momentum in an infinitesimal time interval dt is

$$d\mathbf{L} = \mathbf{N}dt \tag{6.75}$$

The increment $d\mathbf{L}$ is parallel to $\mathbf{N}$ and perpendicular to $\mathbf{L}$, as shown in Fig. 6-6. Since $\mathbf{L}$ and $\mathbf{N}$ are perpendicular, the length of $\mathbf{L}$ is unchanged

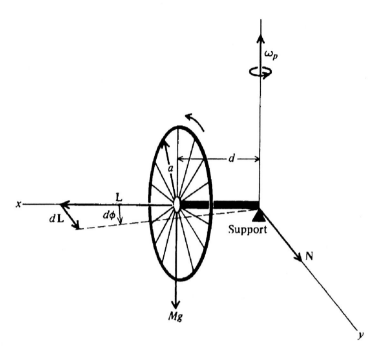

FIGURE 6-6.   Gyroscope effect for a wheel with massive rim.

to first order dt. The direction of $\mathbf{L}$ is rotated counterclockwise, viewed from above in the x, y plane, through an angle $d\phi$ given by

$$d\phi = \frac{dL}{L} = \frac{N}{L}dt \qquad (6.76)$$

If $\mathbf{N}$ remains perpendicular to $\mathbf{L}$ and in the x, y plane, the angular velocity of precession about the z axis is

$$\omega_p = \frac{d\phi}{dt} = \frac{N}{L} \qquad (6.77)$$

This result known as *simple precession*, since we neglected the angular momentum associated with the precession motion. Whenever the applied torque is small or the spin large, simple precession is a good approximation. When the precession angular momentum is taken into account in the description of the motion, an oscillation called *nutation* about the x, y plane may be present, in addition to the precessional motion.

A popular lecture demonstration experiment that illustrates simple precession uses a bicycle rim loaded with lead. The wheel is oriented in a vertical plane as in Fig. 6-6. The suspension point is located a distance

d from the plane of the wheel along the wheel axis. The weight of the wheel then supplies the torque.

$$\mathbf{N} = (Mgd)\hat{\mathbf{y}} \tag{6.78}$$

If the wheel has radius a and mass M and spins with angular velocity ω, the angular momentum is

$$\mathbf{L} = (M\omega a^2)\hat{\mathbf{x}} \tag{6.79}$$

The resulting angular velocity of precession about the z axis is

$$\omega_P = \frac{gd}{\omega a^2} \tag{6.80}$$

where we have used (6.77) to (6.79). For $a = d = 0.3$m and a spin rate of $\omega/2\pi = 200$ r/min, we find precession rate of

$$\frac{\omega_P}{2\pi} = \frac{9.8 \times 3,600}{200(2\pi)^2(0.3)} = 15\text{r/min} \tag{6.81}$$

6.6 The Boomerang

An explanation of why a boomerang returns can be given in terms of the gyroscope effect. The boomerang can take on a variety of shapes. In its most common form it appears as two airfoil-shaped blades meeting at an angle near 90°, as illustrated in Fig. 6-7. However, the characteristic banana-like shape of most boomerangs has little to do with their ability to return. Another version consists of two crossed blades, as shown in Fig. 6-8. The boomerang is thrown overhand in a nearly vertical plane in the manner of Fig. 6-9. As it leaves the hand, the blades are rapidly rotating about the CM, and the CM is moving parallel to the ground. Due to its spin, the boomerang has an angular momentum about the CM that is initially directed to the left, as shown in Fig. 6-10.

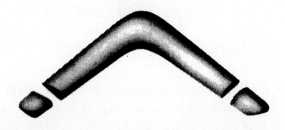

FIGURE 6-7. Common boomerang.

The aerodynamic "lift" forces on the airfoils act perpendicular to the plane of rotation, as indicated in Fig. 6-11. The total aerodynamic force on the boomerang accelerates it perpendicular to the plane of rotation, in the direction of **L**. An upper blade of the boomerang moves more rapidly through the air than a lower blade because the rotation and translation velocities add on the upper blade and subtract on the lower blade. Since the aerodynamic force is larger for higher blade velocities, the upper blade experiences a greater force, and an external torque about the CM is generated by the forces on the airfoils. The torque points opposite the CM velocity direction. Thus the initial directions of **N** and **L** are identical with those for the wheel in Fig. 6-6. From the gyroscope effect discussed in § 6.5 we predict that the plane of rotation precesses counterclockwise about the vertical axis. This precession of the rotational plane accompanied by translational acceleration perpendicular to the rotational plane allows the boomerang to travel in a circular orbit and return to the thrower; see the illustration in Fig. 6-12.

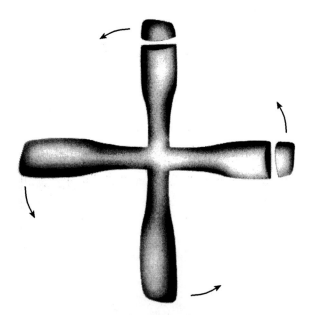

FIGURE 6-8. Cross-blade boomerang.

To discuss the flight of the boomerang in a more quantitative fashion, we consider the crossed-blade boomerang of Fig. 6-8. In this case the CM lies at the blade hub, which simplifies the analysis considerably. We choose the origin of our coordinate system at the hub. Initially, we take

FIGURE 6-9. Proper method of throwing a boomerang.

the CM motion along the negative y axis with velocity $\mathbf{V} = -V\hat{\mathbf{y}}$, as in Fig. 6-10. Rotation occurs around the x axis in a counterclockwise sense with angular velocity ω. One of the blades with length l, mass $\frac{1}{4}M$, and linear mass density $\mu = \frac{1}{4}(M/l)$ is depicted in Fig. 6-13. A point on the blade at a distance r from the CM is specified as a function of time by

$$\mathbf{r} = r(\hat{\mathbf{y}}\cos\omega t + \hat{\mathbf{z}}\sin\omega t) \qquad (6.82)$$

The aerodynamic force is dependent on the transverse component v_t of the air velocity over the airfoil in a direction perpendicular to the long edge of the blade. The force will be approximated by a quadratic dependence on v_t. The force on an element of blade at a distance r is

$$d\mathbf{F} = \hat{\mathbf{x}}cv_t^2\,dr \qquad (6.83)$$

The perpendicular air-velocity component v_t is due to the rotational motion of the blade and to the translation motion of the boomerang CM at velocity V. This tangential velocity component is given by

$$v_t = \omega r + V\sin\omega t \qquad (6.84)$$

Using (6.83) and (6.84), we find

$$d\mathbf{F}(r,t) = \hat{\mathbf{x}}c(\omega^2 r^2 + 2\omega V r\sin\omega t + V^2\sin^2\omega t)\,dr \qquad (6.85)$$

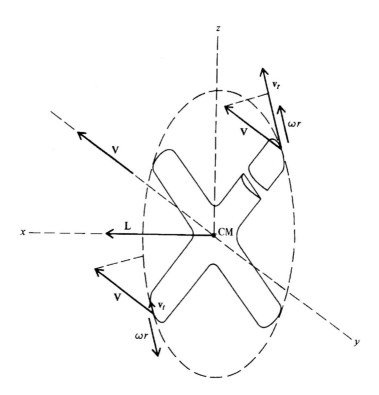

FIGURE 6-10. Boomerang-blade velocities.

As the blade rotates, $d\mathbf{F}$ varies in magnitude. The three remaining blades contribute forces similar to (6.85), but with ωt replaced by $\omega t + \pi/2$, $\omega t + \pi$, and $\omega t + 3\pi/2$, respectively. The net force on the boomerang from all four blades due to the elements of length dr at distance r is

$$d\mathbf{F}(r) = 4\hat{\mathbf{x}}c\left(\omega^2 r^2 + \frac{V^2}{2}\right) \tag{6.86}$$

Adding the elements by integration over r, we find the total force normal to the plane of rotation is

$$\mathbf{F} = 4\hat{\mathbf{x}}cl\left(\frac{\omega^2 l^2}{3} + \frac{V^2}{2}\right) \tag{6.87}$$

From (6.82) and (6.83) the torque about the CM from an element on one of the blades is

$$d\mathbf{N} = \mathbf{r} \times d\mathbf{F} = r(\hat{\mathbf{y}}\cos\omega t + \hat{\mathbf{z}}\sin\omega t) \times \hat{\mathbf{x}}cv_t^2 dr \tag{6.88}$$

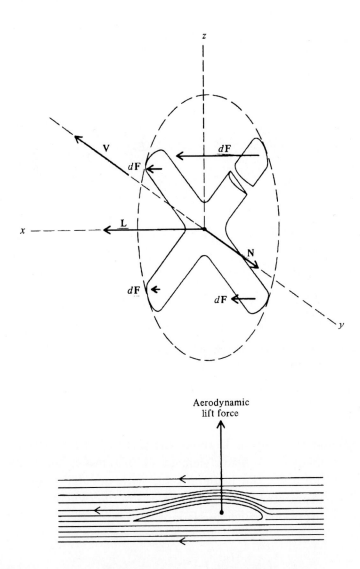

FIGURE 6-11. Aerodynamic forces on a cross-blade boomerang.

Expansion of this result, using (6.84), gives

$$dN = cr(\omega^2 r^2 + 2\omega V r \sin \omega t + V^2 \sin^2 \omega t)(\hat{\mathbf{y}} \sin \omega t - \hat{\mathbf{z}} \cos \omega t)dr \quad (6.89)$$

We add to this the torques from the other three blades' elements to find the net torque

$$dN = 4c\omega V r^2 \hat{\mathbf{y}} dr \quad (6.90)$$

The torque due to all elements is then obtained by integrating (6.90) over

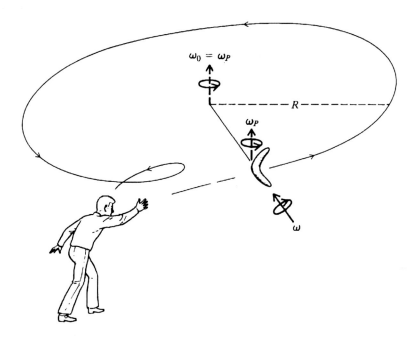

FIGURE 6-12. Typical boomerang orbit.

the length of a blade.

$$\mathbf{N} = \tfrac{4}{3}c\omega V l^3 \hat{\mathbf{y}} \tag{6.91}$$

The angular momentum $\mathbf{L}$ about the CM of the boomerang can be computed as follows. A blade element at distance r has mass $dm = \mu dr = \tfrac{1}{4}(M/l)dr$. As the blade rotates with angular velocity ω, the angular momentum of the element is $d\mathbf{L} = dm r^2 \boldsymbol{\omega}$, where $\boldsymbol{\omega} = \hat{\mathbf{x}}\omega$. The angular momentum of the whole blade is then

$$\mathbf{L} = \frac{M}{4l}\boldsymbol{\omega} \int_0^l r^2 dr = \tfrac{1}{12}Ml^2\boldsymbol{\omega} \tag{6.92}$$

The complete boomerang has angular momentum

$$\mathbf{L} = \tfrac{1}{3}Ml^2\boldsymbol{\omega} \tag{6.93}$$

The constant of proportionality between $\mathbf{L}$ and $\boldsymbol{\omega}$ is called the *moment of inertia*.

$$I = \tfrac{1}{3}Ml^2 \tag{6.94}$$

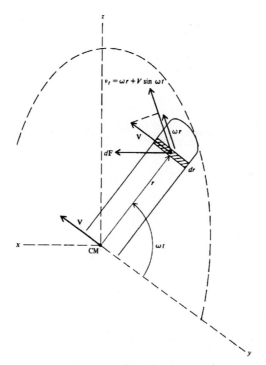

FIGURE 6-13. Diagram of one blade in a four-blade boomerang.

The torque in the $\hat{\mathbf{y}}$ direction induces a precession of the $\mathbf{L}$ vector as given by (6.77). The precession angular velocity is

$$\omega_P = \frac{N}{L} = \frac{\frac{4}{3}c\omega V l^3}{\frac{1}{3}M l^2 \omega} = \frac{4cV l}{M} \tag{6.95}$$

The motion of the CM is influenced by the aerodynamic force normal to the plane of rotation, gravity, and a drag force due to air resistance. For actual boomerangs the gravity and drag forces are not negligible but can be counterbalanced by a small tilt of the boomerang plane from vertical.

Since a boomerang is supposed to return to the thrower, we investigate under what conditions the CM will travel in a circular orbit of radius R with angular velocity ω_0. In a circular orbit the aerodynamic force exactly balances the centrifugal force. In order for the aerodynamic force to be always radial inward toward the center of a circle, the orbital angular velocity ω_0 must match the precession rate ω_P.

$$\omega_0 = \frac{V}{R} = \omega_P \tag{6.96}$$

From (6.95) and (6.96) we can determine the radius of the orbit as

$$R = \frac{M}{4lc} = \frac{\mu}{c} \tag{6.97}$$

where μ is the linear mass density and c is the lift constant determined by the airfoil shape and air properties. By equating the magnitude of the lift force in (6.87) to the mass times the centripetal acceleration,

$$F = 4cl\left(\frac{\omega^2 l^2}{3} + \frac{V^2}{2}\right) = \frac{MV^2}{R} \tag{6.98}$$

we obtain

$$V = \sqrt{\tfrac{2}{3}}\,\omega l \tag{6.99}$$

For a simple circular return flight the CM velocity V and spin ω must be related as in (6.99).

From (6.97) we see that the boomerang has a flight radius which is independent of how hard it is thrown. Of course, if it is thrown very slowly, the effects of gravity will become important, and our theory breaks down. If an indoor boomerang is desired, it should have an exaggerated airfoil shape to obtain a small orbit radius in (6.97). For long flights a boomerang made of dense material is needed, and of course the design should minimize drag. It is said that some native Australians can throw the boomerang 90 m and have it return to their feet. Such a record-setting boomerang would be useless to someone without a very strong arm since it could not be thrown with a smaller radius of orbit. A typical outdoor boomerang orbit may have a diameter of about 25 m. The boomerang starts its flight with a CM velocity of about 25 m/s and a rotation rate of about 100 r/s. It stays in the air for about 5 s.

6.7 Moments and Products of Inertia

The dynamics of rigid-body rotations are contained in (6.48), which relates the time rate of change of the total angular momentum to the external torque. The angular momentum $\mathbf{L}$ about a point O can be computed in terms of the angular velocity $\boldsymbol{\omega}$ in (6.64). We denote the location relative to O of a point mass m_i in the rigid body by $\mathbf{r}_i$. Then from (6.64) the velocity of the mass m_i relative to O is

$$\mathbf{v}_i = \boldsymbol{\omega} \times \mathbf{r}_i \tag{6.100}$$

The angular momentum about O is

$$\mathbf{L} = \sum_i m_i(\mathbf{r}_i \times \mathbf{v}_i) = \sum_i m_i \mathbf{r}_i \times (\boldsymbol{\omega} \times \mathbf{r}_i) \tag{6.101}$$

The summation is over all mass points in the body. Using (2-44a) to expand the triple cross product, we obtain

$$\mathbf{L} = \boldsymbol{\omega}\left(\sum_i m_i|\mathbf{r}_i|^2\right) - \sum_i m_i \mathbf{r}_i(\mathbf{r}_i \cdot \boldsymbol{\omega}) \tag{6.102}$$

We observe that the angular-momentum vector $\mathbf{L}$ will not necessarily be parallel to the angular-velocity vector $\boldsymbol{\omega}$. We can write (6.102) in cartesian components as

$$L_x = \omega_x \sum_i m_i(y_i^2 + z_i^2) - \omega_y \sum_i m_i x_i y_i - \omega_z \sum_i m_i x_i z_i$$

$$L_y = -\omega_x \sum_i m_i y_i x_i + \omega_y \sum_i m_i(x_i^2 + z_i^2) - \omega_z \sum_i m_i y_i z_i \tag{6.103}$$

$$L_z = -\omega_x \sum_i m_i z_i x_i - \omega_y \sum_i m_i z_i y_i + \omega_z \sum_i m_i(x_i^2 + y_i^2)$$

For the coefficients of the angular-velocity components we introduce the notation

$$I_{xx} = \sum_i m_i(y_i^2 + z_i^2) \quad I_{xy} = -\sum_i m_i x_i y_i \quad I_{xz} = -\sum_i m_i x_i z_i$$

$$I_{yx} = -\sum_i m_i y_i x_i \quad I_{yy} = \sum_i m_i(x_i^2 + z_i^2) \quad I_{yz} = -\sum_i m_i y_i z_i$$

$$I_{zx} = -\sum_i m_i z_i x_i \quad I_{zy} = -\sum_i m_i z_i y_i \quad I_{zz} = \sum_i m_i(x_i^2 + y_i^2)$$

$$\tag{6.104}$$

In terms of these quantities we have

$$
\begin{aligned}
L_x &= I_{xx}\omega_x + I_{xy}\omega_y + I_{xz}\omega_z \\
L_y &= I_{yx}\omega_x + I_{yy}\omega_y + I_{yz}\omega_z \\
L_z &= I_{zx}\omega_x + I_{zy}\omega_y + I_{zz}\omega_z
\end{aligned}
\tag{6.105}
$$

The components of angular momentum in (6.105) are then compactly written

$$
L_j = \sum_k I_{jk}\omega_k
\tag{6.106}
$$

In this expression the superscripts j, k take on the values x, y, z. The three quantities I_{jj} are known as *moments of inertia*, and the six I_{jk} with $j \neq k$ are called *products of inertia*. From (6.104) we note that the products of inertia are symmetric.

$$
I_{jk} = I_{kj}
\tag{6.107}
$$

The nine quantities I_{jk} form a *symmetric tensor* and can be written as a 3×3 matrix.

$$
\mathbb{I} \equiv
\begin{pmatrix}
I_{xx} & I_{xy} & I_{xz} \\
I_{yx} & I_{yy} & I_{yz} \\
I_{zx} & I_{zy} & I_{zz}
\end{pmatrix}
\tag{6.108}
$$

In vector notation (6.106) can be written

$$
\mathbf{L} = \mathbb{I} \cdot \boldsymbol{\omega}
\tag{6.109}
$$

From (6.48) and (6.106) the equation of motion for general rotations of a rigid body about its CM point or about a fixed point in space is

$$
N_j = \frac{dL_j}{dt} = \sum_k \frac{d}{dt}(I_{jk}\omega_k)
\tag{6.110}
$$

or more compactly, in vector notation,

$$
\mathbf{N} = \dot{\mathbf{L}} = \frac{d}{dt}(\mathbb{I} \cdot \boldsymbol{\omega})
\tag{6.111}
$$

The kinetic energy of a rigid body can likewise be expressed in terms of the moments and products of inertia. The kinetic energy is given by

$$K = \tfrac{1}{2}\sum_i m_i \mathbf{v}_i \cdot \mathbf{v}_i = \tfrac{1}{2}\sum_i m_i(\boldsymbol{\omega}\times\mathbf{r}_i)\cdot(\boldsymbol{\omega}\times\mathbf{r}_i) = \tfrac{1}{2}\boldsymbol{\omega}\cdot\left[\sum_i m_i\mathbf{r}_i\times(\boldsymbol{\omega}\times\mathbf{r}_i)\right]$$
(6.112)

where we have interchanged dot and cross products in the final step. Using (6.101) and (6.106), we find

$$K = \tfrac{1}{2}\boldsymbol{\omega}\cdot\mathbf{L} = \tfrac{1}{2}\sum_{jk} I_{jk}\omega_j\omega_k \qquad (6.113)$$

In vector notation K can be written

$$K = \tfrac{1}{2}\boldsymbol{\omega}\cdot\mathbb{I}\cdot\boldsymbol{\omega} \qquad (6.114)$$

6.8 Single-Axis Rotations

The equation of motion in (6.110) simplifies considerably for the case of rotation about a single fixed axis. For definiteness we choose the z axis as the axis of rotation, $\boldsymbol{\omega} = \omega_z\hat{\mathbf{z}}$. The components of the angular momentum in (6.105) are then

$$L_x = I_{xz}\omega_z$$
$$L_y = I_{yz}\omega_z \qquad (6.115)$$
$$L_z = I_{zz}\omega_z$$

The equations of motion from (6.110) are

$$N_x = \dot{L}_x = \frac{d}{dt}(I_{xz}\omega_z)$$
$$N_y = \dot{L}_y = \frac{d}{dt}(I_{yz}\omega_z) \qquad (6.116)$$
$$N_z = \dot{L}_z = \frac{d}{dt}(I_{zz}\omega_z)$$

For a rigid body that is symmetrical about the z axis, we find from (6.104) that

$$I_{xz} = I_{yz} = 0 \qquad (6.117)$$

In this case the torques N_x and N_y in (6.116) are zero. On the other hand, if one of the products of inertia, I_{xz} or I_{yz} is nonzero, the body is unbalanced and the bearings must provide the torques N_x or N_y to keep the axis of rotation from moving. For single axis rotation the principal moment of inertia is usually known as the *moment of inertia*.

The rotational motion about the z axis is accelerated by the external torque N_z. From (6.104) we observe that the moment of inertia

$$I_{zz} = \sum_i m_i(x_i^2 + y_i^2) \qquad (6.118)$$

is time independent, since the perpendicular distance $\sqrt{x_i^2 + y_i^2}$ from the rotation axis of the mass point m_i is fixed in a rigid body. Thus the equation of motion for the z component in (6.116) is

$$N_z = \dot{L}_z = I_{zz}\dot{\omega}_z \qquad (6.119)$$

The kinetic energy for rigid-body rotation about the fixed z axis is found from (6.113) to be given by

$$K = \tfrac{1}{2}I_{zz}\omega_z^2 \qquad (6.120)$$

The equation for rotational motion about a fixed axis in (6.119) has the same mathematical structure as the equation for linear motion in one direction.

$$F_z = M\dot{v}_z \qquad (6.121)$$

In fact, the following direct correspondences can be made between the physical quantities of angular and linear motion:

Angular motion	Linear motion
Moment of inertia, I_{zz}	Mass, M
Angular acceleration, $\alpha_z = \dot{\omega}_z = \frac{d^2\phi}{dt^2}$	Linear acceleration, $a_z = \dot{v}_z = \frac{d^2 z}{dt^2}$
Torque, N_z	Force, F_z
Angular velocity, $\omega_z = \frac{d\phi}{dt}$	Linear velocity, $v_z = \frac{dz}{dt}$
Angular position, ϕ	Linear position, z
Angular momentum, $L_z = I_{zz}\omega_z$	Linear momentum, $P_z = Mv_z$
Kinetic energy, $K = \tfrac{1}{2}I_{zz}\omega_z^2$	Kinetic energy, $K = \tfrac{1}{2}Mv_z^2$

As a consequence, we can directly apply the techniques for solving one-dimensional problems in Chapters 1 and 2 to solve (6.113) for rotations about a single axis. For example, if the torque is conservative (function only of the angle ϕ), we can define a potential energy

$$V(\phi) = -\int_{\phi_S}^{\phi} N_z(\phi')d\phi' \qquad (6.122)$$

in correspondence with (2-6).

6.9 Moments-of-Inertia Calculations

The moment of inertia I_O of a rigid body about a given axis through the point O is related to the moment of inertia I_{CM} about a parallel axis which passes the center of mass by the *parallel-axis rule*

$$I_O = I_{CM} + Md^2 \qquad (6.123)$$

where d is the perpendicular distance between the two axes; see Fig. 6-14. In practice I_{CM} is often easier to compute than I_O, making it advantageous to use the parallel-axis rule. In any case only one moment of inertia about a set of parallel axes needs to be computed.

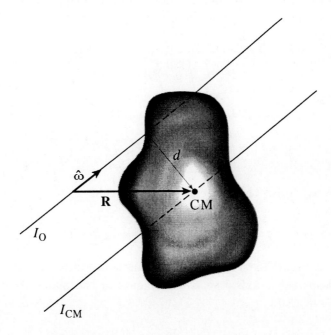

FIGURE 6-14. Parallel-axis rule for moments of inertia.

To prove the parallel axis theorem we use the expression (6.17) for the system angular momentum about O

$$\mathbf{L} = M\mathbf{R} \times \mathbf{V} + \sum_i m_i \mathbf{r}'_i \times \mathbf{v}'_i \tag{6.124}$$

For the moment of inertia we just need the component of $\mathbf{L}$ along $\boldsymbol{\omega}$

$$\mathbf{L} \cdot \hat{\boldsymbol{\omega}} \equiv I_O\,\omega = M\mathbf{R} \times \mathbf{V} \cdot \hat{\boldsymbol{\omega}} + I_{\mathrm{CM}}\,\omega \tag{6.125}$$

The first term on the right side simplifies to

$$\begin{aligned}
M\mathbf{R} \times \mathbf{V} \cdot \hat{\boldsymbol{\omega}} &= M\hat{\boldsymbol{\omega}} \cdot \mathbf{R} \times \mathbf{V} \\
&= M\hat{\boldsymbol{\omega}} \times \mathbf{R} \cdot \mathbf{V} \\
&= M(\hat{\boldsymbol{\omega}} \times \mathbf{R})^2 \omega
\end{aligned} \tag{6.126}$$

By referring to Fig. 6-14 we see that the length of $\hat{\boldsymbol{\omega}} \times \mathbf{R}$ is the perpendicular distance d between the two parallel axes, and $I_O = Md^2 + I_{\mathrm{CM}}$ follows.

Another useful rule for moments of inertia, known as the *perpendicular axis rule*, applies to bodies whose mass is distributed in a single plane. For a body in the x, y plane with mass density per unit area σ, the moments of inertia about the three axes are

$$\begin{aligned}
I_{xx} &= \int y^2 \sigma dA \\
I_{yy} &= \int x^2 \sigma dA \\
I_{zz} &= \int (x^2 + y^2) \sigma dA
\end{aligned} \tag{6.127}$$

From these we derive the perpendicular-axis rule

$$I_{xx} + I_{yy} = I_{zz} \tag{6.128}$$

illustrated in Fig. 6-15. When the mass distribution is azimuthally symmetrical about the z axis, the two moments in the plane are equal and we obtain

$$I_{xx} = I_{yy} = \tfrac{1}{2} I_{zz} \tag{6.129}$$

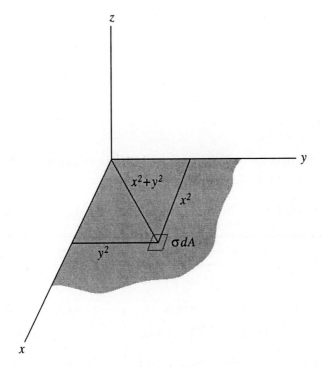

FIGURE 6-15. Perpendicular-axis rule for a body whose mass is distributed only in the x, y plane.

In the applications to be considered in the following two sections, we shall need the moment of inertia of a spherical body about an axis through its center of mass. For a sphere of radius a the mass density is

$$\rho = \frac{M}{\frac{4}{3}\pi a^3} \tag{6.130}$$

where M is the total mass. The integral for the moment of inertia about the z axis,

$$I_{\text{CM}} = \frac{M}{\frac{4}{3}\pi a^3} \int (x'^2 + y'^2) dV' \tag{6.131}$$

can be carried out simply in cylindrical coordinates.

$$x' = r' \cos \phi'$$
$$y' = r' \sin \phi' \tag{6.132}$$
$$dV' = (r' d\phi') dr' dz'$$

After the transformation to cylindrical variables, we find

$$I_{CM} = \frac{M}{\frac{4}{3}\pi a^3} \int\limits_{-a}^{+a} dz' \int\limits_{0}^{\sqrt{a^2 - z'^2}} r'^3 dr' \int\limits_{0}^{2\pi} d\phi' = \frac{M}{\frac{4}{3}\pi a^3} \int\limits_{-a}^{+a} dz' \frac{(a^2 - z'^2)^2}{4}(2\pi)$$

$$= \tfrac{2}{5}Ma^2$$

(6.133)

Some other frequently used moments of inertia of simple uniform bodies are tabulated in Table 6-1.

TABLE 6-1. MOMENTS OF INERTIA OF SOME SIMPLE BODIES

Body	Axis through CM	Moment of inertia
Rod, length l	Perpendicular to rod	$I_{CM} = \frac{1}{12}Ml^2$
Rectangular plate,	Parallel to side b	$I_{CM} = \frac{1}{12}Ma^2$
sides a, b	Perpendicular to plate	$I_{CM} = \frac{1}{12}M(a^2 + b^2)$
Cube, sides a	Perpendicular to face	$I_{CM} = \frac{1}{6}Ma^2$
Hoop, radius a	Perpendicular to plane	$I_{CM} = Ma^2$
Disk, radius a	Perpendicular to plane	$I_{CM} = \frac{1}{2}Ma^2$
	Parallel to plane	$I_{CM} = \frac{1}{4}Ma^2$
Solid cylinder:		
radius a	Along cylinder axis	$I_{CM} = \frac{1}{2}Ma^2$
length l	Perpendicular to cylinder axis	$I_{CM} = \frac{1}{12}M(3a^2 + l^2)$
Spherical shell,		
radius a	Any axis	$I_{CM} = \frac{2}{3}Ma^2$
Solid sphere,		
radius a	Any axis	$I_{CM} = \frac{3}{5}Ma^2$
Solid ellipsoid of		
semi-axes a, b, c	Along axis a	$I_{CM} = \frac{1}{5}M(b^2 + c^2)$

6.10 Impulses and Billiard Shots

For forces that act only during a very short time, it is convenient to use an integrated form of the laws of motion. The translational motion of the center-of-mass point is determined by

$$\dot{\mathbf{P}} = \mathbf{F} \tag{6.134}$$

If we multiply both sides of this equation by dt and integrate over the short time interval $\Delta t = t_1 - t_0$, during which the force acts, we obtain

$$\Delta \mathbf{P} = \mathbf{P}^1 - \mathbf{P}^0 = \int_{t_0}^{t_1} \mathbf{F} dt \tag{6.135}$$

The time integral of the force on the right is called the *impulse*. For angular motion the integrated form of the equation of motion in (6.48) is

$$\Delta \mathbf{L} = \mathbf{L}^1 - \mathbf{L}^0 = \int_{t_0}^{t_1} \mathbf{N} dt \tag{6.136}$$

The time integral of the torque is called the *angular impulse*. For rigid-body rotations about a fixed z axis, we can use (6.119) to rewrite the angular-impulse equation (6.136) as

$$\Delta L_z = I_{zz} \Delta \omega_z = I_{zz}(\omega_z^1 - \omega_z^0) = \int_{t_0}^{t_1} N_z dt \tag{6.137}$$

As an example of the usefulness of the impulse formulation of the equations of motion, we discuss the dynamics of billiard shots. For simplicity we consider only shots in which the cue hits the ball in its vertical median plane in a horizontal direction. In billiard jargon these are shots without "English".

The cue imparts an impulse to the stationary ball at a vertical distance h above the table, as illustrated in Fig. 6-16. The linear impulse from (6.135) is

$$M \Delta V_x = M V_x^1 = \int_{t_0}^{t_1} F_x dt \tag{6.138}$$

where V_x^1 is the velocity of the CM just after impact. The angular impulse of (6.137) about the z axis in Fig. 6-17 which passes through the CM of

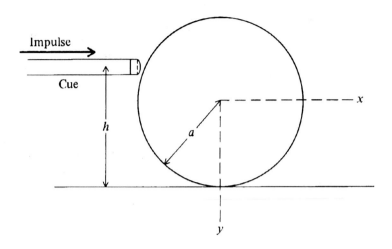

FIGURE 6-16. Impulse imparted to a billiard ball by the cue stick.

the ball is

$$\Delta L_z = I_{zz}\omega_z^1 = \int_{t_0}^{t_1} (h-a)F_x dt \tag{6.139}$$

where a is the radius of the ball. By elimination of the force integral between (6.138) and (6.139) and substitution of the moment of inertia from (6.133), we arrive at the following relation between the spin and velocity of the ball immediately after the impulse:

$$\omega_z^1 = \frac{5}{2}\left(\frac{h-a}{a^2}\right)V_x^1 \tag{6.140}$$

The velocity of the ball at the point of contact with the table is

$$V_c = V_x^1 - a\omega_z^1 = V_x^1\left(\frac{7a-5h}{2a}\right) \tag{6.141}$$

If the ball is to roll without slipping ($V_c = 0$), we find that

$$h = \tfrac{7}{5}a \tag{6.142}$$

Only if the ball is hit exactly at this height does pure rolling take place from the very start. For a *high shot* with $h > \tfrac{7}{5}a$, V_c is opposite in direction to V_x^1. Since the friction at the billiard cloth opposes V_c, it causes V_x to increase and ω_z^1 to decrease until pure rolling sets in. For a *low shot* with $h < \frac{7}{5}a$, the contact velocity V_c is in the direction of V_x^1. In this case the friction force decreases V_x^1 and increases ω_z^1 until rolling occurs. The diagram of Fig. 6-17 summarizes these results. For a ball which is rolling uniformly, the CM moves uniformly, and there is therefore no static frictional force in the direction of the motion.

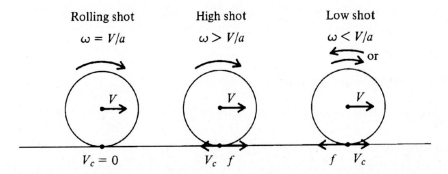

FIGURE 6-17. Rolling, high, and low shots in billiards.

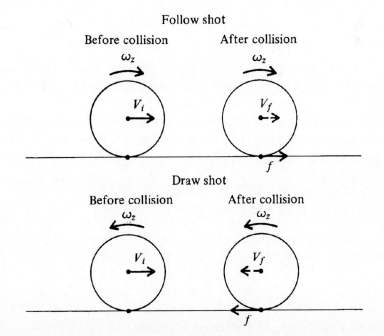

FIGURE 6-18. Motion of the cue ball for follow and draw shots in billiards. V_f is the velocity which will be ultimately acquired by the cue ball.

When the moving cue ball makes a head-on collision with a target ball at rest, the CM of the cue ball momentarily stops, and the target ball moves forward with the CM velocity of the cue ball, as shown earlier, in § 3.4. Since the balls are assumed smooth, the cue ball retains its spin ω_z^1 in the collision. Consequently, the contacting point on the cue ball moves with velocity $V_c = -a\omega_z^1$ immediately after collision. If $\omega_z^1 > 0$ at the moment of collision, the friction force acting opposite to the direction of V_c accelerates the cue ball forward, as illustrated in Fig. 6-18. This is the so-called *follow shot.* If $\omega_z^1 < 0$, the friction force accelerates the cue ball backward until pure rolling motion sets in. This is the *draw shot.* These shots play an important part in the tactics of pool or billiards.

6.11 Super-Ball Bounces

The bizarre behavior observed in bounces of the Wham-O Super-Ball (Registered trademark of Wham-O Corporation, San Gabriel, CA.) can be predicted from the rigid-body equations of motion. The Super-Ball is a hard spherical rubber ball. The bounces of a Super-Ball on a hard surface are almost elastic (*i.e.*, energy-conserving) and essentially nonslip at the point of contact. As an idealization, we shall also neglect gravity in our calculations, though its inclusion does not change the principal results.

We begin with an analysis of a single bounce from the floor. We denote the initial components of the CM velocity by v_x^0 and v_y^0 and the initial spin of the ball about the z axis through the CM by ω_z^0, as pictured in Fig. 6-19. The frictional force f_x and the normal force f_y act on the ball only for a very short time duration, Δt. We can determine the changes in the velocities Δv_x, Δv_y from the linear-impulse equation (6.135), and the change in spin from the angular-impulse equation (6.137). We obtain

$$M\Delta v_x = -\int_{t_0}^{t_1} f_x\, dt$$

$$M\Delta v_y = -\int_{t_0}^{t_1} f_y\, dt \qquad (6.143)$$

$$I_{zz}\Delta\omega_z = a\int_{t_0}^{t_1} f_x\, dt$$

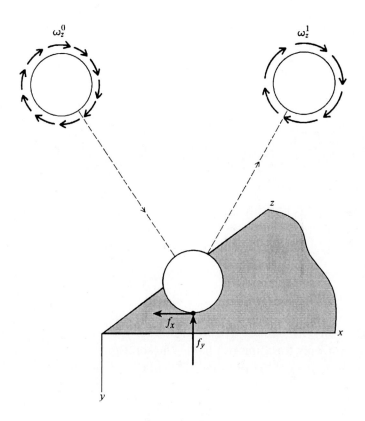

FIGURE 6-19. Super-Ball bounce from a hard surface.

By elimination of the frictional force f_x from the first and third equations, we obtain a relation between Δv_x and $\Delta \omega_z$ caused by f_x.

$$M(v_x^1 - v_x^0) = -\frac{I_{zz}}{a}(\omega_z^1 - \omega_z^0) \tag{6.144}$$

The assumption that f_x and f_y are independent (*i.e.*, that the deformations of the superball result in stresses in the x and y directions, which are independent of one another) requires that the energies associated with the x and y motions be separately conserved. In other words, both f_x and f_y are conservative forces. The conservative nature of f_y leads to

$$\tfrac{1}{2}M(v_y^1)^2 = \tfrac{1}{2}M(v_y^0)^2 \tag{6.145}$$

We conclude that the vertical component of velocity must be reversed by the action of the normal force f_y:

$$v_y^1 = -v_y^0 \tag{6.146}$$

The stipulation that f_x be energy-conserving (*i.e.*, no slipping) yields the condition

$$\tfrac{1}{2}I_{zz}(\omega_z^1)^2 + \tfrac{1}{2}M(v_x^1)^2 = \tfrac{1}{2}I_{zz}(\omega_z^0)^2 + \tfrac{1}{2}M(v_x^0)^2 \tag{6.147}$$

Equations (6.144), (6.145), and (6.147) govern the dynamics of a Super-Ball bounce. One possible solution to (6.144) and (6.147) is

$$\begin{aligned} v_x^1 &= v_x^0 \\ \omega_z^1 &= \omega_z^0 \end{aligned} \tag{6.148}$$

This solution corresponds to zero frictional force in (6.143) and is therefore relevant only for a smooth ball. The solution appropriate for a Super-Ball can be obtained by division of (6.144) and (6.147). We find

$$v_x^1 + v_x^0 = a(\omega_z^1 + \omega_z^0) \tag{6.149}$$

or by rearrangement

$$(v_x^1 - a\omega_z^1) = -(v_z^0 - a\omega_z^0) \tag{6.150}$$

The quantity $(v_x - a\omega_z)$ is just the horizontal velocity at the point on the ball that makes contact with the floor. Hence the velocity at the point of contact is exactly reversed by a bounce. From (6.144) and (6.150) we can solve for the spin and horizontal velocity immediately after the bounce in terms of the initial spin and velocity. With the moment of inertia of the Super-Ball given by (6.133), we arrive at

$$\begin{aligned} v_x^1 &= \tfrac{3}{7}V_x^0 + \tfrac{4}{7}\omega_z^0 a \\ \omega_z^1 &= -\tfrac{3}{7}\omega_z^0 + \tfrac{10}{7}\frac{v_x^0}{a} \end{aligned} \tag{6.151}$$

as the general solution for a Super-Ball bounce.

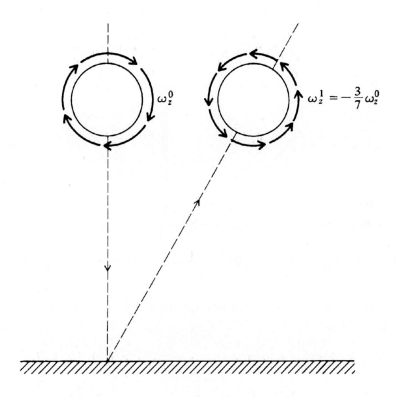

FIGURE 6-20. Deflection of a Super-Ball from a vertical bounce.

As an example of the result in (6.151), a Super-Ball which approaches the floor from a vertical direction ($v_x^0 = 0$) with initial spin ω_z^0 will leave the floor with

$$v_x^1 = \tfrac{4}{7}\omega_z^0 a$$
$$\omega_z^1 = -\tfrac{3}{7}\omega_z^0 \qquad\qquad (6.152)$$
$$v_y^1 = -V_y^0$$

as illustrated in Fig. 6-20. A smooth ball with the same initial velocity and spin would bounce back in the vertical direction.

The unexpected behavior of a Super-Ball is even more dramatically exhibited in successive bounces. As indicated in Fig. 6-21, a Super-Ball thrown to the floor in such a way that it bounces from the underside of a table will return to the hand. We can show this quite simply from repeated applications of (6.151). If the initial spin of the ball is zero, the

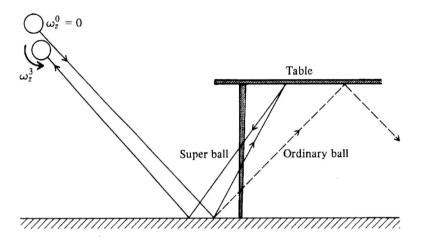

FIGURE 6-21. Return of a Super-Ball when bounced from the underside of a table.

velocity and spin after the first bounce from the floor are

$$v_x^1 = \tfrac{3}{7} v_x^0$$
$$\omega_z^1 = \tfrac{10}{7} \frac{v_x^0}{a}$$

(6.153)

For the bounce off the underside of the table, the angular impulse is opposite in sign to the impulse in (6.143). With the angular impulse reversed, the appropriate modifications of (6.151) for the second bounce are

$$v_x^2 = \tfrac{3}{7} v_x^1 - \tfrac{4}{7} \omega_z^1 a$$
$$\omega_z^2 = -\tfrac{3}{7} \omega_z^1 - \tfrac{10}{7} \frac{v_x^1}{a}$$

(6.154)

When we substitute (6.153) into (6.154), we get

$$v_x^2 = -\tfrac{31}{49} v_x^0$$
$$\omega_z^2 = -\tfrac{60}{49} \frac{v_x^0}{a}$$

(6.155)

Thus the horizontal direction of motion has been reversed. For the final bounce off the floor we again apply (6.151), with (6.155) as initial values. We find

$$v_x^3 = -\tfrac{333}{343} v_x^0$$
$$\omega_z^3 = -\tfrac{130}{343} v_x^0$$

(6.156)

Thus the Super-Ball returns after the three bounces with a slightly lower velocity than when it started, although the total kinetic energy remains the same. A smooth ball would not return, but would continue bouncing between the floor and the table, as indicated by the dashed line in Fig. 6-21.

PROBLEMS

6.1 Center of Mass and the Two-Body Problem

6-1. Find the distance of the center of mass of the earth-moon system from the center of the earth.

6-2. For a two-particle system in a region of uniform gravitational acceleration g, show that the net gravitational torque about the CM point of the system is zero.

6-3. A boat of mass 60 kg and length 4 m is at rest in quiet water. If a man of mass 80 kg walks from the bow to the stern, what distance will the boat move? Neglect water resistance.

6-4. Two particles on a line are mutually attracted by a force

$$F = -fr$$

where f is a constant and r is the distance of separation. At time $t = 0$, particle A of mass M is located at $x = 5$ cm, and particle B of mass $\frac{1}{4}M$ is located at $x = 10$ cm. If the particles are at rest at time $t = 0$, at what value of x do they collide? What is the relative velocity of the two particles at the moment the collision occurs?

6-5. Compare the magnitude of the gravity forces on the moon due to the earth and sun. Despite this result show from (6.22) that the sun is not very important in determining the moon's motion relative to the earth. If the moon's distance from the earth were greater would your conclusion remain valid?

6-6. The two atoms in a diatomic molecule (masses m_1 and m_2) interact through a potential energy

$$V(r) = \frac{a^2}{4r^4} - \frac{b^2}{3r^3}$$

where r is the separation of the atoms.

a) Find the equilibrium separation of the atoms and the frequency of small oscillations about the equilibrium assuming that the molecule does not rotate. How much energy must be supplied to the molecule in order to break it up?

b) Determine the maximum angular momentum which the molecule can have without breaking up, assuming that the motion is in circular orbits. Find the particle separation at the break up angular momentum.

c) Calculate the velocity of each particle in the laboratory system at break up, assuming that the center of mass is at rest. *Hint: break up occurs when V_{eff} no longer has a minimum.*

6-7. Two point masses are connected by a spring with spring constant k but are otherwise free to move in space. The equations of motion are

$$m_1\ddot{\mathbf{r}}_1 = -k(\mathbf{r}_1 - \mathbf{r}_2 - \mathbf{l}) \qquad\qquad m_2\ddot{\mathbf{r}}_2 = k(\mathbf{r}_1 - \mathbf{r}_2 - \mathbf{l})$$

where $\mathbf{l} = l(\mathbf{r}_1 - \mathbf{r}_2)/|\mathbf{r}_1 - \mathbf{r}_2|$.

a) Find the equilibrium separation of the masses and the frequency of oscillation of the masses about equilibrium assuming that the system does not rotate.

b) How will the equilibrium separation of the masses and the frequency of small oscillations about equilibrium change as the system rotates about an axis through the CM perpendicular to the axis of the oscillator?

c) Show that the total energy of the system is conserved.

6-8. Two particles with masses m_1 and m_2 collide head on. Particle 1 has an initial velocity v_1 and particle 2 is initially at rest in the laboratory system. The particles interact through a potential energy

$$V = V_0 \left(\frac{a}{r_{12}}\right)^2$$

where $V_0 > 0$ and $r_{12} = |\mathbf{r}_1 - \mathbf{r}_2|$.

a) Compute the total energy and angular momentum of the two particles in the CM system. Express the results in terms of m_1, m_2, and v_1.

b) Describe the motion qualitatively as it appears in the CM system and in the lab system.

c) Find the distance of closest approach (minimum separation between the particles).

d) Find the velocity of particle 2 in the lab system after the collision.

6-9. Two particles of masses m_1 and m_2 collide. The initial velocity of particle 1 in the lab system is $\mathbf{v}_1$, while particle 2 is initially at rest. The initial impact parameter is b_0, as shown. The particles interact through a repulsive potential $V = V_0/|\mathbf{r}_1 - \mathbf{r}_2|^4$.

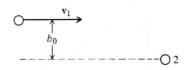

a) Calculate the total energy and angular momentum in the CM system in terms of the particle masses and velocities.

b) Derive the equation of motion for the relative coordinate $\mathbf{r} = \mathbf{r}_1 - \mathbf{r}_2$. Find the distance of closest approach.

c) Show how the angle β between the final velocities of the particles in the lab can be calculated if the magnitude $|\mathbf{v}_{1f}|$ of the final lab velocity of particle 1 is measured.

6.2 Rotational Equation of Motion

6-10. If the force on m_1 by m_2 is $\mathbf{F}_1^{[2]}$ and the force on m_2 by m_1 is $\mathbf{F}_2^{[1]}$, the *extended third law* requires that not only $\mathbf{F}_1^{[2]} + \mathbf{F}_2^{[1]} = 0$ but also that the forces act on a line connecting m_1 and m_2. Show that the total torque due to this pair of forces about a point p at $\mathbf{r}_p$ is zero, thus demonstrating that the extended third law implies zero total internal torque.

6-11. If the potential energy between two particles of a system depends only on their separation show that this potential energy depends on the angle of rotation about a fixed axis only through differences in particle angle coordinates. Then show that the resulting internal torque is zero.

6.3 Rigid Bodies: Static Equilibrium

6-12. A circular tabletop of radius 1 m and mass 3 kg is supported by three equally spaced legs on the circumference. When a vase is placed on the table, the legs support 1, 2, and 3 kg, respectively. How heavy is the vase, and where is it located on the table? What is the lightest vase which might upset the table?

6-13. A spool rests on a rough table as shown. A thread wound on the spool is pulled with force T at angle θ.

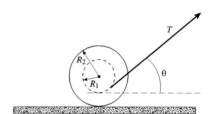

a) If $\theta = 0$ will the spool move to the left or right?

b) Show that there is an angle θ for which the spool remains at rest.

c) At this critical angle find the maximum T for equilibrium to be maintained. Assume a coefficient of friction μ.

6-14. A cylindrical glass full of ice weighs four times as much as when empty. At what intermediate level of filling is the glass least likely to tip? Neglect the mass of the bottom of the glass. Would the result change if the glass contained water (of the same density)? *Hint: show that the maximum angle of tip corresponds to the minimum CM height.*

6.4 Rotations of Rigid Bodies

6-15. Consider again the drag racer of Fig. 1-1 which, while accelerating horizontally, is in vertical and rotational equilibrium. In § 1.3 it was found that the maximum acceleration is μg where μ is the coefficient of friction between the rear tires and the track. Show that in order to realize this optimal acceleration the CM must be located such that the ratio of its height h above the ground to the distance b_2 forward of the rear wheels satisfies $b_2 = \mu h$.

6-16. A vehicle has brakes on all four wheels. At rest the weight supported by each wheel is the same. Find the deceleration which corresponds to maximum possible braking. Calculate the normal forces the on front and back wheels when the brakes are applied. Why are the front brakes the most important in braking?

6.5 Gyroscopic Effect

6-17. A heavy axially symmetric gyroscope is supported at a pivot, as
shown. The mass of the gyroscope is M, and the moment of
inertia about its symmetry axis
is I. The initial angular velocity about its symmetry axis is ω.

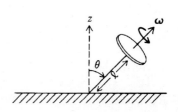

Give a suitable approximate equation of motion for the system,
assuming that ω is very large. Find the angular frequency of the
gyroscopic precession. Show that the above approximation is justi-
fied for

$$\omega \gg \sqrt{\frac{g}{\ell}}$$

where all moments of inertia are taken to be roughly $M\ell^2$.

6.6 The Boomerang

6-18. Show that if the aerodynamic force on a boomerang blade is proportional to v_t (not v_t^2 as in the text), the ratio of spin to CM velocity
must still be related as $V = \sqrt{\frac{2}{3}}\omega l$ for a successful return. Show
that the radius of the orbit is now proportional to the velocity.

6.8 Single-Axis Rotations

6-19. A disk of radius R is oriented in a vertical plane and spinning about
its axis with angular velocity ω. If the spinning disk is set down
on a horizontal surface, with what translational CM velocity will
it roll away? *Hint: friction accelerates the CM and slows rotation
until rolling begins.*

6-20. A spherical asteroid of uniform density 3 g/cm^3 and radius 100 m is
rotating once per minute. It gradually acquires meteoritic material
of the same density until a few billion years later its radius has
doubled. If, on the average, this matter has arrived radially, what
is the final rate of rotation?

6-21. Due to tidal friction the earth's day increases in length by 4.4 $\times$
10^{-8} s each day.

  a) Compute the accumulated error between time based on the earth's
  rotation and absolute time (say by an atomic clock) after one

century and after 3000 years. This accumulated error would be evident in the observation versus prediction of eclipses. *Hint: compare the angle through which an accelerating sphere turns with that of a sphere rotating uniformly.*

b) Estimate the power dissipated assuming a uniform earth. Compare this to a 10^9 W electrical generation facility.

6.9 Moments-of-Inertia Calculations

6-22. A thin, uniform rod of mass M is supported by two vertical strings, as shown. Find the tension in the remaining string immediately after one of the strings is severed.

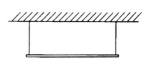

6-23. A physical pendulum consists of a solid cylinder which is free to rotate about a transverse axis displaced by a distance ℓ along the symmetry axis from the center of mass, as illustrated. Find the value of ℓ for which the period is a minimum. Express the result in terms of the mass M and moment of inertia I about a transverse axis through the CM.

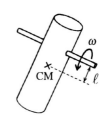

6-24. A pendulum consists of two masses connected by a very light rigid rod, as shown. The pendulum is free to oscillate in the vertical plane about a horizontal axis located a distance a from m_a at a distance b from m_b.

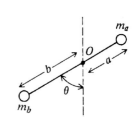

a) Calculate the moment of inertia of the system about 0. Find the location of the center of mass.

b) Set up the equation of motion for the system and derive the potential-energy function.

c) Take $b > a$ and determine the frequency of oscillation for small angles of displacement from the vertical.

d) Derive an exact expression for the period of the pendulum $(|\theta_{max}| < \pi)$.

e) Find the minimum angular velocity which must be given to the system (starting at equilibrium) if it is to continue in rotation instead of oscillating.

6-25. A yo-yo of mass M is composed of two disks of radius R separated by a distance t by a shaft of radius r. A massless string is wound on the shaft, and the loose end is held in the hand. Upon release the yo-yo descends until the string is unwound. The string then begins to rewind, and the yo-yo climbs. Find the string tension and acceleration of the yo-yo in descent and in ascent. Neglect the mass of the shaft and assume the shaft radius is sufficiently small so that the string is essentially vertical.

6-26. Find the inertia tensor components about the origin in terms of the inertial tensor components about the CM. The position of the CM point is $\mathbf{R} = X\hat{\mathbf{x}} + Y\hat{\mathbf{y}} + Z\hat{\mathbf{z}}$.

6-27. A two-dimensional object lies in the x, y plane and is described by the moments of inertia I_{xx}, I_{yy} and the product of inertia I_{xy} for rotations in the x, y plane. In a coordinate system rotated by an angle ϕ the new coordinate components are related to x and y by

$$x' = x \cos\phi + y \sin\phi$$
$$y' = -x \sin\phi + y \cos\phi$$

a) The tensor of inertia elements in the rotated system are $I_{x'x'}$, $I_{y'y'}$ and $I_{x'y'}$. Find the smallest angle ϕ_0 for which $I_{x'y'}$ vanishes. Show that the other solutions are $\phi_0 + n\frac{\pi}{2}$ and are equivalent. The x' and y' axes are known as *principal axes* since for rotations along these axes $\mathbf{L}$ is parallel to $\boldsymbol{\omega}$.

b) If for two rotated systems (rotation angle $\neq \pi/2$) $I_{xy} = I_{x'y'} = 0$, prove that the principal moments are equal. As an example consider an equilateral triangle.

c) If the principal moments are equal and $I_{xy} = 0$ in some coordinate frame show that the moments are equal in any rotated coordinate system. As an example consider a uniform square.

d) Compute the principal moments in terms of a given set I_{xx}, I_{yy} and I_{xy} of inertia elements.

6.10 Impulses and Billiard Shots

6-28. A pencil of length ℓ and mass m lying flat on a frictionless horizontal tabletop receives an impulse on one end at a right angle to the pencil. What is the orientation and position of the center of mass of the pencil at a time t after the impulse?

6-29. A rod of mass m and length ℓ hangs vertically from a horizontal frictionless wire, as shown. Attached to the end of the rod is a small ball also of mass m. The rod is free to move along the wire.

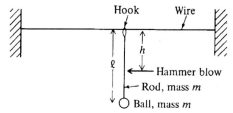

a) Find the location of the center of mass for rod plus ball, taking the hook as the origin of the coordinate system.

b) Find the moment of inertia for rod plus ball about the hook.

c) Use the parallel-axis theorem and the result in part b to find the moment of inertia about the center of mass.

d) The rod-ball system is struck by an impulsive hammer blow a distance h from the hook. Set up equations for the linear and angular motion of the system.

e) Find h such that the hook does not move along the wire at the instant of blow. This point is known as the *center of percussion.*

6-30. A ball of radius a rolling with velocity v on a level surface collides inelastically with a step of height $h < a$, as shown. Find the minimum velocity for which the ball will "trip" up over the step. Assume that no slipping occurs at the impact point. *Hint: compute the total angular momentum about the point of impulsive contact. This angular momentum is conserved and can be used to compute the energy change.*

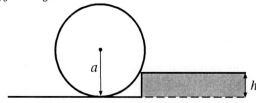

6-31. In the sport of bowling, if the ball is rolled straight down the middle of the alley, pins on the sides will often be left standing (wide splits). A good right-handed bowler will impart a spin to the ball on release, causing it to curve to the left as it goes down the alley and strike the pins somewhat to the side. Describe the required spin and show that the trajectory of the ball is approximately parabolic in shape.

6.11 Super-Ball Bounces

6-32. One of the Super-Ball examples discussed in the text concerned a ball dropped straight down with spin. Discuss the subsequent motion through several bounces.

6-33. Under what conditions will a Super-Ball bounce back and forth as illustrated? How does the spin change?

Chapter 7

ACCELERATED COORDINATE SYSTEMS

The simple form of Newton's second law,

$$\mathbf{F} = m\frac{d^2\mathbf{r_I}}{dt^2} \qquad (7.1)$$

for a particle of mass m holds only in inertial coordinate systems (unaccelerated and not rotating with respect to the distant stars), as denoted by the $\mathbf{I}$ subscript above. On the other hand, physical events are sometimes more simply described with reference to an accelerated or rotating coordinate system. For example, observations of motion on the earth's surface are more simply expressed in terms of a coordinate system fixed on the rotating earth than in terms of an inertial coordinate system. For this reason it is useful to derive the form of the second law which directly applies in accelerated reference frames.

7.1 Transformation to Moving Coordinate Frames

To transform the law of motion to an accelerated reference frame, we first need to relate the time derivatives of vector quantities in moving and fixed coordinate systems. In a moving frame, an arbitrary vector quantity $\mathbf{A}$ can be written

$$\mathbf{A} = A_x\hat{\mathbf{x}} + A_y\hat{\mathbf{y}} + A_z\hat{\mathbf{z}} \qquad (7.2)$$

where the directions of the unit vectors change in time. The time derivative of $\mathbf{A}$ is

$$\frac{d\mathbf{A}}{dt} = \left(\frac{dA_x}{dt}\hat{\mathbf{x}} + \frac{dA_y}{dt}\hat{\mathbf{y}} + \frac{dA_z}{dt}\hat{\mathbf{z}}\right) + \left(A_x\frac{d\hat{\mathbf{x}}}{dt} + A_y\frac{d\hat{\mathbf{y}}}{dt} + A_z\frac{d\hat{\mathbf{z}}}{dt}\right) \qquad (7.3)$$

The first term on the right-hand side of this equation is the time rate of change of $\mathbf{A}$ with reference to the axes of the accelerated frame. We denote this time rate of change of $\mathbf{A}$ in the moving frame by

$$\frac{\delta\mathbf{A}}{\delta t} \equiv \frac{dA_x}{dt}\hat{\mathbf{x}} + \frac{dA_y}{dt}\hat{\mathbf{y}} + \frac{dA_z}{dt}\hat{\mathbf{z}} \qquad (7.4)$$

The second term in (7.3) is due to the rotation of the coordinate system, which causes the direction of the unit vectors to change with time. Since the coordinate system is rigid, we can directly apply the result of (6.64),

$$\frac{d\mathbf{r}}{dt} = \boldsymbol{\omega} \times \mathbf{r} \tag{7.5}$$

to find the time derivatives of the unit vectors

$$\frac{d\hat{\mathbf{x}}}{dt} = \boldsymbol{\omega} \times \hat{\mathbf{x}} \qquad \frac{d\hat{\mathbf{y}}}{dt} = \boldsymbol{\omega} \times \hat{\mathbf{y}} \qquad \frac{d\hat{\mathbf{z}}}{dt} = \boldsymbol{\omega} \times \hat{\mathbf{z}} \tag{7.6}$$

Here $\boldsymbol{\omega}$ is the angular velocity of rotation of the accelerated frame relative to a fixed frame. Upon substitution of (7.4) and (7.6) into (7.3), we have

$$\frac{d\mathbf{A}}{dt} = \frac{\delta \mathbf{A}}{\delta t} + \boldsymbol{\omega} \times (A_x \hat{\mathbf{x}} + A_y \hat{\mathbf{y}} + A_z \hat{\mathbf{z}}) \tag{7.7}$$

or more simply,

$$\frac{d\mathbf{A}}{dt} = \frac{\delta \mathbf{A}}{\delta t} + \boldsymbol{\omega} \times \mathbf{A} \tag{7.8}$$

Accordingly, the time derivative $d\mathbf{A}/dt$ in a fixed reference frame consists of a part $\delta\mathbf{A}/\delta t$ from the time rate of change of $\mathbf{A}$ relative to the axes of the moving frame and a part $\boldsymbol{\omega} \times \mathbf{A}$ from the rotation of these axes relative to the fixed axes. It follows from (7.8) that the time derivative of $\boldsymbol{\omega}$ is independent of the coordinate frame.

$$\frac{d\boldsymbol{\omega}}{dt} = \frac{\delta\boldsymbol{\omega}}{\delta t} = \dot{\boldsymbol{\omega}} \tag{7.9}$$

We next apply the result in (7.8) to the vector $\mathbf{r}$, which specifies the location of a particle with respect to the moving axes. The first time derivative is

$$\frac{d\mathbf{r}}{dt} = \frac{\delta\mathbf{r}}{\delta t} + \boldsymbol{\omega} \times \mathbf{r} \tag{7.10}$$

The second time derivative can likewise be evaluated with the aid of (7.8).

$$\begin{aligned}
\frac{d^2\mathbf{r}}{dt^2} &= \frac{d}{dt}\left(\frac{\delta\mathbf{r}}{\delta t} + \boldsymbol{\omega} \times \mathbf{r}\right) = \frac{\delta}{\delta t}\left(\frac{\delta\mathbf{r}}{\delta t} + \boldsymbol{\omega} \times \mathbf{r}\right) + \boldsymbol{\omega} \times \left(\frac{\delta\mathbf{r}}{\delta t} + \boldsymbol{\omega} \times \mathbf{r}\right) \\
&= \frac{\delta^2\mathbf{r}}{\delta t^2} + \frac{\delta\boldsymbol{\omega}}{\delta t} \times \mathbf{r} + 2\boldsymbol{\omega} \times \frac{\delta\mathbf{r}}{\delta t} + \boldsymbol{\omega} \times (\boldsymbol{\omega} \times \mathbf{r})
\end{aligned} \tag{7.11}$$

To find the law of motion in the accelerated/rotating frame, we relate the location of the particle in the accelerated and inertial frames by the vector $\mathbf{R}$ connecting the origins of the two frames

$$\mathbf{r_I} = \mathbf{r} + \mathbf{R} \tag{7.12}$$

as illustrated in Fig. 7-1. Then, substituting the result in (7.11) into

$$\frac{d^2\mathbf{r_I}}{dt^2} = \frac{d^2\mathbf{r}}{dt^2} + \frac{d^2\mathbf{R}}{dt^2} \tag{7.13}$$

we get

$$\frac{d^2\mathbf{r_I}}{dt^2} = \frac{\delta^2\mathbf{r}}{\delta t^2} + \boldsymbol{\omega} \times (\boldsymbol{\omega} \times \mathbf{r}) + 2\boldsymbol{\omega} \times \frac{\delta\mathbf{r}}{\delta t} + \dot{\boldsymbol{\omega}} \times \mathbf{r} + \frac{d^2\mathbf{R}}{dt^2} \tag{7.14}$$

The form of Newton's law in the non-inertial frame now follows directly from (7.1) and (7.14)

$$m\frac{\delta^2\mathbf{r}}{\delta t^2} = \mathbf{F} - m\left[\boldsymbol{\omega} \times (\boldsymbol{\omega} \times \mathbf{r}) + 2\boldsymbol{\omega} \times \mathbf{v} + \dot{\boldsymbol{\omega}} \times \mathbf{r} + \frac{d^2\mathbf{R}}{dt^2}\right] \tag{7.15}$$

where $\mathbf{v} = \delta\mathbf{r}/\delta t$ is velocity and $\delta^2\mathbf{r}/\delta t^2$ is the acceleration of a particle as observed in the moving coordinate system.

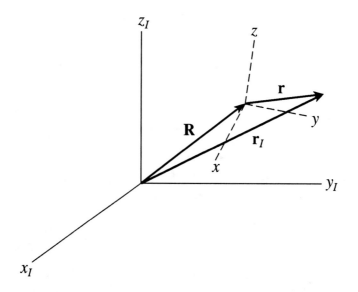

FIGURE 7-1. Inertial and accelerated coordinate frames.

7.2 Fictitious Forces

We can write the equation of motion (7.15) for an accelerated frame in a form similar to (7.1) for an inertial frame.

$$m\frac{\delta^2 \mathbf{r}}{\delta t^2} = \mathbf{F}_{\text{eff}} \qquad (7.16)$$

The acceleration $\delta^2 \mathbf{r}/\delta t^2$ observed in the moving frame is generated by the *effective force*

$$\mathbf{F}_{\text{eff}} = \mathbf{F} - m\left[\boldsymbol{\omega} \times (\boldsymbol{\omega} \times \mathbf{r}) + 2\boldsymbol{\omega} \times \mathbf{v} + \dot{\boldsymbol{\omega}} \times \mathbf{r} + \frac{d^2 \mathbf{R}}{dt^2}\right] \qquad (7.17)$$

The names associated with the so-called *fictitious force* terms on the right-hand side of (7.17) are

Centrifugal force:

$$\mathbf{F}_{cf} = -m\boldsymbol{\omega} \times (\boldsymbol{\omega} \times \mathbf{r}) \qquad (7.18)$$

Coriolis force:

$$\mathbf{F}_{Cor} = -2m\boldsymbol{\omega} \times \mathbf{v} \qquad (7.19)$$

Azimuthal force:

$$\mathbf{F}_{az} = -m\dot{\boldsymbol{\omega}} \times \mathbf{r} \qquad (7.20)$$

Translational force:

$$\mathbf{F}_{tr} = -m\frac{d^2 \mathbf{R}}{dt^2} \qquad (7.21)$$

The centrifugal force of (7.18) is due to the rotational motion of the coordinate system. Since $\boldsymbol{\omega} \cdot \mathbf{F}_{cf} = 0$, the centrifugal force is perpendicular to the rotation axis $\hat{\boldsymbol{\omega}}$. If the angular velocity $\boldsymbol{\omega}$ is chosen to lie along the z axis of the moving frame, as in Fig. 7-2, then

$$\mathbf{F}_{cf} = -m\left[\boldsymbol{\omega}(\boldsymbol{\omega} \cdot \mathbf{r}) - \mathbf{r}\omega^2\right] = m\omega^2(x\hat{\mathbf{x}} + y\hat{\mathbf{y}})$$
$$= m\omega^2 \boldsymbol{\rho} \qquad (7.22)$$

where $\boldsymbol{\rho}$ is the cylindrical-radius vector to the particle from the z axis. The centrifugal force is directed radially outward from the axis of rotation. The result in (7.22) is the same as (5.14).

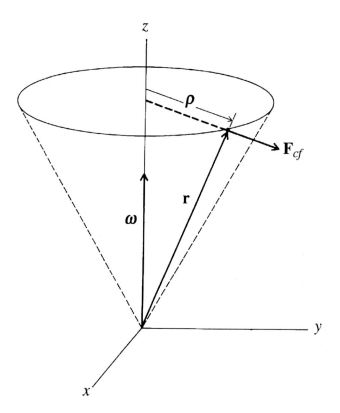

FIGURE 7-2. Centrifugal force $\mathbf{F}_{cf}$ due to rotation with angular velocity $\boldsymbol{\omega}$.

The centrifugal force, along with gravity, accounts for the parabolic shape of the surface of a spinning pail of water. Because of viscous forces the water in a uniformly spinning pail will reach an equilibrium condition where it rotates as if it were a rigid body (*i.e.*, each element has the same angular velocity). Using (7.17), the equation of motion of a small mass of water m on the surface in a frame rotating with the pail is

$$m\frac{\delta^2 \mathbf{r}}{\delta t^2} = \mathbf{F}' + m\mathbf{g} - m\boldsymbol{\omega}\times(\boldsymbol{\omega}\times\mathbf{r}) \qquad (7.23)$$

where the force $\mathbf{F}'$, due to the pressure gradient, is normal to the surface. If the pressure gradient had a component tangent to the surface the pressure would vary along the surface, but this cannot happen since the surface is at atmospheric pressure. Since in equilibrium $\delta^2\mathbf{r}/\delta t^2 = 0$, the *effective-gravity* term

$$\mathbf{g}_{\text{eff}} \equiv \mathbf{g} - \boldsymbol{\omega}\times(\boldsymbol{\omega}\times\mathbf{r}) \qquad (7.24)$$

must also be normal to the surface, by (7.23). In cylindrical coordinates, $\mathbf{g}_{\text{eff}}$ is given by

$$\mathbf{g}_{\text{eff}} = -g\hat{\mathbf{z}} + \omega^2 \rho \hat{\boldsymbol{\rho}} \tag{7.25}$$

From the geometry of Fig. 7-3, the normal requirement on $\mathbf{g}_{\text{eff}}$ can be written

$$\tan\theta = \frac{dz}{d\rho} = \frac{\omega^2 \rho}{g} \tag{7.26}$$

Integration gives

$$z = \frac{\omega^2}{2g}\rho^2 + \text{constant} \tag{7.27}$$

which is a parabolic shape. The solution in (7.27) can alternatively be found from (7.25) by potential-energy methods. The potential energy due to the force $m\mathbf{g}_{\text{eff}}$ is

$$V(z,\rho) = m(gz - \tfrac{1}{2}\omega^2\rho^2) \tag{7.28}$$

as verified by computing $\mathbf{F} = m\mathbf{g}_{\text{eff}} = -\nabla V$. Since there can be no component of force tangential to the surface in equilibrium, the potential energy in (7.28) must be constant on the surface, and the result in (7.27) is thus obtained.

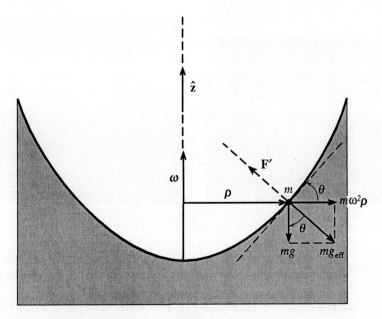

FIGURE 7-3. Parabolic surface of a spinning pail of water.

There are two important aspects of this spinning pail phenomenon which might be mentioned, one of historical and philosophical importance and the other of current practical importance in the construction of telescope mirrors. The *water pail experiment* was of great significance in Isaac Newton's formulation of mechanics. He noted that a spinning water pail achieves a curved surface while one that is not rotating has a flat surface; Newton deduced that a reference frame not rotating with respect to the stars was fundamental.

In the construction of the post-Palomar generation of large land-based optical telescopes short focal lengths must be utilized to minimize the weight and consequent expense. To construct such a mirror by the old technology would require starting with a thick blank disk and then laboriously grinding out the concave shape. An important innovation is to spin a molten pyrex or quartz blank inside a furnace. It obtains precisely the parabolic shape that is needed to focus a distant star into a point image and can be cast with uniform thickness helping to reduce the weight. The surface is then coated with a thin layer of aluminum to make it reflecting. Photos of such a telescope mirror being made by this process are shown in Fig. 7-4.

(a)   **(b)**

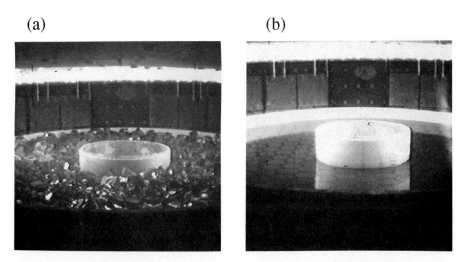

FIGURE 7-4. A parabolic mirror for the WIYN (Wisconsin-Indiana-Yale-NOAO) telescope being made by the spin-casting process. (a) The initial stage of casting and (b) after spin casting. The central plug leaves a hole for light transmission after reflection by a secondary mirror.

The Coriolis force in (7.19) is present when the particle is in motion relative to the rotating coordinate system. Since $\boldsymbol{\omega} \cdot \mathbf{F}_{\text{Cor}} = 0$ and

$\mathbf{v} \cdot \mathbf{F}_{\text{Cor}} = 0$, this force is perpendicular to both $\boldsymbol{\omega}$ and $\mathbf{v}$. The effects of the Coriolis force are important in such problems as calculations of long-range artillery and ballistic missile trajectories and in the description of large-scale atmospheric weather phenomena.

The azimuthal force in (7.20) occurs only when the angular-velocity vector changes with time. Inasmuch as $\mathbf{r} \cdot \mathbf{F}_{\text{az}} = 0$, this force always points in a direction perpendicular to $\mathbf{r}$. If $\boldsymbol{\omega}$ is changing in magnitude but constant in direction, the azimuthal force acts toward maintaining the rotational velocity of the particle.

The translation force in (7.21) is due to the acceleration of the origin of the moving frame relative to an inertial frame. In the special case when the motion of the accelerated coordinate system is purely translational (that is, $\boldsymbol{\omega} = 0$), the equation of motion in (7.16) and (7.17) reduces to

$$m\frac{\delta^2 \mathbf{r}}{\delta t^2} = \mathbf{F} - m\frac{d^2 \mathbf{R}}{dt^2} \tag{7.29}$$

The problem of a pendulum with a moving support provides an interesting application of (7.29). We choose the origin of the moving coordinate system to coincide with the instantaneous location of the support, as illustrated in Fig. 7-5. We shall restrict our discussion to angular motion in the x, y plane defined by $\mathbf{F}$ and $\mathbf{R}$. In terms of the tension $\mathbf{T}$ and the gravitational force $mg\hat{\mathbf{x}}$ acting on the pendulum bob, the equation of motion in the moving frame is

$$m\frac{\delta^2 \mathbf{r}}{\delta t^2} = \mathbf{T} + m(g - A_x)\hat{\mathbf{x}} - mA_y\hat{\mathbf{y}} \tag{7.30}$$

where $\mathbf{A} = d^2\mathbf{R}/dt^2$ is the translational acceleration. Since physical motion occurs along the θ direction, it is advantageous to write (7.30) in polar coordinates using (2.124),

$$\frac{\delta^2 \mathbf{r}}{\delta t^2} = \hat{\mathbf{r}}(\ddot{r} - r\dot{\theta}^2) + \hat{\boldsymbol{\theta}}(r\ddot{\theta} + 2\dot{r}\dot{\theta}) \tag{7.31}$$

and the relations

$$\begin{aligned}
\hat{\mathbf{x}} &= \hat{\mathbf{r}}\cos\theta - \hat{\boldsymbol{\theta}}\sin\theta \\
\hat{\mathbf{y}} &= \hat{\mathbf{r}}\sin\theta + \hat{\boldsymbol{\theta}}\cos\theta
\end{aligned} \tag{7.32}$$

obtained from (2.125). Since $T_\theta = 0$, we find

$$\begin{aligned}
ml\ddot{\theta} &= -m(g - A_x)\sin\theta - mA_y\cos\theta \\
-ml\dot{\theta}^2 &= T_r + m(g - A_x)\cos\theta - mA_y\sin\theta
\end{aligned} \tag{7.33}$$

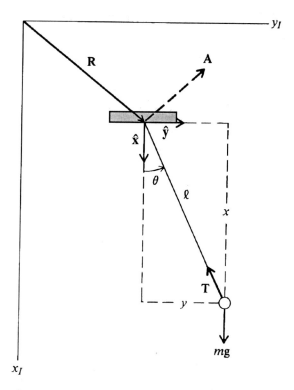

FIGURE 7-5. Pendulum with a support which moves with acceleration $\mathbf{A} \equiv \ddot{\mathbf{R}}$.

The angular motion of the pendulum bob is therefore determined by

$$\ddot{\theta} + \left[\frac{g}{l} - \frac{A_x(t)}{l}\right] \sin \theta = -\frac{A_y(t)}{l} \cos \theta \qquad (7.34)$$

This is the same as obtained more laboriously by the Lagrangian method in Chapter 3 (except with the x and y axes interchanged). For uniform vertical acceleration of the support (A_x = constant and $A_y = 0$), the natural frequency of small oscillations for $A_x < g$ is

$$\omega_0 = \sqrt{\frac{g - A_x}{l}} \qquad (7.35)$$

When $A_x = g$, the pendulum undergoes free-fall motion and it behaves as if the gravity field has vanished. The fact that gravity can be made to disappear (or appear) locally by a coordinate transformation led Einstein to a theory of gravity, the general theory of relativity, in which gravity is linked closely to geometry.

7.3 Motion on the Earth

For the motion of a particle moving near the surface of the earth, it is convenient to choose a coordinate system that is fixed on the earth's surface. We consider in Fig. 7-6 first a reference frame S' rotating with a constant angular velocity $\boldsymbol{\omega}$ relative to S_I (the inertial frame). The frame S' is fixed in the rotating earth and its origin is at the CM of the earth, coinciding with the origin of S_I. The translational vector $\mathbf{R}$ in (7.15) is thus zero. As shown in Fig. 7-6 the location of mass m is given by $\mathbf{r}'$ and its equation of motion from (7.15) is

$$m\frac{\delta^2 \mathbf{r}'}{\delta t^2} = \mathbf{F}' + m\mathbf{g} - m\left[\boldsymbol{\omega}\times(\boldsymbol{\omega}\times\mathbf{r}') - 2\boldsymbol{\omega}\times\mathbf{v}'\right] \qquad (7.36)$$

The net force acting on m in the inertial system has been separated into the gravitational force $m\mathbf{g}$ and any other forces $\mathbf{F}'$.

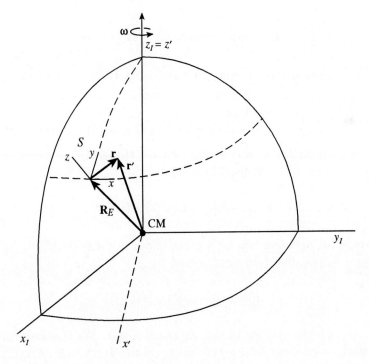

FIGURE 7-6. Inertial reference frame S_I, earth-fixed frame S' whose origin is at the earth CM, and earth-fixed local frame S whose origin is on the earth's surface.

The origin of a local reference frame S fixed on the earth's surface is given by $\mathbf{R}_E$. The location of m in S is specified by the vector $\mathbf{r}$ and hence

$$\mathbf{r}' = \mathbf{R}_E + \mathbf{r} \tag{7.37}$$

Since $\mathbf{R}_E$ is a fixed vector in S', $\delta\mathbf{r}'/\delta t = \delta\mathbf{r}/\delta t$ from (7.37); (7.36) then becomes

$$m\frac{\delta^2\mathbf{r}}{\delta t^2} = \mathbf{F}' + m\mathbf{g} - m\left[\boldsymbol{\omega}\times(\boldsymbol{\omega}\times(\mathbf{R}_E + \mathbf{r})) - 2\boldsymbol{\omega}\times\mathbf{v}\right] \tag{7.38}$$

Due to the earth's large size and its relatively slow rotation the fictitious force corrections are small compared to gravity. The centrifugal force is proportional to the square of the small quantity ω and therefore we can neglect $\mathbf{r}$ compared to $\mathbf{R}_E$ in this term. Thus to a very good approximation the centrifugal force is constant. The Coriolis force is linear in ω and depends on the state of motion relative to S. The equation of motion relative to S is then well approximated by

$$m\frac{\delta^2\mathbf{r}}{\delta t^2} = \mathbf{F}' + m\mathbf{g} - m\boldsymbol{\omega}\times(\boldsymbol{\omega}\times\mathbf{R}_E) - 2m\boldsymbol{\omega}\times\mathbf{v} \tag{7.39}$$

If the earth were perfectly spherical and isotropic, $\mathbf{g}$ would be constant in magnitude and directed toward the center of the earth. In fact, local irregularities, distortions from sphericity, and deviations from uniform density cause slight variations in $\mathbf{g}$ at different points on the earth.

The condition for a particle at rest on the earth ($\mathbf{v} = 0$) to be in equilibrium ($\delta^2\mathbf{r}/\delta t^2 = 0$) from (7.39) is

$$\mathbf{F}' = -m\left[\mathbf{g} - \boldsymbol{\omega}\times(\boldsymbol{\omega}\times\mathbf{R}_E)\right] \tag{7.40}$$

For example, if m is the bob on a plumb line, the tension $\mathbf{F}'$ in the string is opposite to the direction determined by

$$\mathbf{g}_{\text{eff}} = \mathbf{g} - \boldsymbol{\omega}\times(\boldsymbol{\omega}\times\mathbf{R}_E) \tag{7.41}$$

The plumb bob thus points in the direction of $\mathbf{g}_{\text{eff}}$. We conclude that $\mathbf{g}_{\text{eff}}$ is the effective gravitational acceleration on the earth. The magnitude of the correction term to $\mathbf{g}$ is

$$\left|\boldsymbol{\omega}\times(\boldsymbol{\omega}\times\mathbf{R}_E)\right| = \omega^2 R_E \sin\theta \tag{7.42}$$

where θ is the colatitude angle between $\boldsymbol{\omega}$ and $\mathbf{R}_E$. For the earth's angular velocity of rotation,

$$\omega = \frac{2\pi}{\tau} = \frac{2\pi}{24 \times 3{,}600} = 0.727 \times 10^{-4}\,\text{rad/s} \tag{7.43}$$

we find that the correction term is small,

$$\omega^2 R_E \sin\theta = (0.727 \times 10^{-4})^2 (6{,}371 \times 10^3) \sin\theta \simeq 0.03 \sin\theta \ \text{m/s}^2 \tag{7.44}$$

This correction is less than 0.3% of g, but nonetheless measurable. The direction of the correction is radially outward from the rotation axis, as illustrated in Fig. 7-7.

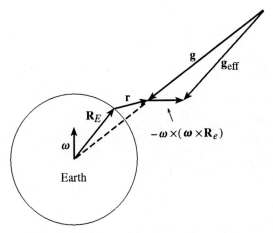

FIGURE 7-7. Effective gravitational acceleration $\mathbf{g}_{\text{eff}}$ (with the relative magnitude of the centrifugal acceleration exaggerated for clarity).

The differential equation (7.39) which describes the motion of a particle on the earth can be expressed in terms of $\mathbf{g}_{\text{eff}}$ in (7.41) as

$$m\frac{\delta^2 \mathbf{r}}{\delta t^2} = \mathbf{F}' + m\mathbf{g}_{\text{eff}} - 2m\boldsymbol{\omega} \times \mathbf{v} \tag{7.45}$$

For convenience we choose the z axis of the coordinate system so that

$$\mathbf{g}_{\text{eff}} = -g_{\text{eff}}\hat{\mathbf{z}} \tag{7.46}$$

The y axis is taken to point north and the x axis east, as pictured in Fig. 7-8. The components of $\boldsymbol{\omega}$ along these axes are

$$\boldsymbol{\omega} = 0\hat{\mathbf{x}} + \omega \sin\theta\,\hat{\mathbf{y}} + \omega \cos\theta\,\hat{\mathbf{z}} \tag{7.47}$$

where θ is the colatitude angle as measured from the north polar axis.

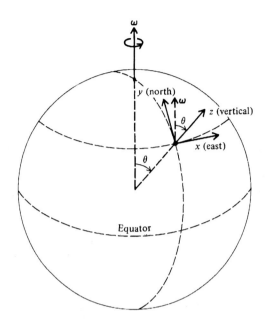

FIGURE 7-8. Coordinate frame fixed on the surface of the earth at colatitude angle θ.

Substituting (7.47) into the Coriolis force term,

$$\mathbf{F}_{\text{Cor}} = -2m\boldsymbol{\omega} \times \mathbf{v} \qquad (7.48)$$

of (7.45), we have

$$\mathbf{F}_{\text{Cor}} = 2m\omega[(v_y \cos\theta - v_z \sin\theta)\hat{\mathbf{x}} - v_x \cos\theta\hat{\mathbf{y}} + v_x \sin\theta\hat{\mathbf{z}}] \qquad (7.49)$$

The direction of deflection of the particle from its direction of motion due to the Coriolis force follows directly from (7.49). In the Northern Hemisphere, $0 \leq \theta \leq \pi/2$, we find

Velocity direction	Deflection direction
North ($v_y > 0$)	East
East ($v_x > 0$)	South and up
South ($v_y < 0$)	West
West ($v_x < 0$)	North and down
Up ($v_z > 0$)	West
Down ($v_z < 0$)	East

For motion parallel to the earth's surface ($v_z = 0$), the particle is always deflected to the right in the Northern Hemisphere and to the left in the Southern Hemisphere.

The trade winds and weather circulations of high- and low-pressure areas are striking examples of Coriolis force effects. The equatorial region of the earth generally receives more heat from the sun. The warm air rises and is replaced by a flow of air from the temperate regions. The air moving south from the Northern Hemisphere is deflected westward by the Coriolis effect. This accounts for the steady prevailing winds to the west and south, known as the *trade winds*.

On a smaller scale a low-pressure region in the Northern Hemisphere on the order of 200 km across is associated with a counterclockwise circulation of the air because of the Coriolis force effect on the air flowing in. The pressure gradient is largely balanced by the Coriolis force. Under certain circumstances this cyclonic motion builds up to great intensity and destructive power in the form of a hurricane, cyclone, or typhoon. High-pressure areas force air outward. This airflow deflects to the right and produces clockwise circulation in the Northern Hemisphere. Vortices on a still smaller scale such as tornados, dust devils, water spouts, and the bathtub vortex are not directly influenced by Coriolis effects to any great extent. Nevertheless, some of these vortices often have a counterclockwise motion because of general counterclockwise movements which spawn them.

7.4 Foucault's Pendulum

In 1851 Jean Foucault exhibited a pendulum at the Pantheon in Paris which through Coriolis force dramatically illustrated the rotation of the earth. Today Foucault pendulums are on exhibit in many public buildings and planetariums. One of the most famous hangs in the United Nations Building in New York, as illustrated in Fig. 7-9. The Foucault pendulum is a simple plane pendulum which can oscillate a long time without being appreciably damped by friction. Its oscillation plane is observed to rotate slowly with time, confirming in a dramatic way that a reference frame in which the distant stars appear fixed is more fundamental than one in which the earth is fixed and the stars rotate about the earth.

The motion of the Foucault pendulum can be determined from (7.45). We take **r** to represent the distance of the bob of mass m from its equilibrium position. At rest the pendulum hangs along the direction $\mathbf{g}_{\text{eff}}$, and

FIGURE 7-9. Foucault pendulum which hangs in the United Nations Building in New York City. (*Photo courtesy of United Nations.*)

the tension in the string is $\mathbf{F}' = -m\mathbf{g}_{\text{eff}}$. If the earth did not rotate, the Coriolis force term in (7.45) would not be present and the motion would occur in a fixed plane. With Coriolis force present the bob will deflect to the right, out of its plane as shown in Fig. 7-10. On the return swing the pendulum bob again deflects to the right and after one period the pendulum plane has rotated clockwise as viewed from above.

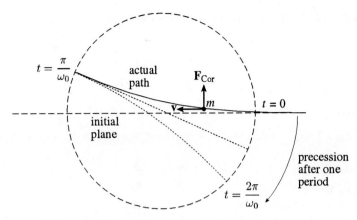

FIGURE 7-10. Deflection of Foucault pendulum bob by Coriolis force as viewed from above. The precession angle is greatly exaggerated in the figure.

A way to obtain the Foucault precession frequency is to view the system from a new frame S_F which rotates with the angular velocity $\boldsymbol{\omega}_F$ of the pendulum relative to our local earth fixed frame S. Starting from the equation of motion (7.45) in S, we transform it to frame S_F using (7.11)

$$m\frac{\bar{\delta}^2\mathbf{r}}{\bar{\delta}t^2} = (\mathbf{F}' + m\mathbf{g}_{\text{eff}} - 2m\boldsymbol{\omega}\times\mathbf{v}) - m\boldsymbol{\omega}_F\times(\boldsymbol{\omega}\times\mathbf{r}) - 2m\boldsymbol{\omega}_F\times\mathbf{v}_F \quad (7.50)$$

Here $\bar{\delta}^2\mathbf{r}/\bar{\delta}t^2$ is the second time derivative of $\mathbf{r}$ relative to the axes of the S_F frame. The particle velocity in the S_F frame is

$$\mathbf{v}_F = \mathbf{v} + \boldsymbol{\omega}_F\times\mathbf{r} \quad (7.51)$$

Substituting (7.51) into (7.50) gives

$$m\frac{\bar{\delta}^2\mathbf{r}}{\bar{\delta}t^2} = \mathbf{F}' + m\mathbf{g}_{\text{eff}} + m\boldsymbol{\omega}_F\times(\boldsymbol{\omega}_F\times\mathbf{r}) - 2m(\boldsymbol{\omega}+\boldsymbol{\omega}_F)\times\mathbf{v} \quad (7.52)$$

To see the advantage of viewing the Foucault pendulum from a frame rotating with angular velocity $\boldsymbol{\omega}_F$ we observe

1. The Foucault pendulum precesses about the vertical axis. Thus we take

$$\boldsymbol{\omega}_F = \omega_F\hat{\mathbf{z}} \quad (7.53)$$

where $\hat{\mathbf{z}}$ is the vertical direction at the earth's surface.

2. For small displacements the pendulum motion is nearly perpendicular to $\hat{z}$.

3. The $\boldsymbol{\omega}_F \times (\boldsymbol{\omega}_F \times \mathbf{r})$ term is small compared to the already small centrifugal term in $\mathbf{g}_{\mathrm{eff}}$. It can be neglected.

4. For small pendulum displacements v_z is negligible; using (7.47)

$$(\boldsymbol{\omega}+\boldsymbol{\omega}_F) \times \mathbf{v} = -\hat{\mathbf{x}} v_y (\omega \cos\theta + \omega_F) + \hat{\mathbf{y}} v_x (\omega \cos\theta + \omega_F) - \hat{\mathbf{z}} v_x (\omega \sin\theta) \tag{7.54}$$

Thus the pendulum motion will remain in its initial plane in this frame if

$$\omega_F = -\omega \cos\theta \tag{7.55}$$

An earth-fixed observer in S thus sees the pendulum plane precessing slowly clockwise (in the northern hemisphere) with angular frequency $\omega_F = -\omega \cos\theta$. We note that in the southern hemisphere $\cos\theta$ is negative and ω_F automatically adjusts in sign. The time required for the pendulum plane to precess by 2π is

$$\tau_P = \frac{2\pi}{\omega_F} = \frac{(1 \text{ day})}{\cos\theta} \tag{7.56}$$

The precession vanishes at the equator and is a maximum at the north pole, where the pendulum precesses clockwise through a complete revolution every 24 h. From the viewpoint of an observer in space, the oscillation plane at the north pole remains fixed, while the earth turns counterclockwise beneath it.

7.5 Dynamical Balance of a Rigid Body

The formulation of the equations of motion in a rotating reference system is also quite valuable in the description of rigid-body motion. As an introduction to the general treatment of rigid-body rotational motion, we discuss a simple example of a dumbbell formed by two point masses m at the ends of a massless rod of length l. The dumbbell rotates at a fixed inclination θ with constant angular velocity $\boldsymbol{\omega}$ about a pivot at the center of the rod, as shown in Fig. 7-11. The equation of motion (6.48) in a fixed reference frame for rotation about the pivot of the rod is

$$\mathbf{N} = \frac{d\mathbf{L}}{dt} \tag{7.57}$$

where $\mathbf{N}$ is the external torque on the rod applied at the pivot. The angular momentum is given by

$$\mathbf{L} = m(\mathbf{r}_1 \times \mathbf{v}_1 + \mathbf{r}_2 \times \mathbf{v}_2) \qquad (7.58)$$

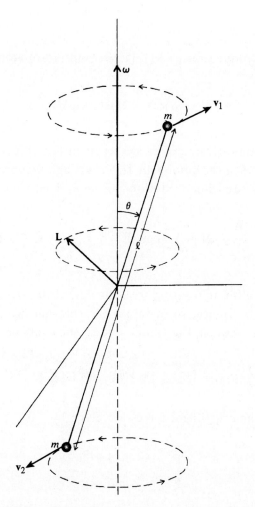

FIGURE 7-11. Dumbbell rotating about a pivot at center of the rod at a fixed inclination angle θ.

Since $\mathbf{r}_2 = -\mathbf{r}_1$ and $\mathbf{v}_1 = \boldsymbol{\omega} \times \mathbf{r}_1$ and $\mathbf{v}_2 = \boldsymbol{\omega} \times \mathbf{r}_2$ we have

$$\mathbf{v}_2 = -\mathbf{v}_1 \qquad (7.59)$$

Thus $\mathbf{L}$ can be expressed in terms of $\boldsymbol{\omega}$ and $\mathbf{r}_1$ as

$$\mathbf{L} = 2m\mathbf{r}_1 \times (\boldsymbol{\omega} \times \mathbf{r}_1) = 2m\left[\boldsymbol{\omega} r_1^2 - \mathbf{r}_1(\boldsymbol{\omega} \cdot \mathbf{r}_1)\right] \qquad (7.60)$$

Since $\mathbf{L}$ is perpendicular to $\mathbf{r}_1$ and lies in the plane determined by $\mathbf{r}_1$ and $\boldsymbol{\omega}$, it also rotates with angular velocity $\boldsymbol{\omega}$. From (7.8), we then have

$$\frac{d\mathbf{L}}{dt} = \boldsymbol{\omega} \times \mathbf{L} \tag{7.61}$$

The external torque from (7.57), (7.60), and (7.61) necessary to maintain the rotation is

$$\mathbf{N} = \boldsymbol{\omega} \times \mathbf{L} = 2m(\mathbf{r}_1 \times \boldsymbol{\omega})(\mathbf{r}_1 \cdot \boldsymbol{\omega}) \tag{7.62}$$

We can alternatively derive the result in (7.62) in a coordinate frame which rotates with the dumbbell. In a rotating reference frame the following rigid-body equation of motion can be derived from (7.16) to (7.21):

$$\frac{\delta\mathbf{L}}{\delta t} = \mathbf{N} + \sum_i \mathbf{r}_i \times (\mathbf{F}_{cf}^i + \mathbf{F}_{Cor}^i + \mathbf{F}_{az}^i + \mathbf{F}_{tr}^i) \tag{7.63}$$

In a coordinate frame rotating with the dumbbell, $\delta\mathbf{L}/\delta t = 0$ and $\mathbf{F}_{Cor}^i = \mathbf{F}_{az}^i = \mathbf{F}_{tr}^i = 0$. Hence, to maintain the rotation, the torque applied at the pivot must balance the torque due to the centrifugal forces.

$$\mathbf{N} = -(\mathbf{r}_1 \times \mathbf{F}_{cf}^1 + \mathbf{r}_2 \times \mathbf{F}_{cf}^2) \tag{7.64}$$

From (7.18) the centrifugal forces are given by

$$\begin{aligned} \mathbf{F}_{cf}^1 &= -m\boldsymbol{\omega} \times (\boldsymbol{\omega} \times \mathbf{r}_1) \\ \mathbf{F}_{cf}^2 &= -m\boldsymbol{\omega} \times (\boldsymbol{\omega} \times \mathbf{r}_2) \end{aligned} \tag{7.65}$$

Using $\mathbf{r}_1 = -\mathbf{r}_2$, the torque reduces to

$$\begin{aligned} \mathbf{N} &= 2m\mathbf{r}_1 \times [\boldsymbol{\omega} \times (\boldsymbol{\omega} \times \mathbf{r}_1)] \\ &= 2m\mathbf{r}_1 \times [\boldsymbol{\omega}(\boldsymbol{\omega} \cdot \mathbf{r}_1) - \mathbf{r}_1\omega^2] \\ &= 2m(\mathbf{r}_1 \times \boldsymbol{\omega})(\mathbf{r}_1 \cdot \boldsymbol{\omega}) \end{aligned} \tag{7.66}$$

in agreement with the result in (7.62).

In terms of the angle θ between ω and $\mathbf{r}_1$, the angular momentum and torque in (7.60) and (7.62) of the rotating dumbbell can be written

$$\mathbf{L} = \tfrac{1}{2}ml^2\omega(\hat{\omega} - \cos\theta\hat{\mathbf{r}}_1) \tag{7.67}$$

$$\mathbf{N} = \tfrac{1}{2}ml^2\omega\sin\theta\cos\theta\hat{\mathbf{n}} \tag{7.68}$$

where $\hat{\mathbf{n}} = \mathbf{r}_1\times\omega/|\mathbf{r}_1\times\omega|$. For $\theta = \pi/2$, we find

$$\begin{aligned} \mathbf{L} &= (\tfrac{1}{2}ml^2)\omega \\ \mathbf{N} &= 0 \end{aligned} \tag{7.69}$$

In this orientation the motion does not require an imposed torque.

From (7.67) and (7.68), we see that torques on the rod are present whenever the angular momentum $\mathbf{L}$ does not lie along the axis of rotation ω. This result is generally true for rigid-body rotations.

A practical application in which it is important that $\mathbf{L}$ and ω are parallel is the dynamic balance of automobile tires. If a wheel is not balanced, noise and vibration result in the car and excessive wear occurs on the tire. There are two criteria for complete balance of a wheel:

(1) *Static balance:* Unless the CM of the wheel lies on the rotation axis, a time-varying centrifugal force is present. This acts to make the axle oscillate and imparts vibration to the car. In a static balance the wheel is removed from the car and mounted on a vertical axis. Weights are attached around the rim of the wheel until the wheel is in equilibrium in a horizontal plane.

(2) *Dynamic balance:* Even when the CM lies on the wheel axis, it is possible that in rotation the angular momentum does not lie along the axis. If we specify the x axis as the rotation axis, $\omega = \omega\hat{\mathbf{x}}$, the angular-momentum vector from (6.105) is

$$\mathbf{L} = (I_{xx}\hat{\mathbf{x}} + I_{yx}\hat{\mathbf{y}} + I_{zx}\hat{\mathbf{z}})\omega \tag{7.70}$$

Unless the products of inertia I_{yx} and I_{zx} vanish, $\mathbf{L}$ does not lie along ω. The time variation of $\mathbf{L}$ then leads to a time-varying torque, causing the wheel to wobble. A dynamic balance consists of the application of weights until the wheel spins smoothly with no wobble. Since modern tires are usually very nearly symmetrical, a static balance alone is often sufficient to ensure good driving results.

7.6 Principal Axes and Euler's Equations

For a rigid body of arbitrary shape, the rotational equation of motion
(6.48) in a fixed coordinate system or with origin at the center-of-mass
point is

$$N_j = \frac{dL_j}{dt} = \frac{d}{dt}(I_{jk}\omega_k) \tag{7.71}$$

where a sum over the index k is implied. Since the moments and products
of inertia I_{jk} relative to the fixed coordinate system change as a function
of time as the body rotates, the description of the motion through (7.71)
can be cumbersome and difficult. The analysis of the motion can often be
greatly simplified by choosing instead a *body-fixed* coordinate system that
rotates with the body. In this reference frame the moments and products
of inertia are time-independent. Using (7.8) and (7.71), the equation of
motion with respect to the moving body axes is

$$N_j = \frac{\delta L_j}{\delta t} + (\boldsymbol{\omega} \times \mathbf{L})_j \tag{7.72}$$

A further simplification can be made by a judicious choice of the orientations of the rotating axes with respect to the rigid body. As we shall
shortly prove, it is always possible to make a choice of axes in the body
for which all the products of inertia vanish.

$$I_{ij} = 0 \qquad \text{for } i \neq j \tag{7.73}$$

The axes for which (7.73) holds are called the *principal axes* of the rigid
body. For these axes the angular-momentum components in (6.105) reduce to

$$L_1 = I_{11}\omega_1 \equiv I_1\omega_1$$
$$L_2 = I_{22}\omega_2 \equiv I_2\omega_2 \tag{7.74}$$
$$L_3 = I_{33}\omega_3 \equiv I_3\omega_3$$

where I_1, I_2, I_3 denote the principal moments of inertia. From (7.72),
expressing the cross product in cartesian coordinates, we obtain *Euler's
equations of motion* for a rigid body in terms of the coordinate system
aligned with the principal axes of the body.

$$N_1 = I_1\dot{\omega}_1 + (I_3 - I_2)\omega_3\omega_2$$
$$N_2 = I_2\dot{\omega}_2 + (I_1 - I_3)\omega_1\omega_3 \tag{7.75}$$
$$N_3 = I_3\dot{\omega}_3 + (I_2 - I_1)\omega_2\omega_1$$

It should be emphasized that the angular velocity and torque components appearing above refer to the ω and $\mathbf{N}$ vectors of the inertial system projected onto the principal body axes. The Euler equations are a convenient starting point for many discussions of rigid body rotations.

To illustrate the application of Euler's equations, we return to the rotating rod of the preceding section. The principal axes of the body lie along and perpendicular to the rod, as illustrated in Fig. 7-12. With the z axis along the rod and the x axis in the plane of the rod and ω, the components of ω are

$$
\begin{aligned}
\omega_1 &= \omega \sin \theta \\
\omega_2 &= 0 \\
\omega_3 &= \omega \cos \theta
\end{aligned}
\tag{7.76}
$$

where θ is the angle between ω and the rod. The principal moments of inertia are

$$
I_1 = I_2 = m \left(\frac{\ell}{2}\right)^2 + m \left(\frac{\ell}{2}\right)^2 = \tfrac{1}{2} m\ell^2
\tag{7.77}
$$

$$
I_3 = 0
$$

Using (7.76) and (7.77) in (7.75), we find

$$
\begin{aligned}
N_1 &= 0 \\
N_2 &= (\tfrac{1}{2} m\ell^2)\, \omega^2 \sin \theta \cos \theta \\
N_3 &= 0
\end{aligned}
\tag{7.78}
$$

where $\dot{\omega} = 0$ has been used. This result obtained from Euler's equations is the same as (7.68).

In the derivation of (7.75) we have used the diagonal property in (7.73) of the inertia tensor in the principal-axes coordinate system. We will now establish this property. Suppose that there exists a direction in space ω for which $\mathbf{L}$ is parallel to ω

$$
\mathbf{L} = I\omega
\tag{7.79}
$$

If such a direction can be found it will by definition be a principal axis since the products of inertia vanish and the principal moment is I. In the original coordinate system, ω will in general have three components:

$$
\omega = \omega_1 \hat{\mathbf{x}} + \omega_2 \hat{\mathbf{y}} + \omega_3 \hat{\mathbf{z}}
\tag{7.80}
$$

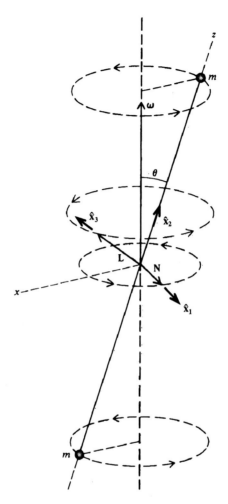

FIGURE 7-12. Principal axes $\hat{x}_1, \hat{x}_2, \hat{x}_3$ of the dumbbell.

From (7.79) the components of **L** along the inertial axes are

$$
\begin{aligned}
L_1 &= I\omega_1 \\
L_2 &= I\omega_2 \\
L_3 &= I\omega_3
\end{aligned}
\tag{7.81}
$$

These components of **L** must be equivalent to the expression for **L** given in (6.105), namely,

$$L_1 = I_{11}\omega_1 + I_{12}\omega_2 + I_{13}\omega_3$$
$$L_2 = I_{21}\omega_1 + I_{22}\omega_2 + I_{23}\omega_3 \qquad (7.82)$$
$$L_3 = I_{31}\omega_1 + I_{32}\omega_2 + I_{33}\omega_3$$

Equating the components in (7.81) and (7.82), we find

$$(I_{11} - I)\omega_1 + I_{12}\omega_2 + I_{13}\omega_3 = 0$$
$$I_{21}\omega_1 + (I_{22} - I)\omega_2 + I_{23}\omega_3 = 0 \qquad (7.83)$$
$$I_{31}\omega_1 + I_{32}\omega_2 + (I_{33} - I)\omega_3 = 0$$

which in vector notation is

$$\sum_{j=1}^{3} I_{ij}\omega_j = I\omega_i \qquad (7.84)$$

or

$$\mathbb{I} \cdot \boldsymbol{\omega} = I\boldsymbol{\omega} \qquad (7.85)$$

For $\boldsymbol{\omega} \neq 0$, this system of homogeneous equations for $(\omega_1, \omega_2, \omega_3)$ has solutions only if the determinant of the coefficients of the $\boldsymbol{\omega}$ components vanishes.

$$\begin{vmatrix} (I_{11} - I) & I_{12} & I_{13} \\ I_{21} & (I_{22} - I) & I_{23} \\ I_{31} & I_{32} & (I_{33} - I) \end{vmatrix} = 0 \qquad (7.86)$$

This leads to a cubic equation in I_1 of the form

$$I^3 + aI^2 + bI + c = 0 \qquad (7.87)$$

where a, b, c are products of the inertia tensor elements I_{ij}. There are three real solutions for I from (7.87), any is appropriate to (7.79). The other two solutions refer to the principal moments about two other orthogonal principal axes. The detailed proofs are outlined in the exercises. By construction we have therefore shown that it is always possible to find a principal-axis system for any rigid body. In many applications the choice of principal axes is obvious from the symmetry of the body. If two of the principal moments are equal the body is called *rotation symmetric* about the third axis in that plane. If all three principal moments are equal the body is *rotation isotropic*.

7.7 The Tennis Racket Theorem

The solution of Euler's equations for a rigid body with unequal principal moments of inertia can be beautifully illustrated with a tennis or badminton racket. The three principal axes of a tennis racket are readily identified to be (1) along the handle, (2) perpendicular to the handle in the plane of the strings, and (3) perpendicular to the handle and strings. When a tennis racket is tossed into the air with a spin about one of the principal axes, a curious phenomenon is observed. If the initial spin is about either axis (1) or axis (3), the racket continues to spin uniformly about the initial axis and can easily be recaught. On the other hand, if the initial spin is about axis (2), the motion quickly becomes irregular, with spin developing about all three principal axes, which makes it difficult to catch the falling racket. The explanation of the observed behavior follows from Euler's equations. To apply Euler's equations to the tennis racket, we choose the origin of the principal-axes coordinate system at the CM of the racket, as illustrated in Fig. 7-13.

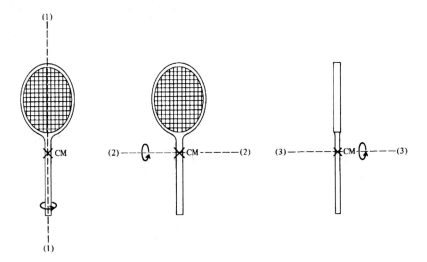

FIGURE 7-13. Principal axes of the tennis racket.

Since gravity is a uniform force in the vicinity of the earth's surface, there are no gravitational torques about the CM of the racket. If we neglect torques due to wind resistance, Euler's equations (7.75) simplify to

$$I_1\dot{\omega}_1 + (I_3 - I_2)\omega_3\omega_2 = 0$$
$$I_2\dot{\omega}_2 + (I_1 - I_3)\omega_1\omega_3 = 0 \qquad (7.88)$$
$$I_3\dot{\omega}_3 + (I_2 - I_1)\omega_2\omega_1 = 0$$

These equations can be written as

$$\dot{\omega}_1 + r_1\omega_2\omega_3 = 0; \quad r_1 = \frac{I_3 - I_2}{I_1} \tag{7.89}$$

$$\dot{\omega}_2 - r_2\omega_3\omega_1 = 0; \quad r_2 = \frac{I_3 - I_1}{I_2} \tag{7.90}$$

$$\dot{\omega}_3 + r_3\omega_1\omega_2 = 0; \quad r_3 = \frac{I_2 - I_1}{I_3} \tag{7.91}$$

If we assume the ordering

$$I_1 \leq I_2 \leq I_3 \tag{7.92}$$

we consequently have all the r_i positive. We note the sign asymmetry in the three Euler equations (7.89)–(7.91).

The tennis racket theorem concerns stability of spin about the principal axes. We assume that the racket is initially spun nearly about one of the principal axes and we will determine if this spin state is stable. Assume first that the spin is initially nearly along the intermediate axis (2) or

$$\boldsymbol{\omega} \simeq \omega_2\hat{\mathbf{y}}; \quad \omega_1, \omega_3 \ll \omega_2 \tag{7.93}$$

If ω_1 and ω_3 are small as hypothesized their product is negligible and (7.90) implies that

$$\omega_2 \simeq \text{constant} \tag{7.94}$$

Equations (7.89) and (7.91) then comprise a set of coupled linear equations in ω_1 and ω_3

$$\begin{aligned} \dot{\omega}_1 + (r_1\omega_2)\omega_3 = 0 \\ \dot{\omega}_3 + (r_3\omega_2)\omega_1 = 0 \end{aligned} \tag{7.95}$$

For the trial exponential solution

$$\begin{aligned} \omega_1 = a_1 e^{\lambda t} \\ \omega_3 = a_3 e^{\lambda t} \end{aligned} \tag{7.96}$$

two algebraic equations must be satisfied

$$\begin{aligned} a_1\lambda + r_1\omega_2 a_3 = 0 \\ a_3\lambda + r_3\omega_2 a_1 = 0 \end{aligned} \tag{7.97}$$

Solving both equations for the ratio a_3/a_1, we obtain

$$\frac{a_3}{a_1} = -\frac{\lambda}{r_1\omega_2} = -\frac{r_3\omega_2}{\lambda} \tag{7.98}$$

The second equality determines λ to be

$$\lambda = \pm\omega_2\sqrt{r_1r_3} \tag{7.99}$$

Then from (7.98) the ratio of amplitudes

$$a_3 = \mp a_1\sqrt{\frac{r_3}{r_1}} \tag{7.100}$$

The general solution is a superposition of these two solutions

$$
\begin{aligned}
\omega_1(t) &= ae^{\omega_2\sqrt{r_1r_3}\,t} + be^{-\omega_2\sqrt{r_1r_3}\,t} \\
\omega_3(t) &= \sqrt{\frac{r_3}{r_1}}\left[-ae^{\omega_2\sqrt{r_1r_3}\,t} + be^{-\omega_2\sqrt{r_1r_3}\,t}\right]
\end{aligned}
\tag{7.101}
$$

where a and b are constants determined by the initial conditions. Since r_1 and r_3 are positive the solution is a superposition of increasing and decreasing exponentials in time. The increasing exponential term will make ω_1 and ω_3 large even if they started small and hence *rotation about the intermediate axis (2) is unstable.* Our solution for axis (2) is of course strictly valid only at times for which the product $\omega_1\omega_3$ is small.

Next take the initial angular velocity nearly along one of the extreme principal moment axes, say axis (1)

$$\boldsymbol{\omega} = \omega_1\hat{\mathbf{x}}, \qquad \omega_2, \omega_3 \ll \omega_1 \tag{7.102}$$

By (7.89) we obtain

$$\omega_1 \simeq \text{constant} \tag{7.103}$$

and ω_2 and ω_3 satisfy

$$
\begin{aligned}
\dot{\omega}_2 - r_2\omega_1\omega_3 &= 0 \\
\dot{\omega}_3 + r_3\omega_1\omega_2 &= 0
\end{aligned}
\tag{7.104}
$$

Proceeding as before, we know the exponential solutions analogous to

(7.96) must satisfy (7.104) except that in this case

$$\lambda = \pm\omega_1 \sqrt{-r_2 r_3} = \pm i\omega_1 \sqrt{r_2 r_3}$$
$$a_3 = \pm a_2 \sqrt{r_3/r_2}$$

(7.105)

The solutions now are oscillatory and can be written in the form

$$\omega_1(t) = a \sin\left(\omega_1 \sqrt{r_2 r_3}\, t + \alpha\right)$$
$$\omega_3(t) = \sqrt{\frac{r_3}{r_2}}\, a \cos\left(\omega_1 \sqrt{r_2 r_3} + \alpha\right)$$

(7.106)

where a and α are constants determined by the initial conditions. Thus if ω_2 and ω_3 are initially small, they will remain small. *Rotation about axis (1) is thus stable.*

For an initial spin about axis (3) the solution to Euler's equations is similar to the case of axis (1). The exponential factor λ is again purely imaginary and the solutions are oscillatory. We conclude that *rotations along extreme axes are stable while the intermediate axis is unstable.*

For spin about axis (1) the angular velocity vector $\omega = \omega_1 \hat{\mathbf{x}} + \omega_2 \hat{\mathbf{y}} + \omega_3 \hat{\mathbf{z}}$ precesses in a small cone about principal axis (1) as shown in Fig. 7-14. A similar picture applies to the largest principal axis (3). For rotation along principal axis (2) the angular velocities about axes (1) and (3) grow rapidly with time and the racket tumbles.

To calculate the principal moments of inertia for our specific case of the tennis racket we use a grossly simplified model for the racket. We represent the mass distribution of the racket by a circular hoop of radius a and mass m_a connected to a thin rod of length ℓ and mass m_ℓ. The total mass of the racket is $M = m_a + m_\ell$. The CM of the racket is located on principal axis (1) at a distance R from the center of the hoop, where

$$MR = m_a(0) + m_\ell \left(a + \frac{\ell}{2}\right)$$

(7.107)

or

$$R = \frac{m_\ell}{M}\left(a + \frac{\ell}{2}\right)$$

(7.108)

The moment of inertia of the racket about principal axis (1) comes entirely from the hoop. We use the perpendicular-axis rule in (6.128) to obtain

$$I_1 = \tfrac{1}{2} m_a a^2$$

(7.109)

FIGURE 7-14. Stable precession of the angular velocity $\boldsymbol{\omega}$ about principal axis (1) of the tennis racket.

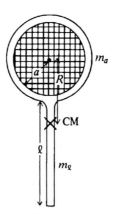

FIGURE 7-15. Dimensions of tennis racket model.

To compute the moment of inertia about principal axis (2), we will make use of the parallel-axis rule of (6.123). The moment of inertia of the hoop about an axis through its CM and parallel to principal axis (2) is $\frac{1}{2}m_a a^2$. By (6.123), the hoop makes a contribution to I_2 of

$$I_2^{hoop} = \tfrac{1}{2}m_a a^2 + m_a R^2 \tag{7.110}$$

since R is the perpendicular distance between the two parallel axes. The moment of inertia of the handle about an axis parallel to principal axis (2) passing through the CM of the handle is $\frac{1}{12}m_\ell\ell^2$. Again using (6.123), we find that the contribution of the handle to I_2 is

$$I_2^{handle} = \tfrac{1}{12}m_\ell\ell^2 + m_\ell\left(a + \frac{\ell}{2} - R\right)^2 \tag{7.111}$$

where $a + \ell/2 - R$ is the distance between these parallel axes. Combining (7.110) and (7.111) and substituting for R from (7.108), we obtain

$$I_2 = \tfrac{1}{2}m_a a^2 + m_a\left(\frac{m_\ell}{M}\right)^2\left(a + \frac{\ell}{2}\right)^2 + \tfrac{1}{12}m_\ell\ell^2 + m_\ell\left(\frac{m_a}{M}\right)^2\left(a + \frac{\ell}{2}\right)^2$$

This can be further simplified to

$$I_2 = \tfrac{1}{2}m_a a^2 + \tfrac{1}{12}m_\ell\ell^2 + \frac{m_a m_\ell}{M}\left(a + \frac{\ell}{2}\right)^2 \tag{7.112}$$

Finally, for principal axis (3), the racket lies in a plane perpendicular to the axis, and we can use the perpendicular-axis rule of (6.128) to obtain

$$I_3 = I_1 + I_2 \tag{7.113}$$

By comparison of (7.109), (7.112), and (7.113), we see that the principal moments of inertia are ordered as $I_1 < I_2 < I_3$. Characteristic parameters for our tennis racket model are

$$
\begin{aligned}
a &= 0.13\,\text{m} & \ell &= 0.38\,\text{m} \\
R &= 0.18\,\text{m} & M &= 0.33\,\text{kg} \\
m_\ell &= 0.18\,\text{kg} & m_a &= 0.15\,\text{kg}
\end{aligned}
\tag{7.114}
$$

The principal moments of inertia from (7.109), (7.112), (7.113), and (7.114) are

$$
\begin{aligned}
I_1 &= 0.13 \times 10^{-2}\text{kg} \cdot \text{m}^2 \\
I_2 &= 1.24 \times 10^{-2}\text{kg} \cdot \text{m}^2 \\
I_3 &= 1.37 \times 10^{-2}\text{kg} \cdot \text{m}^2
\end{aligned}
\tag{7.115}
$$

The condition of stability of the motion about a principal axis which has either the largest or the smallest moment of inertia and the instability about the other principal axis is often called the *tennis racket theorem*. The conclusions of this theorem can be readily demonstrated by throwing this book (with a rubber band around it) or other oblong object into the air with a spin about one of the principal axes. The detailed nature of the spin about the stable axes is similar to the free symmetric top discussed in the next section.

7.8 The Earth as a Free Symmetric Top

Since the earth is nearly spherical in shape, the gravitational torques exerted on the earth by the sun and the moon are quite small. To a good approximation the rotational motion can therefore be described by Euler's equations with no external torques. Since the earth is nearly axially symmetric, the principal moments of inertia for the two axes in the equatorial plane are equal.

$$I_1 = I_2 = I \tag{7.116}$$

The third principal axis with moment of inertia I_3 is along the polar symmetry axis. From (7.88) the differential equations for the earth's motion in an earth-based coordinate frame are

$$\dot{\omega}_1 + \frac{I_3 - I}{I}\omega_3\omega_2 = 0$$
$$\dot{\omega}_2 - \frac{I_3 - I}{I}\omega_1\omega_3 = 0 \tag{7.117}$$
$$\dot{\omega}_3 = 0$$

Any rigid body which obeys this set of torque-free equations is called a *free axially symmetric top*. The exact solution to this coupled set of equations is easily obtained. The last equation above implies that ω_3 is constant.

$$\omega_3(t) = \omega_3(0) = \omega_3 \tag{7.118}$$

The equations (7.117) can be solved using the method of (7.97)–(7.101). The solution is

$$\omega_1(t) = a\cos(\Omega t + \alpha)$$
$$\omega_2(t) = a\sin(\Omega t + \alpha) \tag{7.119}$$

where

$$\Omega = \omega_3\left(\frac{I_3 - I}{I}\right) \tag{7.120}$$

The magnitude of the angular-velocity vector ω is

$$\omega = \sqrt{\omega_1^2 + \omega_2^2 + \omega_3^3} = \sqrt{a^2 + \omega_3^2} \qquad (7.121)$$

Since the components ω_1 and ω_2 in (7.119) trace out a circle of radius a while ω_3 and ω remain constant, an observer on the earth sees the angular-velocity vector precesses uniformly about the symmetry axis with angular velocity Ω, as shown in Fig. 7-16.

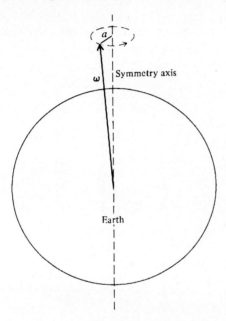

FIGURE 7-16. Precession of the earth's spin about the symmetry axis.

The period of precession of ω about the earth's symmetry axis is

$$\tau = \frac{2\pi}{\Omega} = \left(\frac{I}{I_3 - I}\right) \frac{2\pi}{\omega_3} \qquad (7.122)$$

For the earth, since $2\pi/\omega_3 = 1$ day, the period of precession in days is determined by the moment-of-inertia ratio. For an earth of uniform density and oblate spheroidal shape, the value of this ratio, calculated from the measured radii of the earth, is

$$\frac{I}{I_3 - I} \approx 300 \qquad (7.123)$$

Although the earth becomes more dense toward its center, the moment-of-inertia ratio is not appreciably changed from the uniform-density result.

Thus the expected precessional period is about 300 days. The precession of $\boldsymbol{\omega}$ about the symmetry axis of the earth is known as the *Chandler wobble*.

FIGURE 7-17. Star trails in the night sky above the Mauna Kea Observatory in Hawaii, photographed with a nine hour exposure camera. The stars appear as circular arcs due to the earth's rotation. The smallest bright arc is that of Polaris.

The direction of the earth's axis of rotation (*i.e.*, the direction of $\boldsymbol{\omega}$) can be experimentally determined by location of the point in the night sky which appears to remain stationary as the earth rotates, as illustrated in Fig. 7-17. The direction of the earth's rotational axis is observed to precess about the symmetry axis with a period of about 440 days. The angle between $\boldsymbol{\omega}$ and the symmetry axis is quite small. In fact, at the north pole, $\boldsymbol{\omega}$ never moves more than about 10 m from the symmetry axis. The actual motion of $\boldsymbol{\omega}$ is rather irregular, being strongly affected by earthquakes and seasonal changes. In fact, it is only due to these effects that the motion has a nonvanishing amplitude. On a quiet earth, viscous effects would damp out such a motion, and $\boldsymbol{\omega}$ would soon lie along the symmetry axis (this minimizes energy for fixed $\mathbf{L}$). The discrepancy between the expected period of 300 days and the observed value of about 440 days is primarily due to the nonrigidity of the earth.

7.9 The Free Symmetric Top: External Observer

The description of the earth's rotational motion as a free symmetric top in § 7.8 was appropriate for an observer at rest in the rotating reference frame. In this section we concentrate on the motion of a free symmetric top as viewed by an external observer in an inertial frame. Since any object tossed into the air is basically a free top, the inertial description has a wide range of applications.

For a symmetric top the angular momentum and angular velocity projected onto the principal axes $(\hat{\mathbf{x}}, \hat{\mathbf{y}}, \hat{\mathbf{z}})$ in the top are

$$
\begin{aligned}
\mathbf{L} &= I(\omega_1\hat{\mathbf{x}} + \omega_2\hat{\mathbf{y}}) + I_3\omega_3\hat{\mathbf{z}} \\
\boldsymbol{\omega} &= (\omega_1\hat{\mathbf{x}} + \omega_2\hat{\mathbf{y}}) + \omega_3\hat{\mathbf{z}}
\end{aligned}
\tag{7.124}
$$

where $\hat{\mathbf{z}}$ is in the direction of the symmetry axis. By eliminating $(\omega_1\hat{\mathbf{x}} + \omega_2\hat{\mathbf{y}})$ in these equations, the angular-velocity vector $\boldsymbol{\omega}$ can be expressed in terms of $\hat{\mathbf{L}}$ and $\hat{\mathbf{z}}$ as

$$
\boldsymbol{\omega} = \frac{L}{I}\hat{\mathbf{L}} - \Omega\hat{\mathbf{z}}
\tag{7.125}
$$

where

$$
\Omega \equiv \left(\frac{I_3 - I}{I}\right)\omega_3
$$

as before, in (7.120). Since (7.125) is a linear relation among $\boldsymbol{\omega}$, $\mathbf{L}$, and $\hat{\mathbf{z}}$, these three vectors must lie in a plane. The absence of torques on the top implies that $\mathbf{L}$ is constant in the inertial system. Thus the $\boldsymbol{\omega}, \hat{\mathbf{z}}$ plane rotates (precesses) around the direction of $\mathbf{L}$. According to (7.125), the motion of the top as viewed from the inertial frame can be resolved into non-orthogonal components ω_L along $\hat{\mathbf{L}}$ and ω_3 along $\hat{\mathbf{z}}$ as

$$
\begin{aligned}
\omega_L &= \frac{L}{I} \\
\omega_3 &= -\Omega
\end{aligned}
\tag{7.126}
$$

Since $\hat{\mathbf{z}}$ is a vector fixed in the body (*i.e.*, it rotates with the body), we have from (7.6) and (7.125)

$$
\dot{\hat{\mathbf{z}}} = \boldsymbol{\omega}\times\hat{\mathbf{z}} = (\omega_L\hat{\mathbf{L}})\times\hat{\mathbf{z}}
\tag{7.127}
$$

Hence the symmetry axis $\hat{\mathbf{z}}$ rotates (or precesses) with fixed angular velocity $\omega_L\hat{\mathbf{L}}$ about the fixed inertial axis $\hat{\mathbf{L}}$. If we were riding on the top what would we experience? The angular velocity $\boldsymbol{\omega}^*$ of the top as observed from the *precessing frame* which rotates with angular velocity $\omega_L\hat{\mathbf{L}}$ is

$$
\boldsymbol{\omega}^* = \boldsymbol{\omega} - \omega_L\hat{\mathbf{L}} = -\Omega\hat{\mathbf{z}}
$$

The motion of the top as seen from this body-fixed frame is a rotation about the symmetry axis $\hat{\mathbf{z}}$ at the angular rate $-\Omega$. Since this is the rate that the top rotates with respect to $\boldsymbol{\omega}$ (which is a fixed vector in the precessing frame), we conclude that $+\Omega$ is the rate that $\boldsymbol{\omega}$ rotates with respect to the body, in agreement with the result (7.120) found from Euler's equations.

In the motion of the top, the angles that the symmetry axis $\hat{\mathbf{z}}$ makes with the vectors $\mathbf{L}$ and $\boldsymbol{\omega}$ remain constant, as can be shown from (7.125) and (7.127) or from energy- and angular-momentum conservation. Since there are no torques, both the angular momentum $\mathbf{L}$ and the rotational kinetic energy K are constant. From (7.124) we can write $\mathbf{L}$ as

$$
\mathbf{L} = I\omega_n\hat{\mathbf{n}} + I_3\omega_3\hat{\mathbf{z}}
\tag{7.128}
$$

where

$$
\omega_n\hat{\mathbf{n}} \equiv \omega_1\hat{\mathbf{x}} + \omega_2\hat{\mathbf{y}}
\tag{7.129}
$$

is orthogonal to $\hat{\mathbf{z}}$. In terms of the components ω_n and ω_3, we have

$$
\begin{aligned}
L^2 &= I^2\omega_n^2 + I_3^2\omega_3^2 \\
2K &= \mathbf{L}\cdot\boldsymbol{\omega} = I\omega_n^2 + I_3\omega_3^2
\end{aligned}
\tag{7.130}
$$

where we used the expression (6.113) for the rotational kinetic energy. The constancy of L^2 and K in (7.130) requires in turn that ω_n and ω_3 be constant. The magnitude of $\boldsymbol{\omega}$ in (7.124),

$$\omega = \sqrt{\omega_n^2 + \omega_3^3}$$

is then also constant. From the geometry of Fig. 7-18, the angles of interest are determined by

$$\tan \alpha = \frac{\omega_n}{\omega_3}$$

$$\tan \theta = \frac{L_n}{L_3} = \frac{I\omega_n}{I_3\omega_3} \tag{7.131}$$

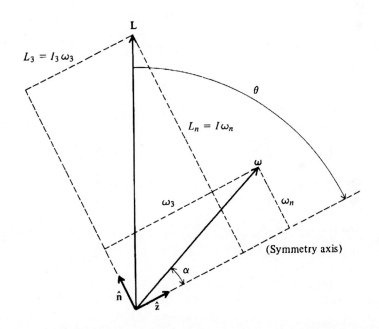

FIGURE 7-18. Components of angular velocity and angular momentum along the symmetry axis $\hat{\mathbf{z}}$ of the top and an axis $\hat{\mathbf{n}}$ perpendicular to the symmetry axis in the plane of $\boldsymbol{\omega}$ and $\mathbf{L}$.

The fixed relative orientation of $\mathbf{L}$, $\boldsymbol{\omega}$, and $\hat{\mathbf{z}}$ follows immediately from these results. If we eliminate ω_n/ω_3 in (7.131), we obtain

$$\tan \alpha = \frac{I_3}{I} \tan \theta \tag{7.132}$$

For an oblate top (pancake or coinlike), $I_3 > I$, and the angle α is larger than θ. For a prolate top (football or cigar-shape), $\alpha < \theta$, which is the case illustrated in Fig. 7-18.

A simple geometric construction can be made to illustrate symmetrical free-top motion in an inertial reference frame. This construction is based on the constancy of the angles θ and α. As the plane containing the vectors $\boldsymbol{\omega}$ and $\hat{\mathbf{z}}$ precesses about $\mathbf{L}$, the vector $\boldsymbol{\omega}$ sweeps out a cone (the space cone) of half-angle $(\theta - \alpha)$ about the fixed direction $\mathbf{L}$. In the coordinate system fixed in the top, the vector $\boldsymbol{\omega}$ sweeps out a cone (the body cone) of half-angle α. Since $\boldsymbol{\omega}$ sweeps out both the space and body cones, the line of contact between the two cones is simply the vector $\boldsymbol{\omega}$. The points on the body cone which lie on the vector $\boldsymbol{\omega}$ are instantaneously at rest with respect to the fixed-space cone because $\boldsymbol{\omega}$ is the instantaneous axis of rotation of the top. As a consequence, the body cone must roll on the fixed-space cone without slipping. Thus we have a qualitative picture of the top's motion as the body cone rolling on the space cone. This is illustrated in Fig. 7-19(a) for a prolate top, and in Fig. 7-19(b) for an oblate top.

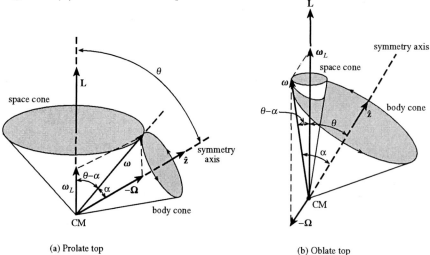

(a) Prolate top   (b) Oblate top

FIGURE 7-19. Space and body cones for (a) prolate top, (b) oblate top.

7.10   The Heavy Symmetric Top

Untold generations of children have been fascinated by the precessing, rising, sleeping, and dying of spinning tops. The theory of spinning tops plays an important role in a wide variety of disciplines ranging from astronomy to applied mechanics to nuclear physics. In this section we discuss the motion of a symmetric top in a gravity field for a special case in which the point of contact of the top with supporting surface, the pivot, is fixed.

To analyze the motion of the top it is convenient to introduce the *Euler angle coordinates* ϕ, θ, ψ shown in Fig. 7-20. In this figure the origin of the inertial coordinate system x_I, y_I, z_I is at the fixed point of contact of the top. A coordinate system x, y, z is obtained by the rotation through an angle ϕ about $\hat{\mathbf{z}}_I$, followed by a rotation through θ about $\hat{\mathbf{x}}$. The $\hat{\mathbf{x}}$ axis is called the *line of nodes*. Then a coordinate system x', y', z' is obtained by a rotation through an angle ψ about the $\hat{\mathbf{z}}$ axis. The angles ϕ, θ, ψ uniquely specify the x', y', z' system relative to x_I, y_I, z_I and are useful in describing the orientation of a symmetric top as illustrated in Fig. 7-21, where x', y', z' are taken to be the body-fixed axes of the top.

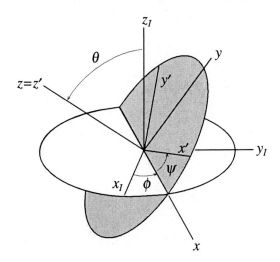

FIGURE 7-20. Euler angle coordinates describing rotations.

The angular velocity of the top is given in terms of the Euler angles by

$$\boldsymbol{\omega} = \dot{\phi}\hat{\mathbf{z}}_I + \dot{\theta}\hat{\mathbf{x}} + \dot{\psi}\hat{\mathbf{z}} \tag{7.133}$$

Using the geometry of Fig. 7-20

$$\hat{\mathbf{z}}_I = \cos\theta\hat{\mathbf{z}} + \sin\theta\hat{\mathbf{y}} \tag{7.134}$$

and thus

$$\boldsymbol{\omega} = \dot{\theta}\hat{\mathbf{x}} + \dot{\phi}\sin\theta\hat{\mathbf{y}} + (\dot{\psi}\dot{\phi}\cos\theta)\hat{\mathbf{z}} \tag{7.135}$$

Because the top is symmetric, the moments of inertia $I_{xx} = I_{yy} \equiv I$ are the same about any set of orthogonal axes in the $\hat{\mathbf{x}}, \hat{\mathbf{y}}$ plane. Thus the

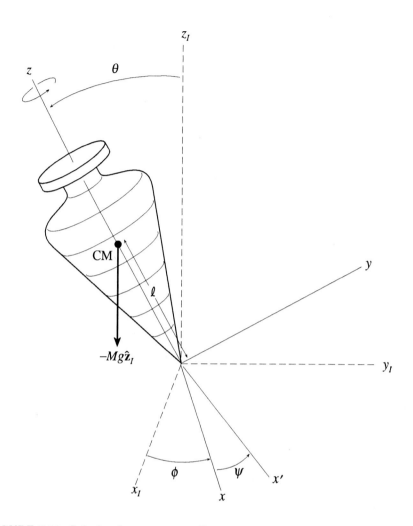

FIGURE 7-21. Spinning heavy top coordinates.

angular momentum about the origin is

$$\mathbf{L} = I\dot{\theta}\hat{\mathbf{x}} + I\dot{\phi}\sin\theta\hat{\mathbf{y}} + I_3(\dot{\psi} + \dot{\phi}\cos\theta)\hat{\mathbf{z}} \tag{7.136}$$

where $I_{zz} \equiv I_3$. Using these expressions for $\boldsymbol{\omega}$ and $\mathbf{L}$ and (6.113), the kinetic energy of the top is

$$K = \frac{1}{2}\boldsymbol{\omega} \cdot \mathbf{L} = \frac{1}{2}I(\dot{\theta}^2 + \dot{\phi}\sin^2\theta) + \frac{1}{2}I_3(\dot{\psi} + \dot{\phi}\cos\theta)^2 \tag{7.137}$$

The gravitational potential energy is

$$V = Mg\ell\cos\theta \tag{7.138}$$

where ℓ is the distance from the point of contact to the CM of the top.

The Lagrangian $L = K - V$ for the top is a function of θ, $\dot{\theta}$, $\dot{\phi}$, $\dot{\psi}$. Since there is no dependence on ϕ or ψ there are two conserved general momenta. The first is

$$p_\psi = \frac{\partial L}{\partial \dot{\psi}} = I_3(\dot{\psi} + \dot{\phi}\cos\theta) \tag{7.139}$$

$$\equiv I_3\omega_3$$

Thus the angular velocity ω_3 along the symmetry (z) axis is constant. The second conserved momentum is

$$p_\phi = \frac{\partial L}{\partial \dot{\phi}} = I\dot{\phi}\sin^2\theta + I_3\omega_3\cos\theta \tag{7.140}$$

The ϕ equation of motion is then $\dot{p}_\phi = 0$ which gives

$$I\ddot{\phi}\sin\theta = \dot{\theta}(I_3\omega_3 - 2I\dot{\phi}) \tag{7.141}$$

The Lagrange equation $\dot{p}_\theta = \frac{\partial L}{\partial \theta}$ gives the other equation of motion

$$I\ddot{\theta} = (Mg\ell - I_3\omega_3\dot{\phi} + I\dot{\phi}^2\cos\theta)\sin\theta \tag{7.142}$$

Motion in ϕ corresponds to precession and variations in θ are known as nutation.

We now use this last equation of motion to investigate the conditions under which the motion is pure precession. For pure precession the angle θ is constant. For constant ϕ θ the right-hand side of (7.141) must vanish and we can solve for $\dot{\phi}$ in terms of θ.

$$\dot{\phi} = \frac{I_3\omega_3}{2I\cos\theta}\left(1 \pm \sqrt{1 - \frac{4Mg\ell I\cos\theta}{I_3^2\omega_3^2}}\right) \tag{7.143}$$

For physical solutions, the quantity under the radical sign must not be negative. Since $\cos\theta > 0$ for a top on a table, the spin ω_3 must satisfy

$$\omega_3 \geq \sqrt{\frac{4Mg\ell I\cos\theta}{I_3^2}} \tag{7.144}$$

Only if the top has at least this minimum value of spin is pure precession possible. For a spin which is much greater than this minimum value, we can approximate the square root in (7.143) by the first two terms in a binomial expansion. We then find two possible approximate solutions for the precessional rate $\dot{\phi}$.

Slow precession:

$$\dot{\phi} = \frac{Mg\ell}{I_3\omega_3} \equiv \omega_p \tag{7.145}$$

Fast precession:

$$\dot{\phi} = \frac{I_3\omega_3}{I\cos\theta} \tag{7.146}$$

For the first solution, $\dot{\phi} \ll \omega_3$, and the angular momentum vector $\mathbf{L}$ lies nearly along the $\hat{\mathbf{z}}$ axis. This solution corresponds to slow gyroscopic precession, as discussed in § 6.5. For the second solution, $\dot{\phi} \approx \omega_3$, and $\mathbf{L}$ lies nearly along the z axis. In this case, $L\cos\theta \approx I_3\omega_3$ and $\dot{\phi} \approx L/I$, which is just the angular frequency ω_L in the force-free-top limit of (7.126). This solution with rapid precession about the vertical direction is independent of gravity in the limit $\omega_3 \gg (\omega_3)_{\min}$.

For a rapidly spinning top, slow precession and small nutation are frequently observed. To find an approximate solution for the motion with this condition, the quadratic terms in $\dot{\phi}$ and $\dot{\theta}$ in the differential equations (7.141) and (7.142) can be neglected. We then obtain

$$\ddot{\phi}\sin\theta = \frac{I_3\omega_3}{I}\dot{\theta}$$

$$\ddot{\theta} = \left(\frac{Mg\ell}{I} - \frac{I_3\omega_3}{I}\dot{\phi}\right)\sin\theta \tag{7.147}$$

In terms of ω_p from (7.145) and ω_L defined as

$$\omega_L \equiv \frac{I_3\omega_3}{I} \tag{7.148}$$

these equations can be written

$$\ddot{\phi}\sin\theta = \omega_L\dot{\theta} \tag{7.149}$$

$$\ddot{\theta} = \omega_L(\omega_p - \dot{\phi})\sin\theta \tag{7.150}$$

If we time-differentiate (7.149) and substitute (7.150), we find

$$\frac{d^2\dot{\phi}}{dt^2} + \omega_L^2\dot{\phi} = \omega_L^2\omega_p \tag{7.151}$$

where again we have dropped a quadratic term (of order $\ddot{\phi}\dot{\theta}$). We can immediately write down the solution to this equation for $\dot{\phi}$.

$$\dot{\phi}(t) = \omega_p + a\cos(\omega_L t + \alpha) \tag{7.152}$$

For the initial conditions $\dot{\phi} = \omega_0$, $\dot{\theta} = 0$, $\theta = \theta_0$ at $t = 0$, we find $\ddot{\phi}_0 = 0$ from (7.149) and

$$\dot{\phi}(t) = \omega_p - (\omega_p - \omega_0)\cos\omega_L t \tag{7.153}$$

The solution for $\phi(t)$ follows by integration.

$$\phi(t) = \omega_p t - \left(\frac{\omega_p - \omega_0}{\omega_L}\right)\sin\omega_L t \tag{7.154}$$

To solve for θ, we plug (7.153) into (7.149). This gives

$$\dot{\theta} = (\omega_p - \omega_0)\sin\omega_L t\sin\theta \tag{7.155}$$

Since ω_p and ω_0 are small quantities, we can make the approximation $\sin\theta \approx \sin\theta_0$ on the right-hand side of this equation. The solution for θ is then found by integration.

$$\theta(t) = \theta_0 + \left(\frac{\omega_p - \omega_0}{\omega_L}\right)\sin\theta_0(1 - \cos\omega_L t) \tag{7.156}$$

This completes the formal solution of the equations of motion in the approximation of slow precession and small nutation.

The solution for $\theta(t)$ in (7.150) exhibits nutation of the top between the angular limits θ_0 and $\theta_0 + 2[(\omega_p - \omega_0)/\omega_L]\sin\theta_0$. The sign of $(\omega_p - \omega_0)$ determines which is the upper and which is the lower bound on θ. The precession $\phi(t)$ in (7.154) has a sinusoidal motion associated with the nutation which is superimposed on the steady precession. When the initial precession ω_0 equals ω_p, the top undergoes steady precessional motion with no nutation. In Fig. 7-22 the curves traced out by the symmetry axis of the top are shown for various initial values ω_0.

The nutation frequency of the top from (7.156) is ω_L. We see from (7.145) and (7.148) that as the spin ω_3 of the top increases, the nutation frequency ω_L increases, while the precession frequency ω_p decreases. Furthermore, the nutation amplitude is inversely proportional to ω_L, so that nutation of a fast top is not so visible. When a fast top is spun on a hollow surface, however, a buzzing tone can often be heard with a frequency corresponding to the nutation frequency.

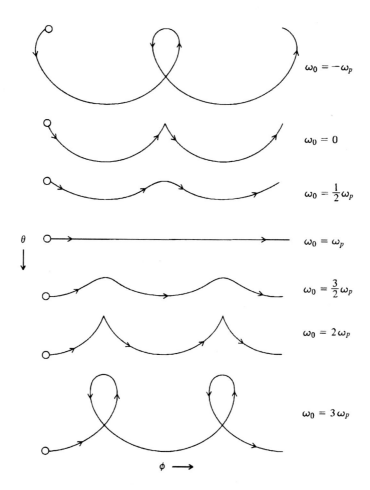

FIGURE 7-22. Nutation curves traced out by the symmetry axis of the top for various initial conditions. The top is started at the same value of θ in each case and the resulting curves are trochoids.

The phenomenon of nutation exhibited by our formal solution above can be understood from a more elementary viewpoint. For a top which is spinning rapidly, the angular momentum **L** is nearly along the symmetry axis $\hat{\mathbf{z}}$ of the top. The gravitational torque is obtained from (7.138)

$$\mathbf{N} = -\frac{dV}{d\theta}\hat{\boldsymbol{\theta}} = Mg\ell\sin\theta\hat{\mathbf{x}} \tag{7.157}$$

N is perpendicular to the z_I axis and causes the precession about the vertical direction. The angular velocity of precession $\boldsymbol{\omega}_p = \omega_p\hat{\mathbf{z}}_I$ can be found by equating

$$\frac{d\mathbf{L}}{dt} = \mathbf{N} = -Mg\ell \sin\theta \,\hat{\mathbf{x}}$$

to

$$\frac{d\mathbf{L}}{dt} = \boldsymbol{\omega}_p \times \mathbf{L} = \omega_p L(\hat{\mathbf{z}}_I \times \hat{\mathbf{z}})$$

Since $\hat{\mathbf{z}}_I \times \hat{\mathbf{z}} = -\hat{\mathbf{x}} \sin\theta$ and $L \approx I_3 \omega_3$, we obtain the expected result

$$\omega_p \approx \frac{Mg\ell}{I_3\omega_3} \qquad (7.158)$$

for the angular frequency of steady precession about the vertical z_I axis. For ω_3 very large, the precession rate ω_p is quite slow and the symmetry axis is nearly stationary. If the top is now given a slight push, it instantaneously acquires a small angular-momentum component $\Delta\mathbf{L}$ perpendicular to $\hat{\mathbf{z}}$. The resulting total angular momentum $\bar{\mathbf{L}} \equiv \mathbf{L} + \Delta\mathbf{L}$ points in a direction slightly different from the symmetry axis $\hat{\mathbf{z}}$. The ensuing motion is like that of a free top with the symmetry axis precessing around $\bar{\mathbf{L}}$ in a small circle. The angular frequency of this circular motion is found from (7.136) to be

$$\omega_L = \frac{\bar{\mathbf{L}}}{I} \approx \frac{I_3\omega_3}{I} \qquad (7.159)$$

The complete motion of the symmetry axis is a superposition of this rapid free-top circular motion about the direction $\bar{\mathbf{L}}$ on the slow precession of $\hat{\bar{\mathbf{L}}}$ about the vertical direction.

7.11 Slipping Tops: Rising and Sleeping

When a spinning top similar to that in Fig. 7-21 is set down on a rough surface, the top usually slips initially. A frictional force directed opposite to the instantaneous skidding velocity acts to accelerate the CM of the top until the velocity of slipping is reduced to zero and pure rolling motion sets in. If the top is spinning rapidly when it is set down, it tends to maintain a fixed angle θ with the vertical, since the nutation is small. The normal force at the point of contact is then essentially the entire weight of the top, and so the frictional force is

$$|\mathbf{f}| \approx \mu Mg \qquad (7.160)$$

where μ is the coefficient of friction. This friction force will cause the top to move in the direction of the force but that effect will not be considered here. The friction force will also cause a torque

$$|\mathbf{N}| \approx \mu M g \ell$$

about the CM of the top which is roughly perpendicular to the peg for a thin peg, as illustrated in Fig. 7-23. For a rapidly spinning top, $\mathbf{L}$ lies nearly along the symmetry axis. Since the frictional torque is perpendicular to the symmetry axis, $\mathbf{L}$ precesses toward the vertical. The angular velocity of this precession is

$$\dot{\theta} = -\frac{|\mathbf{N}|}{|\mathbf{L}|} \approx -\frac{\mu M g \ell}{I_3 \omega_3}$$

or

$$\dot{\theta} \approx -\mu \omega_p \tag{7.161}$$

by (7.158). The angular velocity of the rising motion is just the product of the precessional angular velocity and the coefficient of skidding friction.

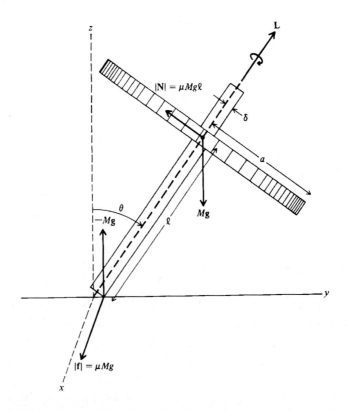

FIGURE 7-23. Forces on a rising top.

As the top rises, kinetic energy is converted into potential energy and the spin of the top decreases. In addition, some of the energy is dissipated by friction. The rate of frictional dissipation of energy,

$$\frac{dE}{dt} = -fv \tag{7.162}$$

where f is the frictional force and v is the velocity at the point of contact, can be quite small for a thin peg. Nevertheless, the effects of friction in causing the top to rise are dramatic.

Once the top has risen to a vertical position, the point of contact is the symmetry axis, and the frictional force is much smaller. From (7.142), the equation of motion for very small θ is then approximately

$$I\ddot{\theta} - (Mg\ell - I_3\omega_3\dot{\phi} + I\dot{\phi}^2)\theta = 0 \tag{7.163}$$

provided that the dissipation of energy by friction is neglected. In terms of the quantities ω_p and ω_L defined in (7.145) and (7.148), this equation can be written as

$$\ddot{\theta} + \omega_L \left(-\omega_p + \dot{\phi} - \frac{1}{\omega_L}\dot{\phi}^2\right)\theta = 0 \tag{7.164}$$

For a given $\dot{\phi}$, the motion in θ will be stable about $\theta = 0$ if

$$-\omega_p + \dot{\phi} - \frac{1}{\omega_L}\dot{\phi}^2 > 0 \tag{7.165}$$

The corresponding requirement on $\dot{\phi}$ is

$$\frac{\omega_L}{2}\left(1 - \sqrt{1 - \frac{4\omega_p}{\omega_L}}\right) < \dot{\phi} < \frac{\omega_L}{2}\left(1 + \sqrt{1 - \frac{4\omega_p}{\omega_L}}\right) \tag{7.166}$$

For a high value of the spin ω_3,

$$\frac{\omega_p}{\omega_L} = \frac{Mg\ell/I_3\omega_3}{I_3\omega_3/I} \ll 1 \tag{7.167}$$

and the condition in (7.166) is satisfied. The spinning top *sleeps* in the vertical position until friction slows down the spin to the value

$$\omega_3 = \sqrt{\frac{4Mg\ell I}{I_3^2}} \tag{7.168}$$

for which

$$\omega_L = 4\omega_p$$

and (7.166) becomes unphysical (*i.e.*, complex). At this point, the θ motion of the top becomes unstable and the top wobbles and goes down as θ increases from zero.

To develop a feeling for the motion of a typical top, we consider as an example a top made of a thin disk of radius a and mass M which is supported by a narrow peg of length $\ell = a/2$ and negligible mass, as illustrated in Fig. 7-23. The moments of inertia about the point of contact of the peg with the table are

$$I_3 = \tfrac{1}{2}Ma^2$$
$$I = \tfrac{1}{4}Ma^2 + M\ell^2 = \tfrac{1}{2}Ma^2 \tag{7.169}$$

If the top has a radius $a = 3$ cm and is set down with an initial spin of $\omega_3 = 300$ rad/s (about 50 r/s), the angular velocity of precession from (7.145) is

$$\omega_p = \frac{Mg(a/2)}{\tfrac{1}{2}Ma^2\omega_3} = \frac{g}{a\omega_3} = \frac{980}{3(300)} \simeq 1 \text{ rad/s} \tag{7.170}$$

For a coefficient of friction $\mu = 1/10$, the angular velocity from (7.161) of the top's rise toward the vertical is

$$\dot{\theta} = -\mu\omega_p = -0.1 \text{ rad/s} \tag{7.171}$$

If the top is started at its maximum angle of inclination,

$$\theta = \arctan\frac{a/2}{a} = 0.46 \text{ rad} \tag{7.172}$$

the time to rise to the vertical is

$$t = \frac{\theta}{|\dot{\theta}|} = \frac{0.46}{0.1} \simeq 5 \text{ s} \tag{7.173}$$

In this length of time the axis of the top has made

$$\frac{\omega_p t}{2\pi} = \frac{1(5)}{6.28} = 0.8 \text{ revolution} \tag{7.174}$$

about the vertical and

$$\frac{\omega_3 t}{2\pi} = \frac{300(5)}{6.28} = 240 \text{ revolutions} \tag{7.175}$$

about the symmetry axis. From (7.168), the condition for the motion at $\theta = 0$ to be stable is

$$\omega_3 > \sqrt{\frac{4Mg\ell I}{I_3^2}} = \sqrt{\frac{4Mg(a/2)(\frac{1}{2}Ma^2)}{(\frac{1}{2}Ma^2)^2}} = \sqrt{\frac{4g}{a}} \tag{7.176}$$

From the parameters of our top, we find

$$\sqrt{\frac{4g}{a}} = \sqrt{\frac{4(980)}{3}} = 36 \text{ rad/s} \tag{7.177}$$

Since the inequality $\omega_3 > \sqrt{4g/a}$ is satisfied for the initial spin $\omega_3 = 300$ rad/s, the top will *sleep* in its vertical position.

7.12   The Tippie-Top

When a tippie-top is spun on a smooth table, it turns itself upside-down, as pictured in Fig. 7-24. The usual high school and college rings likewise flip over to spin with their heavy ends upward. This fascinating behavior is due to a small frictional force at the point of contact with the table.

The frictional force is parallel to the table, opposing the velocity of slipping, as illustrated in Fig. 7-25. Since the horizontal direction of this force rotates rapidly with the angular frequency ω of the top, the time average of the force is zero, resulting in little effect on the motion of the CM.

The CM of a tippie-top is located close to the center of curvature, as indicated in Fig. 7-25. The gravitational torque can therefore be neglected. Furthermore, the frictional torque is nearly horizontal. This horizontal torque also rotates with angular frequency ω and time averages to zero. As a result, $d\mathbf{L}/dt \approx 0$, on the average, and the angular momentum of the top is nearly conserved. If the tippie-top is initially spun with the spin upward, as in Fig. 7-24, the approximately fixed direction of $\mathbf{L}$ is vertical with respect to the table.

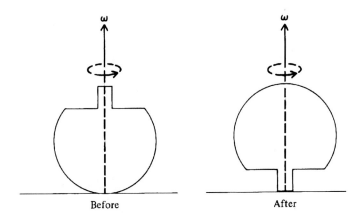

FIGURE 7-24. Flipping of a tippie-top.

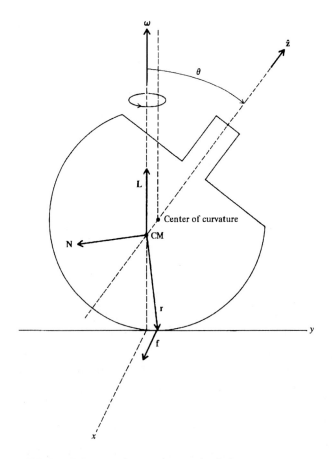

FIGURE 7-25. Frictional force and torque on a tippie-top.

The tipping motion is readily analyzed in a coordinate system which rotates with the top. In this reference frame the time average of the torque $\mathbf{N}$ is nonzero. The equation of motion (7.72) is

$$\mathbf{N} = \frac{\delta \mathbf{L}}{\delta t} + \boldsymbol{\omega} \times \mathbf{L} \qquad (7.178)$$

As a simplifying approximation, we take the three principal moments of inertia about the CM as equal (the shape of the tippie-top is nearly spherical). Then

$$\mathbf{L} \approx I\boldsymbol{\omega} \qquad (7.179)$$

and the $\boldsymbol{\omega} \times \mathbf{L}$ term in (7.178) vanishes. $\mathbf{L}$ remains vertical and $\mathbf{N}$ horizontal throughout the motion. Since $\mathbf{L}$ and $\mathbf{N}$ are perpendicular, the torque causes the angular momentum to precess uniformly in the body frame, according to (7.178). Taking the component of (7.178) along the symmetry axis $\hat{\mathbf{z}}$, we find

$$\hat{\mathbf{z}} \cdot \mathbf{N} = \hat{\mathbf{z}} \cdot \frac{\delta \mathbf{L}}{\delta t} = \frac{\delta(\hat{\mathbf{z}} \cdot \mathbf{L})}{\delta t} \qquad (7.180)$$

where θ is the angle between $\mathbf{L}$ and $\hat{\mathbf{z}}$. From (7.179) and (7.180), we obtain

$$\dot{\theta} = \frac{N}{L} \approx \frac{\mu M g R}{I \omega} \qquad (7.181)$$

where R is the radius of the top. Thus θ increases with time, and the tippie-top flips over. Once the stem scrapes the table, the subsequent rise to the vertical is almost the same as an ordinary rising top, as treated in §7.11. We can estimate the time required for the tippie-top to flip over by use of (7.181). A spin of $\omega = 300$ rad/s is easily imparted to a tippie-top of radius $R = 1.5$ cm and moment of inertia $I \approx \frac{2}{3}MR^2$ (hollow sphere). For a coefficient of friction $\mu = 1/10$, we obtain

$$\dot{\theta} = \frac{3\mu g}{2R\omega} = \frac{3(0.1)(980)}{2(1.5)(300)} = 0.3 \text{ rad/s} \qquad (7.182)$$

The flip time is, roughly,

$$t = \frac{\theta}{\dot{\theta}} \approx \frac{\pi}{0.3} \approx 10 \text{ s} \qquad (7.183)$$

PROBLEMS

7.2 Fictitious Forces

7-1. A particle of mass m moves in a smooth straight horizontal tube which rotates with constant velocity ω about a vertical axis which intersects the tube. Set up the equations of motion in polar coordinates and derive an expression for the distance of the particle from the rotation axis. If the particle is at $r = r_0$ at $t = 0$, what velocity must it have along the tube in order that it will be very close to the rotation axis after a very long time?

7-2. The WIYN telescope mirror blank was cast on a spinning platform in an oven. By this spin-casting technique the desired parabolic shape can be achieved without having to remove much glass by grinding. Light rays parallel to the symmetry axis of the mirror are focused to a point on the axis. This focal distance f is one-half the radius of curvature of the mirror.

a) If the mirror spins with angular velocity ω show that the focal distance is $f = g/2\omega^2$.

b) The mirror is to have a f/diameter ratio of 1.75 and a diameter of 3.5 m. Find the required spin in revolutions/min.

7-3. A bug of mass 1 g crawls out along a radius of a phonograph record turning at $33\frac{1}{3}$ r/min. If the bug is 6 cm from the center and traveling at the rate of 1 cm/s, what are the forces on the bug? What added torque must the motor supply because of the bug?

7-4. When ice skaters spin in place while pulling in their arms and legs, the striking increase in angular velocity is a consequence of angular momentum conservation. The fictitious forces which act to spin the skater are the Coriolis and azimuthal forces. For simplicity assume the spinning skater holds dumbbells initially at arm's length and by internal body forces draws them toward the rotation axis. Neglect any mass other than that of the dumbbells. Analyze the situation from the point of view of a rotating coordinate system in which the dumbbells are at rest except for radial motion. Show that the resultant forces imply that angular momentum is conserved.

7-5. For problem 7-1 use the Lagrangian method to

a) Find the radial equation of motion of the mass m.

b) Find the general constraint force Q'_θ.

c) Interpret this constraint force in terms of the Coriolis force.

7-6. A bead of mass m is constrained to move frictionlessly on a hoop of radius R. The hoop rotates with constant angular velocity w about a vertical axis which coincides with a diameter of the hoop.

a) Obtain the equation of motion by applying Newton's law in the rest frame of the hoop.

b) Find the critical angular velocity Ω below which the bottom of the hoop provides a stable-equilibrium position for the bead.

c) Find the stable-equilibrium position for $w > \Omega$.

7.3 Motion on the Earth

7-7. A spherical planet of radius R rotates with a constant angular velocity w. The effective gravitational acceleration g_{eff} is some constant, g, at the poles and $0.8\,g$ at the equator. Find g_{eff} as a function of the polar angle θ and g. With what velocity must a rocket be fired vertically upward from the equator to escape completely from the planet?

7-8. A particle of mass m is constrained to move in a vertical plane which rotates with constant angular velocity w. Find the equations of motion of the particle, including the force of gravity.

7-9. A particle moves with velocity v on a smooth horizontal plane. Show that the particle will move in a circle due to the rotation of the earth; find the radius of the circle.

7-10. A ball is thrown vertically upward with velocity v_0 on the earth's surface. If air resistance is neglected, show that the ball lands a distance $(4w \sin \theta v_0^3 / 3g^2)$ to the west, where w is the angular velocity of the earth's rotation and θ is the colatitude angle.

7.4 Foucault's Pendulum

7-11. Show using (7.45) that the Foucault pendulum equations of motion in cartesian coordinates are

$$\ddot{x} + w_0^2 x - 2w \cos \theta \dot{y} = 0$$
$$\ddot{y} + w_0^2 y + 2w \cos \theta \dot{x} = 0$$

where w is the earth's angular velocity and θ is the colatitude angle. Solve this system of coupled equations retaining only the leading order in w/w_0 using the trial solutions $x = c_x e^{i\Omega t}$ and $y = c_y e^{i\Omega t}$.

Show that the two allowed angular frequencies are $\Omega_\pm = \omega_0 \pm \omega \cos\theta$ and that $(c_y/c_x)_\pm = \pm i$. Impose the initial values $x(0) = a$, $\dot{x}(0) = 0$, $y(0) = 0$, $\dot{y}(0) = 0$. Use the trigonometric identity in (2.168) to find the Foucault rotation in the x, y plane and determine the period of the rotation.

7.6 Principal Axes and Euler's Equations

7-12. Prove that if $I_1 \neq I_2 \neq I_3$ the principal axes are orthogonal. *Hint: start with (7.85) for $\omega^{(a)}$ and I_a and dot with $\omega^{(b)}$. Then do the same with $\omega^{(b)}$ and I_b and dot with $\omega^{(a)}$. Subtract the two equations and use the fact that $\mathbb{I}$ is a symmetric tensor.*

7-13. Show that the principal moments of inertia are real. *Hint: starting with (7.85) dot with ω^* (the complex conjugate) and subtract the resulting equation from its complex conjugate. Then use the fact that $\mathbb{I}$ is real and symmetric.*

7-14. A flat rectangular plate of mass M and sides a and $2a$ rotates with angular velocity ω about an axle through two diagonal corners, as shown. The bearings supporting the plate are mounted just at the corners. Find the force on each bearing.

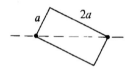

7-15. A point particle of mass $m = 1\,\text{kg}$ is located at the point $(x_0,\ y_0,\ 0)$

a) Calculate the tensor of inertia.

b) Find the principal axes and interpret the result.

7-16. For a prism mass distribution show that if any two axes perpendicular to the axis of the prism have equal moments of inertia, then all the axes in this plane are principal axes. *Hint: consider the expression for kinetic energy K in the ω_1, ω_2 plane where ω_1 and ω_2 lie in the plane perpendicular to the prism axes. The curve of constant K is an ellipse. The result follows from the geometry of the ellipse.*

7.7 The Tennis Racket Theorem

7-17. Show that the tennis racket in §7.7 is properly designed so that a hard stroke to the ball at the center of the racket head does not jar the player's hand by causing impulsive torques.

7-18. A tennis racket is swung underhand and released so that it rises vertically with an initial spin about the unstable principal axis. At the instant of release the end of the handle is at rest. The racket subsequently rises to a height of 5 m.

a) Determine the time of rise to maximum height.

b) Find the initial angular velocity $\omega_2(0)$ about the CM.

c) For an initial spin axis $\omega_3(0)$ that is 1 percent of $\omega_2(0)$, compute the time at which the racket begins to tumble.

7-19. Write $2K$ and L^2 for a general rigid body, in terms of the principal-axis components of $\boldsymbol{\omega}$. From this, demonstrate the *tennis racket theorem* for a free rigid body using conservation of K and L^2.

7.9 The Free Symmetric Top: External Observer

7-20. A coin in a horizontal plane is tossed into the air with angular velocity components ω_1 about a diameter through the coin and ω_3 about the principal axis perpendicular to the coin. If ω_3 were equal to zero, the coin would simply spin around its diameter. For ω_3 nonzero, the coin will precess. What is the minimum value of ω_3/ω_1 for which the wobble is such that the same face of the coin is always exposed to an observer looking from above? With a little practice, this is a clever way to arrange the outcome of a coin flip!

7-21. A satellite with three distinct principal axes tumbles as it orbits the earth. A flexible antenna is deployed which slowly dissipates energy as the spacecraft tumbles. Eventually the spacecraft stabilizes without further action. Find the final regular rotational state of the satellite. *Hint: consider what happens to the energy and angular momentum in the principal axes coordinate system.*

7.10 The Heavy Symmetric Top

7-22. Why is it difficult to spin a pencil on its point? Illustrate quantitatively with a pencil of length 10 cm and diameter 0.5 cm.

7-23. Give an expression, in terms of rotations about the CM, for the kinetic energy of a heavy top whose point is fixed but pivots freely. When the kinetic energy of the CM motion is included show explicitly that the total kinetic energy is the same as when calculated relative to the pivot point.

7-24. A spinning heavy top is placed on a smooth horizontal table. The top point will now trace out a complex curve and the CM will repeatedly rise and fall. Find the Lagrangian and discuss how it differs from the fixed-point case.

7-25. A hollow conical segment of half
angle α, mass M and side length
ℓ spins on a sharp pivot at its
apex as shown. The cone is made
from a uniform thin sheet and has
an open base.

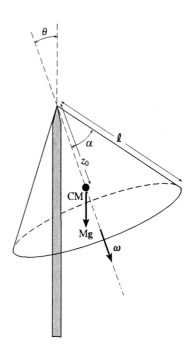

   a) The cone is initially rapidly
   spun clockwise as viewed from
   above with angular velocity ω_3
   about the symmetry axis. Find
   the direction and rate of the
   slow precession in terms of
   M, g, I_3 and the distance z_0
   of the CM from the apex.

   b) Calculate the principal moments, the location of the CM
   and the rate of slow precession.

7-26. A disk is spun about a vertical diameter. As it looses energy
through friction it begins to wobble with slowly decreasing angle θ.
Assume that the only motion of the CM is to fall slowly and that
the disk rolls without slipping.

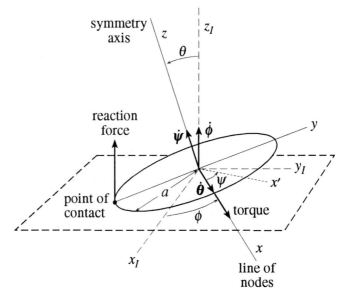

   a) Show that the rolling condition is $\omega_3 = 0$.

b) Show that the precession of the line of nodes $\dot{\phi}$ is given by $\dot{\phi} = \omega_0/\sqrt{\sin\theta}$, where $\omega_0 = 2\sqrt{g/a}$. The wobble rate thus increases as the disk lies down. *Hint: Use part a) to establish that* $\mathbf{L}$ *lies along the* $\hat{\mathbf{y}}$ *axis. The torque lies along the* $\hat{\mathbf{x}}$ *axis. Use* $\mathbf{N} = \dot{\mathbf{L}} = \dot{\phi} \times \mathbf{L}$.

c) Show that the component of spin along the vertical x-axis is given by $\omega_{\text{vert}} = \omega_0(\sin\theta)^{3/2}$. Thus even though the wobble rate increases without limit as the disk lies down, the spin as seen from above actually comes to a stop. This effect can be impressively demonstrated using a heavy disk, with a mark on top, spinning on a cement floor.

7-27. Apply Euler's equations to obtain the equations of motion for the symmetric heavy top. Use the $\hat{\mathbf{x}}, \hat{\mathbf{y}}, \hat{\mathbf{z}}$ coordinate system of Fig. 7-20, where $\hat{\mathbf{x}}$ is the line of nodes. Because of the symmetry of the top these coordinates form a principal system even though the top is not at rest. Show that the equations of motion are equivalent to those of § 7.10. *Hint: The* $\hat{\mathbf{x}}, \hat{\mathbf{y}}, \hat{\mathbf{z}}$ *system has angular velocity* $\boldsymbol{\omega} - \dot{\psi}\hat{\mathbf{z}}$ *relative to the inertial system.*

7.11 Slipping Tops: Rising and Sleeping

7-28. For a top with a spherical peg end, show that the effective peg radius is $\delta = \delta_0\sin\theta$, where θ is the inclination angle. Would you expect this top to rise higher than a similar top with a cut-off peg?

7.12 The Tippie-Top

7-29. Analyze the motion of a tippie-top using an inertial frame on the table. *Hint: Find the implications for* $\boldsymbol{\omega}$ *of the precession of* $\mathbf{L}$ *about the vertical direction.*

Chapter 8

GRAVITATION

According to Newton's law of universal gravitation each pair of particles in the universe is mutually attracted with a force proportional to the product of their masses, inversely proportional to the square of the distance between them, and directed along the line joining them. The proportionality constant G in the gravitational force law is known as Newton's constant. Although G is the least precisely measured fundamental constant, known only to one part in 10^4, its constancy is very well checked by careful analyses of solar system motions to better than one part in 10^{12} per year, which corresponds to a variation of no more than one percent over the age of the universe. Newtonian physics provides a nearly complete understanding of the motions of the planets, satellites, stars, galaxies and the universe as a whole. Indeed, it has only been in this century that a few tiny discrepancies have been uncovered whose explanation requires the more complete theory of gravity provided by Einstein's general relativity.

8.1 Attraction of a Spherical Body: Newton's Theorem

The statement of Newton's law of gravity applies to the attraction between two point masses, whereas celestial bodies are roughly spherical collections of particles. The theorem, first shown by Newton, that a spherically symmetric body acts as if its mass is concentrated at its center, is an essential step in the application of the law of gravitation to celestial mechanics. A corollary is that a particle located in a spherical mass distribution at a radius r from the center of the distribution experiences a net gravitational force only from the mass $M(r)$ within the radius r and the net force is as if $M(r)$ were located at $r = 0$. We give a proof of Newton's theorem using the concept of potential energy.

The gravitational potential energy between two point masses m and M separated by a distance r is

$$V(r) = -\frac{GMm}{r} \tag{8.1}$$

The corresponding force on m due to M is given by

$$\mathbf{F} = -\boldsymbol{\nabla}V(r) = GMm\frac{d}{dr}\left(\frac{1}{r}\right)\hat{\mathbf{r}} = -\frac{GMm}{r^2}\hat{\mathbf{r}} = -\frac{GMm}{r^3}\mathbf{r} \qquad (8.2)$$

where $\mathbf{r} = \mathbf{r}_m - \mathbf{r}_M$. It is convenient to define the gravitational force on m as $\mathbf{F} = m\mathbf{g}$, where $\mathbf{g}$ is the *acceleration of gravity* at the position of m, independent of the value of m. (The fact that any mass at a given position in a gravitational field has the same acceleration $\mathbf{g}$ is known as the *equivalence principle*.) Correspondingly the gravitational potential energy is defined as $V = m\Phi$, where Φ is the gravitational *potential*, so

$$\mathbf{g} = -\boldsymbol{\nabla}\Phi \qquad (8.3)$$

From (8.2) the gravitational *potential* due to a mass M at a distance r is

$$\Phi(\mathbf{r}) = -\frac{GM}{r} \qquad (8.4)$$

We first calculate the gravitational potential due to a uniform spherical shell of mass M at a distance R from the center of the shell, as illustrated in Fig. 8-1. To begin, we evaluate the potential energy of a circular ring element of mass dM shown in Fig. 8-1. If the radius of the shell is a, the surface mass density is

$$\sigma = \frac{M}{4\pi a^2} \qquad (8.5)$$

The circular ring element has differential area $dA = 2\pi(a\sin\theta)(ad\theta)$ and mass

$$\begin{aligned} dM &= (2\pi a^2 \sin\theta d\theta)\sigma \\ &= \frac{M}{2}\sin\theta d\theta \end{aligned} \qquad (8.6)$$

The distance r from dM to the point where the potential is being evaluated is given by the law of cosines

$$r^2 = a^2 + R^2 - 2aR\cos\theta \qquad (8.7)$$

By differentiation we obtain

$$rdr = aR\sin\theta d\theta = 2aR\frac{dM}{M}$$

so that dM can be expressed as

$$dM = \frac{Mrdr}{2aR} \qquad (8.8)$$

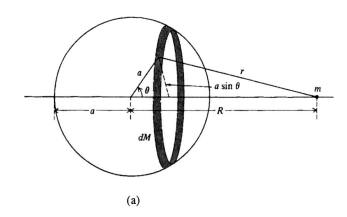

(a)

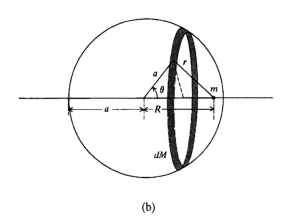

(b)

FIGURE 8-1. Gravitational attraction of a point mass m and a differential ring element dM on a spherical shell of mass M, with (a) m outside the shell and (b) m inside the shell.

The potential due to the ring mass dM is

$$d\Phi(r) = -\frac{G\,dM}{r} = -\frac{GM}{2aR}dr \tag{8.9}$$

The contributions of all ring elements on the shell are obtained by integration over r

$$\Phi(r) = \int_{r_{\min}}^{r_{\max}} d\Phi(r) = -\frac{GM}{2aR}(r_{\max} - r_{\min}) \tag{8.10}$$

We see from Fig. 8-1 that $r_{\max} = R + a$ and $r_{\min} = |R - a|$ and thus when

r is outside the shell

$$r_{\text{max}} - r_{\text{min}} = (R + a) - (R - a) = 2a \qquad (8.11)$$

whereas when m is inside the shell,

$$r_{\text{max}} - r_{\text{min}} = (R + a) - (a - R) = 2R \qquad (8.12)$$

Thus the potential is

$$\Phi(R) = \begin{cases} -\dfrac{GM}{R} & R > a \qquad\qquad (8.13) \\[2ex] -\dfrac{GM}{a} & R < a \qquad\qquad (8.14) \end{cases}$$

Since $\Phi(r)$ is constant inside the shell, $\mathbf{g}$ vanishes there. When r is outside the shell, the potential in (8.13) is as if the mass M of the shell were concentrated at the center of the shell. Since a spherically symmetric solid body can be represented as a collection of concentric spherical shells, the gravitational force on m due to a spherical body is as if the total mass M were concentrated at the center of the sphere. Newton's theorem follows: the gravitational force of any spherically symmetric distribution of matter at a distance R from the center is the same as if all the mass within the sphere of radius R were concentrated at the center.

8.2 The Tides

When a body moves in a non-uniform gravitational field, it is subjected to tide-generating forces. These shearing forces may even tear the body apart—this is a possible origin of the rings of Saturn.

The acceleration of the body $\mathbf{a}_B$ is the total gravitational force on its component masses divided by its total mass. (If the body is spherically symmetric, then the result of Newton's theorem and the "action equals reaction" principle is that $\mathbf{a}_B$ is simply the value of $\mathbf{g}(\mathbf{r})$ at the center of the body.) If we use coordinates centered on the body (*i.e.*, "falling with the body") the gravitational field becomes $\mathbf{g}(\mathbf{r}) - \mathbf{a}_B$. If we separate $\mathbf{g}$ into the part due to the body itself $\mathbf{g}_{\text{self}}$ (which vanishes at the center of the body) and to the part due to external masses $\mathbf{g}_{\text{ext}}$, then the gravitational field in the frame fixed on the body is $\mathbf{g}_{\text{self}} + (\mathbf{g}_{\text{ext}} - \mathbf{a}_B)$. The second term, $(\mathbf{g}_{\text{ext}} - \mathbf{a}_B)$, is the tidal field.

Tidal forces on a planet are maximum along a line to the external force center and give two high tides on opposite sides of the planet. For a planet in a circular orbit about the sun the origin of the double tide is easily explained by the following argument. The forces acting on a mass m are the attractive gravitational force GmM/r^2 and the repulsive centrifugal force $m\omega^2 r$ due to the revolution of the planet about the sun. At the CM of the planet the gravity force exactly balances the centrifugal force since there is no radial acceleration in a circular orbit. At the point closest to the sun, the sun's gravitational attraction is larger than at the CM and the centrifugal force is smaller, giving a net tidal force in the direction of the sun. At the farthest point on the planet from the sun the centrifugal force exceeds that of gravity and there is a tidal force directed away from the sun.

The ocean tides on earth are caused by the variation from place to place of the gravitational attraction due to the moon and the sun. The atmosphere, the ocean, and the solid earth all experience tidal forces, but only the effects on the ocean are commonly observed. To estimate the gross features of the midocean tides, we begin with a static theory in which the rotation of the earth about its axis is neglected. The daily rotation of the earth will be invoked later to explain the propagation of the tides.

To calculate the tide-generating force, we consider the acceleration of a small mass m on the ocean's surface under the combined influence of the gravitational attraction of the earth and a distant mass M, as shown in Fig. 8-2. The coordinates of the masses m, M_E, M in an inertial frame are represented by the vectors $\mathbf{r}_1$, $\mathbf{r}_2$, $\mathbf{r}_3$, respectively. For convenience, we denote the relative coordinates of the masses by

$$\mathbf{r} = \mathbf{r}_1 - \mathbf{r}_2$$
$$\mathbf{R} = \mathbf{r}_2 - \mathbf{r}_3 \tag{8.15}$$
$$\mathbf{d} = \mathbf{r}_1 - \mathbf{r}_3 = \mathbf{R} + \mathbf{r}$$

With this notation, the motion of m and M_E due to gravitational forces is determined by

$$m\ddot{\mathbf{r}}_1 = -\frac{GmM_E\hat{\mathbf{r}}}{r^2} - \frac{GmM}{d^2}\hat{\mathbf{d}} \tag{8.16}$$

$$M_E\ddot{\mathbf{r}}_2 = -\frac{GM_E M}{R^2}\hat{\mathbf{R}} \tag{8.17}$$

By dividing the first equation by m, the second equation by M_E, and then

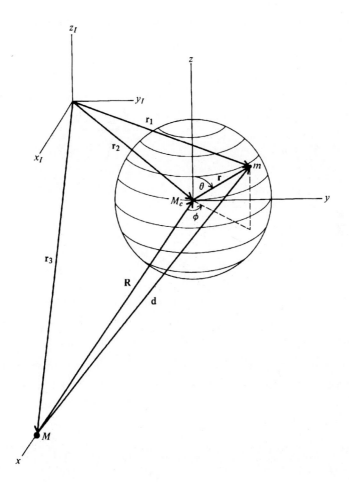

FIGURE 8-2. Location of a point on the earth's surface and a distant mass M in an inertial frame and an earth-centered frame.

subtracting, we find the equation of motion for the relative coordinate $\mathbf{r}$.

$$\ddot{\mathbf{r}} = -\frac{GM_E\hat{\mathbf{r}}}{r^2} - GM\left(\frac{\hat{\mathbf{d}}}{d^2} - \frac{\hat{\mathbf{R}}}{R^2}\right) \tag{8.18}$$

This result could have been directly obtained from (6.22). The first term on the right-hand side of (8.18) is the central gravity force of the earth on a particle of unit mass. The second term is the tide-generating force per unit mass due to the presence of the distant mass M. The tide-generating force is the difference between the forces on the surface of the earth and at the center of the earth. The direction and relative magnitude of the tide-generating force due to M are plotted in Fig. 8-3 for points around the earth's equator. The effect of this force is to produce the two tidal bulges which, as the earth rotates, are observed twice daily as high tides.

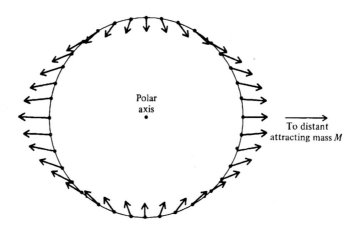

FIGURE 8-3. Tide-generating force on the surface of the earth at the equator due to a distant mass.

If the tidal forces are small compared to the gravitational force on the CM and the distance to the external force center is large compared to the planetary radius we can approximate (8.18) as follows. By (8.15) we can express the second factor of (8.18) as

$$
\frac{\hat{\mathbf{d}}}{d^2} - \frac{\hat{\mathbf{R}}}{R^2} = \frac{\mathbf{d}}{d^3} - \frac{\mathbf{R}}{R^3} = \frac{\mathbf{R} + \mathbf{r}}{d^3} - \frac{\mathbf{R}}{R^3}
$$
$$
= \mathbf{R}\left(\frac{1}{d^3} - \frac{1}{R^3}\right) + \frac{\mathbf{r}}{d^3}
$$

(8.19)

We form the square of d

$$
d^2 = R^2 + r^2 + 2\mathbf{R}\cdot\mathbf{r}
$$
$$
d = R\left(1 + \frac{2\mathbf{R}\cdot\mathbf{r}}{R^2} + \frac{r^2}{R^2}\right)^{1/2}
$$

(8.20)

Then for $R \gg r$ we apply the binomial expansion $(1+\beta)^n \simeq 1+n\beta+\cdots$, with $\beta = \mathbf{R}\cdot\mathbf{r}/R^2$ and $n = 1/2$, and retain only leading terms

$$
d \simeq R\left(1 + \frac{\mathbf{R}\cdot\mathbf{r}}{R^2} + \cdots\right)
$$

(8.21)

The quantity d^{-3} in (8.19) can be approximated by

$$\frac{1}{d^3} \simeq \frac{1}{R^3}\left(1 - \frac{3\mathbf{R}\cdot\mathbf{r}}{R^2}\right)$$

$$= \frac{1}{R^3} - \frac{3\hat{\mathbf{R}}\cdot\mathbf{r}}{R^4} \tag{8.22}$$

where the binomial expansion with $n = -3$ has been applied. To first order in $\mathbf{r}$ (8.19) becomes

$$\frac{\hat{\mathbf{d}}}{d^2} - \frac{\hat{\mathbf{R}}}{R^2} = \mathbf{R}\left(\frac{1}{d^3} - \frac{1}{R^3}\right) + \frac{\mathbf{r}}{R^3}$$

$$\simeq -\frac{3(\mathbf{R}\cdot\mathbf{r})}{R^4} + \frac{\mathbf{r}}{R^3} \tag{8.23}$$

$$\simeq \frac{1}{R^3}\left[-3\hat{\mathbf{R}}\left(\hat{\mathbf{R}}\cdot\mathbf{r}\right) + \mathbf{r}\right]$$

In our choice of coordinate system in Fig. 8-2, $\hat{\mathbf{R}} = -\hat{\mathbf{x}}$ and thus

$$\frac{\hat{\mathbf{d}}}{d^2} - \frac{\hat{\mathbf{R}}}{R^2} \simeq \frac{1}{R^3}\left(-3x\hat{\mathbf{x}} + \mathbf{r}\right) \tag{8.24}$$

In this approximation the tidal acceleration of (8.18) is

$$\ddot{\mathbf{r}} = -\frac{GM_E\hat{\mathbf{r}}}{r^2} + \frac{GM}{R^3}(3x\hat{\mathbf{x}} - \mathbf{r}) \tag{8.25}$$

Since gravitational forces are conservative this force per unit mass can be derived from a potential and we may write

$$\ddot{\mathbf{r}} \equiv -\nabla_\mathbf{r}\Phi \tag{8.26}$$

where $\nabla_\mathbf{r}$ means the gradient with respect to the vector $\mathbf{r} = x\hat{\mathbf{x}} + y\hat{\mathbf{y}} + z\hat{\mathbf{z}}$ whose origin is at the center of the earth. It is easy to guess that the potential whose negative gradient is the right side of (8.25) is

$$\Phi = -\frac{GM_E}{r} - \frac{GM}{R^3}\left(\frac{3}{2}x^2 - \frac{1}{2}r^2\right) \tag{8.27}$$

Since $x = r\sin\theta\cos\phi$, we have

$$\Phi = -\frac{GM_E}{r} - \frac{GM}{r}\left(\frac{r}{R}\right)^3\left(\frac{3}{2}\sin^2\theta\cos^2\phi - \frac{1}{2}\right) \tag{8.28}$$

For equilibrium of the ocean surface, the net tangential force on m must vanish. Equivalently, the potential at any point on the ocean's surface must be constant. We choose the constant to be $\Phi(\mathbf{r}) = -GM_E/R_E$, where R_E is the undistorted spherical radius of the earth (i.e., when the distant M is absent). Using this condition in (8.28) gives

$$r - R_E = \frac{M}{M_E} \frac{r^3 R_E}{R^3} \left(\frac{3}{2} \sin^2 \theta \cos^2 \phi - \frac{1}{2} \right) \tag{8.29}$$

Since the height of the tidal displacement

$$h(\theta, \phi) \equiv r - R_E \tag{8.30}$$

is quite small compared with R_E, (8.29) gives

$$h(\theta, \phi) \simeq \frac{M}{M_E} \frac{R_e^4}{R^3} \left(\frac{3}{2} \sin^2 \theta \cos^2 \phi - \frac{1}{2} \right) \tag{8.31}$$

For a given colatitude angle θ in (8.31), the high tides occur at $\phi = 0$ and $\phi = \pi$, and low tides occur at $\phi = \pi/2$ and $\phi = 3\pi/2$. The difference in height between high and low tide, known as the *tidal range*, is

$$\Delta h = \frac{3}{2} \frac{M}{M_E} \frac{R_e^4}{R^3} \sin^2 \theta \tag{8.32}$$

The tidal displacement h is largest at $\theta = 90°$ (on the equator). The tidal distortion is illustrated in Fig. 8-4. The tide for an ocean devoid of continents has a prolate spheroid shape (football-like), with the major axis in the direction of the distant mass. The calculation of such an ideal tide was first made by Newton in 1687.

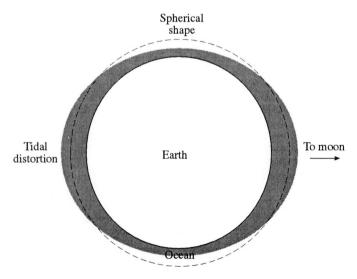

FIGURE 8-4. Tidal distortion at the earth's equator on an exaggerated scale.

The preceding discussion applies to the tidal forces induced by a single astronomic body. If there are two tide-producing bodies the net tide is the superposition of the separate tides. (If the bodies are not collinear with the planet, the total tidal shape is not axially symmetric but a triaxial ellipsoid instead.) From (8.31) the ratio of the maximum heights of the lunar (L) and solar ($\odot$) tides on earth is

$$\frac{h_L}{h_\odot} = \left(\frac{M_L}{M_\odot}\right)\left(\frac{a_E}{a_L}\right)^3 \tag{8.33}$$

where a_L is the earth-moon distance and a_E is the earth-sun distance. The numerical value of this ratio is

$$\frac{h_L}{h_\odot} = \frac{(1/81.5)M_E}{\left(\frac{1}{3}\times 10^6\right)M_E}\left(\frac{1.5\times 10^8 \text{ km}}{3.8\times 10^5 \text{ km}}\right)^3 = 2.2 \tag{8.34}$$

Thus the sun's tidal effect is smaller than the moon's, but it is not negligible. When the sun and moon are lined up (new or full moon), an especially large tide results (spring tide), and when they are at right angles (first or last quarter moon), their tidal effects partially cancel (neap tide). The diagram in Fig. 8-5 illustrates these orientations of the moon relative to the earth and sun.

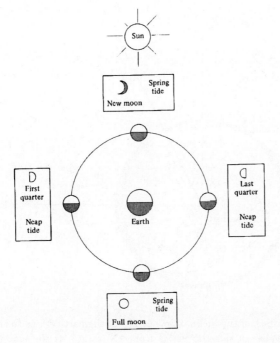

FIGURE 8-5. Relation of the phases of the moon to the tides on earth.

The tidal range due to the moon at a point on the earth-moon axis can be calculated from (8.32). We get

$$\Delta h \left(\theta = \frac{\pi}{2} \right) = \frac{3}{2} \left(\frac{1}{81.5} \right) \left(\frac{6,371}{384,000} \right)^3 (6,371 \times 10^3) = 0.56 \, \text{m}$$

This figure agrees roughly with the measured tidal difference in midocean. As the earth rotates about its own axis, the tidal maxima, which lie on the earth-moon axis, will pass a given point on the earth's surface approximately two times a day. More precisely, since the orbital rotation of the moon about the earth (with period of $27\frac{1}{3}$ days) is in the same sense as the earth's own rotation (with period 24 h), two tidal maxima pass a given spot on earth every $(24 + 24/27\frac{1}{3})$ h. Thus high tide occurs every 12 h and 26.5 min, and high tide is observed about 53 min later each day.

The two high tides are not of the same height because of the inclination of the earth's axis to the normal of the moon's orbital plane about the earth. In the Northern Hemisphere the high tide which occurs closest to the moon is higher, as illustrated in Fig. 8-6.

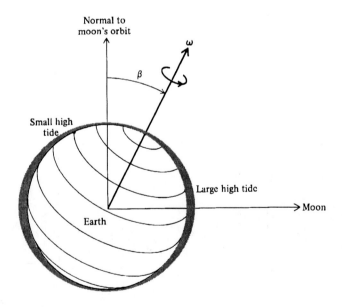

FIGURE 8-6. Effect of the inclination angle β of the earth's axis to the moon's orbital plane on the heights of tides. β varies from 17° to 29° as the moon's elliptical orbit precesses slowly about the normal to the plane of the earth's heliocentric orbit.

The tides are in reality more complicated than described above. Along coastal regions the configuration of the land masses and the ocean bottom cause considerable amplification or suppression of the tidal range. Over the world, tidal ranges vary as much as twenty meters.

The friction of the moving tidal waves against ocean bottoms and the continental shorelines dissipates energy at a rate estimated at 7 billion horsepower. To supply this energy, the earth's rotation about its axis slows down at the rate of 4.4×10^{-8} s per day. The cumulative time over a century is about 28 s. This gradual lengthening of the day is confirmed by the observation that various astronomical events such as eclipses seem to run systematically ahead of calculations based on observations over preceding centuries.

8.3   Tidal Evolution of a Planet-Moon System

The earth-moon system has very little external torque acting upon it on the average. The total angular momentum of the system is thus nearly constant. The consequence of angular momentum conservation is that the moon spirals outward about a half a centimeter each month as the earth's rotation is slowed by tidal friction. Ultimately the moon's distance will increase by over forty percent of its present value and our day will lengthen by a factor of about 50. The moon will then remain stationary above one spot on the earth.

To see this, we make the following simplifications, which are sufficiently accurate to represent the physical situation.

1. The spin angular momentum $\mathbf{S} = I\omega$ of the earth is parallel to the orbital angular momentum $\mathbf{L}$ of the moon about the earth. (The earth's spin precesses about the normal to the ecliptic plane with a period of 26,000 years and the plane of the moon's orbit about the earth precesses similarly with a period of about 19 years, so the average values of both $\mathbf{S}$ and $\mathbf{L}$ are perpendicular to the ecliptic plane—the plane of earth's orbit around the sun).

2. The total angular momentum

$$\mathbf{J} = \mathbf{L} + \mathbf{S} = (L + S)\hat{\mathbf{L}} \tag{8.35}$$

   is constant (we are neglecting the solar tidal drag).

3. The moon's orbit about the earth is circular and lies in the ecliptic plane (point 1 above).

4. The moon is much less massive than the earth and the moon's spin angular momentum is negligible.

In a reference frame with the earth at rest at the origin, the energy of the earth-moon system is

$$E = \frac{1}{2}mv^2 - \frac{\alpha}{r} + \frac{1}{2}I\omega^2 \tag{8.36}$$

where m is the mass of the moon, v is its velocity, r is its distance from the earth, $\alpha = GmM_E$, and I is the moment of inertia of the earth about its spin axis. It is useful to express E in terms of the angular momenta. The last term in (8.36) is the spin energy of the earth $S^2/(2I)$, where $S = I\omega$. The first two terms in (8.36) are the orbital energy of the moon, which can be expressed in terms of $L = mvr$ by using the circular-orbit balance of gravitational and centrifugal forces

$$m\frac{v^2}{r} = \frac{\alpha}{r^2} \tag{8.37}$$

We obtain

$$E = -\frac{m\alpha^2}{2L^2} + \frac{S^2}{2I} \tag{8.38}$$

Because the total angular momentum $J = L + S$ is conserved, we can express S as $J - L$ and thus get E expressed in terms of one independent variable quantity, L

$$E = -\frac{m\alpha^2}{2L^2} + \frac{(J-L)^2}{2I} \tag{8.39}$$

If tidal friction is present the energy E (kinetic plus potential) of the system as well as L and S are not constant. The ultimate state of this system will be the state of lowest energy. The extreme values of E with J held fixed are determined by

$$0 = \frac{dE}{dL} = \frac{M_L\alpha^2}{L^3} - \frac{(J-L)}{I} \tag{8.40}$$

Using (8.38) and $S = J - L$ this condition can be expressed as

$$\frac{L}{M_L r^2} = \frac{S}{I} \tag{8.41}$$

The left-hand side is the orbital angular velocity Ω and the right-hand side is the spin angular velocity ω so the condition (8.41) of extreme energy at fixed total angular momentum is simply *corotation*

$$\Omega = \omega \tag{8.42}$$

In general, for fixed total angular momentum J about the CM, the state of minimum energy of an isolated system is rigid rotation. (Another example is the state of water in an isolated spinning bucket. Eventually the water rotates as a rigid body with the same angular velocity as the bucket.)

At present $\Omega < \omega$ for the earth-moon system. In Fig. 8-7 we plot Ω and ω as a function of r. There are two solutions for corotation.

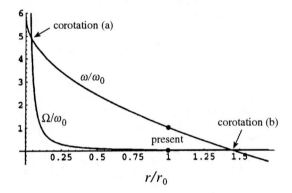

FIGURE 8-7. The spin angular velocity ω and the orbital angular velocity Ω for the earth-moon system as a function of orbital angular momentum. The subscript 0 denotes present value.

In Fig. 8-8 the energy of the earth-moon-system is plotted versus r. The two extrema correspond to the corotation points of Fig. 8-7. Case (a) is an unstable equilibrium; the bulk of the angular momentum is in the spin of the earth. Case (b) is a stable equilibrium; the bulk of the angular momentum is in the orbit of the moon. In Fig. 8-9 a more detailed plot of the energy is shown for the more immediate past and future. In the past the spin angular momentum S was larger and the orbital angular momentum L was smaller, corresponding to a higher energy for the system.

The earth's day is lengthening by 4.4×10^{-8} s/day, which corresponds to an angular acceleration of

$$\dot{\omega} = -0.85 \times 10^{-21} \text{ rad/s}^2 \tag{8.43}$$

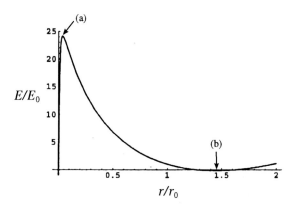

FIGURE 8-8. The energy of the earth-moon system versus the moon's orbital angular momentum for constant total angular momentum J. Here E_0 and r_0 are the present values. The labels (a) and (b) refer to the corotation points of Fig. 8-7.

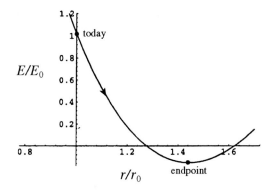

FIGURE 8-9. A blow-up of Fig. 8-8 near the present time. As the energy of the earth-moon system decreases due to tidal drag the moon's distance r increases.

Using (8.38), $L + S$ constant, and $S = I\omega$, we obtain

$$\frac{\dot{r}}{r} = \frac{2\dot{L}}{L} = -\frac{2\dot{S}}{L} = -\frac{2\dot{\omega}}{\omega}\frac{S}{L} \tag{8.44}$$

Then using the present values we find

$$\dot{r} \simeq 0.4 \text{ cm/month} \tag{8.45}$$

Thus the moon is spiraling outward roughly one-half centimeter per revolution. This process will continue until the energy reaches minimum at $r = 1.44r_0$, where r_0 is the present earth-moon separation. At this point corotation is achieved and lunar tidal drag vanishes. (From then on, solar tidal drag evolves the system.)

The torque between the earth and the moon that transfers S to L is caused by the tidal friction. The earth's rotation acts to drag the tidal bulge ahead of the line between the earth and moon, as shown in Fig. 8-10. The lead angle Δ can be calculated by equating the torque applied by the moon to the tidal bulge (which depends on Δ; it obviously vanishes for $\Delta = 0°$ or $90°$) to the torque implied by $\dot{\omega}$,

$$N = I\dot{\omega} \qquad (8.46)$$

The tidal torque on a volume element of water is

$$dN_{\text{tide}} = \rho_{\text{H}_2\text{O}} \left(R_E^2 \sin\theta \, d\theta \, d\phi \right) h(\theta, \phi) \left(-\frac{\partial \Phi_{\text{tide}}}{\partial \phi} \right) \qquad (8.47)$$

where $\rho_{\text{H}_2\text{O}}$ is the density of water and $-\frac{\partial \Phi_{\text{tide}}}{\partial \phi}$ is the torque per unit mass. From (8.31) for a tide displaced by an angle Δ as in Fig. 8-10,

$$h(\theta, \phi) = \frac{M_L}{2M_c} \left(\frac{R_E}{R} \right)^2 R_c \left[3 \sin^2\theta \, \cos^2(\phi - \Delta) - 1 \right] \qquad (8.48)$$

We then integrate (8.47) over the surface of the earth to obtain

$$N_{\text{tide}} = -\frac{6}{5} \frac{M_L G}{M_c} \rho_{\text{H}_2\text{O}} R_c^2 \left(\frac{R_c}{R} \right)^2 \sin 2\Delta \qquad (8.49)$$

Equating the two torques (8.46) and (8.49) using (8.43) gives the angle Δ that the tide leads the direction to the moon

$$\Delta \simeq 10 \text{ degrees} \qquad (8.50)$$

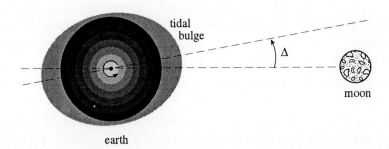

FIGURE 8-10. Earth-moon system as seen from above the north pole. Friction drags the tidal bulge ahead of the moon. This drag is opposed by a torque due to the moon's attraction.

In the past the moon was closer to the earth. At the present recession rate of 0.4 cm per month, two billion years ago the moon would have been at about three quarters of its present distance. The tidal height would have been double that at present and the increased tidal bulge would have caused larger tidal friction. On the other hand, differences in continental configurations and ocean levels might have decreased tidal drag in the distant past.

The moon could never have been closer than the *Roche limit.* According to this limit a moon having the same density as the planet will be pulled apart by tidal forces at distances closer than $R = 2.44R_E$. Most astronomical bodies are held together by their self-gravity, which is stronger for large bodies than the chemical forces that hold rocks together. As a satellite comes within the Roche limit tidal forces overcome the self-gravity and the satellite falls apart.

8.4   General Relativity: The Theory of Gravity

Einstein's theory of general relativity is a theory of gravity. At this level we do not have the mathematical tools to completely discuss the theory because it is expressed most naturally in the language of metric differential geometry. We can however illustrate some of the physical ideas which underlie general relativity and explore a few instances in which it differs from Newtonian gravity. These differences can be dramatic in very intense gravity fields.

A. The Principle of Equivalence

There are two aspects of mass: inertia as it appears in the second law and a proportionality constant in the gravity force. The equivalence of the two has the important consequence that all objects fall equally in a gravity field. Newton tested this hypothesis by verifying that pendulum bobs made of different materials have the same period to roughly 1 part in a thousand. Modern tests of the equivalence principle have improved this limit to one part in 10^{12}.

This remarkable equivalence led Albert Einstein to propose that locally (*i.e.*, at any given point) one cannot distinguish between the acceleration of a reference frame (*e.g.*, in an elevator) and gravitational force. Free fall is indistinguishable from being located in a gravity-free region. The value of $\mathbf{g} = -\boldsymbol{\nabla}\Phi$ is a frame-dependent quantity. The general principle of relativity requires that in free fall all physical laws reduce to those in an inertial frame.

B. Gravitational Frequency Shift

One of the most direct implications of the principle of equivalence is that "higher clocks run faster." According to Einstein's theory, if waves are emitted on the earth with frequency ν as in Fig. 8-11(a) they arrive a distance h below with frequency ν' where

$$\nu' = \nu \left(1 + \frac{gh}{c^2}\right) \tag{8.51}$$

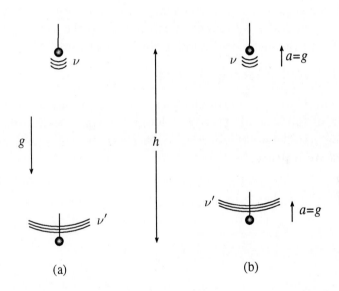

(a)   (b)

FIGURE 8-11.   Gravitational frequency shift. In (a) the lower observer receives waves at higher frequency. In (b) the equivalence principle relates this shift to a Doppler shift. The clocks are supported against gravity in (a), and accelerated in (b), by strings.

To see how this comes about we invoke the principle of equivalence. The same frequency shift will occur in the situation of Fig. 8-11(b), where both emitter and receiver are being accelerated upward with $a = g$ in a region far from the earth. Consider waves emitted at $t = 0$ with frequency ν. When these waves reach the receiver a time h/c later, the receiver is moving faster by velocity $\Delta v = g(h/c)$ than the emitter at the time of emission. The waves will therefore be Doppler shifted to a higher frequency by the familiar formula

$$\nu' = \nu \left(1 + \frac{\Delta v}{c}\right) = \nu \left(1 + \frac{gh}{c^2}\right) \tag{8.52}$$

which is the desired result (8.51). On the earth, for $h = 100\,\text{m}$, the Doppler velocity is

$$\Delta v = gh/c \simeq 9.8(100)/(3 \times 10^8)\,\text{m/s} \\ \simeq 1\,\text{cm/hour} \tag{8.53}$$

This very small gravitational frequency shift has been verified to 1% accuracy. If ν and ν' are thought of as the rate of the ticking of a clock, a higher clock runs faster and a lower clock runs slower.

C. Gravitational Bending of Light

Light appears to travel in straight lines, but this is only because we live in a region of fairly weak gravity. The gravitational deflection of light propagating parallel to the earth's surface can be computed using the principle of equivalence.

Referring to Fig. 8-12(a), we imagine a laboratory set up in a rocket ship. While the rocket is on the earth, Einstein's theory predicts that a light beam traveling a distance ℓ across the ship is deflected by a distance

$$d = \tfrac{1}{2}g(\ell/c)^2 \tag{8.54}$$

Now we imagine the rocket ship to be far from the earth, as in Fig. 8-12(b); in the absence of gravity the deflection is zero. If the rockets are then fired to give an acceleration $a = g$, the ship, during the light-transit time ℓ/c, moves forward by a distance $\tfrac{1}{2}g(\ell/c)^2$ compared to where it would have been in free fall, so the light hits the other side of the ship at a distance $d = \tfrac{1}{2}g(\ell/c)^2$ below the receiving point of (b). This is just the result (8.54).

In a non-uniform gravity field the equivalence principle implies a deflection of light which can be straightforwardly found by integrating up the contributions to the deflection from each part of the light's path. However, there is an additional contribution, not predictable from the equivalence principle, from the curvature of space. In Einstein's theory, for light passing a massive body these two contributions happen to be

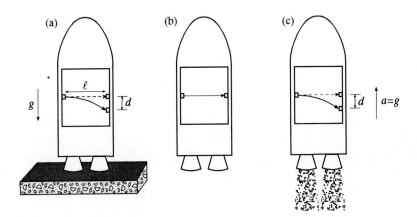

FIGURE 8-12. Bending of light in uniform non-inertial fields. In (a) the rocket ship rests on the earth and a deflection d due to gravity is observed. In (b) the ship is moving at constant velocity with no gravity present and no deflection is seen. When the rockets are fired in (c) the acceleration g of the ship results in the same deflection d.

just equal. The result for a light ray passing a spherical mass M at an impact parameter b [see Fig. 8-13] is a deflection angle

$$\phi = 2r_S/b \tag{8.55}$$

if $b \gg r_S$, where r_S is the *Schwarzschild radius*

$$r_S = 2GM/c^2 \tag{8.56}$$

One of the original empirical successes of the general relativity theory was the observation of the deflection of starlight by the sun during a total solar eclipse. In this case the maximum deflection angle (for light just grazing the sun) is 1.8 arc seconds.

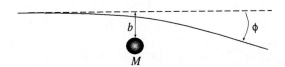

FIGURE 8-13. Light ray passing a mass M at impact parameter b. The light is deflected by the angle ϕ.

As the impact parameter b of the incident light is decreased to a critical value

$$b_{\text{crit}} = \sqrt{\frac{27}{4}} r_S \tag{8.57}$$

the deflection angle increases without limit. The light spirals into an unstable circular orbit around the mass. For $b < b_{\text{crit}}$ the light goes in but never comes back out.

At distances r approaching r_S the gravitational acceleration becomes infinite and the gravitational redshift becomes infinite. If the radius of an object of mass M decreases to the Schwarzschild radius, nothing can prevent its further gravitational collapse to a point singularity (a *black hole*). We cannot directly observe the singularity because the gravitational attraction is so strong that nothing, not even light, can emerge from it. Table 1 gives r_S for some familiar astronomical objects.

Table 1: Schwarzschild radius of our earth, sun and galaxy

	earth	sun	our galaxy
r_S	0.9 cm	3 km	3×10^{11} km
			(2000 AU or 1/100 ly)

Once a black hole has formed it is a one way street—things can go in but nothing can come out. Black holes can in principle be of any mass but the most obvious mechanism for producing them starts with a sufficiently massive star. The gravity of a spent star of sufficient core mass ($\gtrsim 3M_\odot$) is so strong that nothing can prevent collapse to a point. Super-massive black holes apparently form at the centers of galaxies (including our own) as a normal by-product of galactic formation. The high velocity of stars near the center of many galaxies and a hot gas disk are evidence for black holes containing millions of solar masses.

D. Gravitational Lenses

Massive point sources of gravity deflect light and can act as lenses that distort the appearance of objects behind them and produce characteristic images. The simplest effect occurs when light rays (or radio waves) from a very distant point source pass near a galaxy and continue toward the earth as shown in Fig. 8-14. From the geometry of this figure and the deflection (8.55) we obtain

$$\phi = \frac{b}{d} = \frac{2r_S}{\phi d} \tag{8.58}$$

Thus ϕ is given by

$$\phi = \sqrt{\frac{2r_S}{d}} \tag{8.59}$$

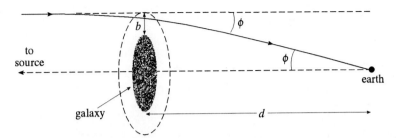

FIGURE 8-14. A light ray from a very distant source, usually a quasar, passes near a distant galaxy and is deflected to an observer on earth. The scale of the transverse dimension in this drawing is greatly exaggerated.

If the source lies directly behind the lensing galaxy an *Einstein ring* of angular radius ϕ would be observed. Only a few such rings have been seen. If the source is near to being collinear with the lensing galaxy, its image forms a partial arc. An example of an Einstein ring is shown in Fig. 8-15. If the mass of lensing object is small the ring will not be resolvable but a brightening of the source will be seen as the lensing object, the earth and the source become collinear. This *microlensing* effect has provided direct evidence for dark matter in our galaxy in the form of small stars called *brown dwarfs*.

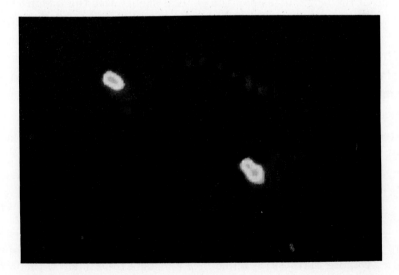

FIGURE 8-15. Einstein ring produced by a galaxy serving as a gravitational lens (not visible). The object MG 113+0456 consists of a hot core and produces two luminous lobes on either side. The lens images one lobe into an Einstein ring and the core into two bright spots. The other lobe is far enough away that it is not affected by the lens. This picture is taken with 15 GHz (2 cm) waves by the Very Large Array radio telescope.

8.5 Planetary Motion—Perihelion Advance

The orbits of planets as predicted from general relativity are almost identical to the Newtonian orbits discussed in Chapter 5. The effective potential energy in general relativity that governs the shape of the orbit of a small mass m in the presence of a large mass M is

$$V_{\text{eff}}(r) = -\frac{\alpha}{r} + \frac{L^2}{2mr^2} - \frac{r_S L^2}{2mr^3} \tag{8.60}$$

The last two terms of (8.60) can be written as

$$V_{cf} = \frac{L^2}{2mr^2}\left(1 - \frac{r_S}{r}\right) \tag{8.61}$$

Under normal circumstances r_S/r is very small. For example, for the planet Mercury $r_S/r \simeq 5 \times 10^{-8}$. Extraordinary lengths must therefore be taken to test general relativity using planetary motion. On the other hand, when matter is so compressed that r can be comparable to r_S, orbits can be very different from Newtonian.

One of the original tests of the general theory was the *advance of the orbit of Mercury*. In Newtonian theory a planet moves in a fixed elliptical orbit. The extra r_S term in (8.60) causes the major axis of an elliptical orbit to precess in the direction of motion, or *advance*. The effect is cumulative so that observations of Mercury over centuries allowed even this small effect to be seen. In what follows we briefly indicate how this advance is calculated.

Referring to § 5.2 we repeat the same steps using the potential energy (8.60) to find the relativistic orbit equation

$$\left(\frac{du}{d\theta}\right)^2 = f(u) = -u^2 + \frac{2m\alpha}{L^2}u + \frac{2mE}{L^2} + r_S u^3 \tag{8.62}$$

where $u = 1/r$. Except for the $r_S u^3$ term, this equation is the same as (5.50). We note that $f(u)$ is cubic in u and by a fundamental theorem of algebra can be factored as

$$f(u) = r_S(u - u_1)(u - u_2)(u - u_3) \tag{8.63}$$

We assume the roots are ordered as $u_1 < u_2 < u_3$. Comparing powers of

u between (8.62) and (8.63) we obtain

$$u_1 + u_2 + u_3 = \frac{1}{r_S}$$

$$u_1 u_2 + u_1 u_3 + u_2 u_3 = \frac{2m\alpha}{L^2 r_S} \qquad (8.64)$$

$$u_1 u_2 u_3 = -\frac{2mE}{L^2 r_S}$$

In the Newtonian limit $r_S = 0$ the largest root u_3 becomes infinite.

As the planet orbits, u varies between u_1 and u_2 and by integrating (8.62) using the factored form (8.63) the angle swept out during a complete revolution is

$$\theta = 2 \int_{u_1}^{u_2} \frac{du}{\sqrt{r_S(u - u_1)(u - u_2)(u - u_3)}} \qquad (8.65)$$

This is an elliptic integral but for small r_S we can approximate the integral since $u \ll u_3$. Using binomial expansion and retaining the leading correction of order r_S the turning angle after one radial period is by (8.65)

$$\theta = \frac{2}{\sqrt{r_S u_3}} \int_{u_1}^{u_2} \frac{du \left(1 + \frac{1}{2} u/u_3\right)}{\sqrt{(u - u_1)(u_2 - u)}} \qquad (8.66)$$

Using

$$\int_{u_1}^{u_2} \frac{du}{\sqrt{(u - u_1)(u_2 - u)}} = \pi, \qquad \int_{u_1}^{u_2} \frac{u\,du}{\sqrt{(u - u_1)(u_2 - u)}} = \frac{\pi}{2}(u_1 + u_2) \qquad (8.67)$$

(8.66) becomes

$$\theta = \frac{2\pi}{\sqrt{r_S u_3}} \left(1 + \frac{u_1 + u_2}{4 u_3}\right) \qquad (8.68)$$

In the Newtonian limit $r_S = 0$, the roots of $f(u)$ in (8.62) are

$$u_{1,2} = \frac{m\alpha}{L^2}(1 \pm \epsilon) \qquad (8.69)$$

where $\epsilon = \sqrt{1 + 2EL^2/m\alpha^2}$. For small r_S, the roots u_1 and u_2 are approximately given by (8.69) and

$$u_1 + u_2 = \frac{2m\alpha}{L^2} + \mathcal{O}(r_S) \tag{8.70}$$

and by the first equation of (8.64) we have

$$r_S u_3 = 1 - \frac{2m\alpha r_S}{L^2} + \mathcal{O}(r_S) \tag{8.71}$$

The orbit turning angle formula (8.68) becomes

$$\theta - 2\pi = \frac{3\pi m\alpha}{L^2} r_S \tag{8.72}$$

Since θ exceeds 2π the orbit "advances." In terms of the orbit major axis

$$2a = r_1 + r_2 = \frac{1}{u_1} + \frac{1}{u_2}$$
$$= \frac{L^2}{m\alpha} \frac{2}{1 - \epsilon^2} \tag{8.73}$$

the orbital advance formula (8.72) becomes

$$\theta - 2\pi = \frac{3\pi}{1 - \epsilon^2} \frac{r_S}{a} \text{ rad/orbit} \tag{8.74}$$

For Mercury $\epsilon = 0.2056$, $a = 0.579 \times 10^8$ km, and the orbit period is 0.241 yr. With $r_S = 2.95$ km for the sun, the advance in arc seconds per century is then

$$\theta - 2\pi \simeq 43.0 \text{ arc seconds/century} \tag{8.75}$$

The observed advance of Mercury's orbit is 5597.7 arc-sec/century but almost all is due to perturbations of the other planets, leaving 43.1 ± 0.21 arc seconds unaccounted for by Newtonian physics. The correct explanation for the extra observed perihelion advance was the first confirmation of Einstein's general relativity theory.

An astonishing astronomical object (PSR 1913+16), which was discovered in 1974 by Hulse and Taylor, has significant general relativistic effects. This system consists of a neutron star *pulsar* of mass $1.4M_\odot$ orbiting with a period of roughly 8 hours about another neutron star of similar mass with a separation not much greater than the Sun's diameter. A pulsar is a rotating neutron star that emits radiation which arrives as periodic pulses. It is thought that the pulsing is due to the radiation being emitted in a narrow beam which rotates with the star and so sweeps past us periodically like a lighthouse beacon. Because the pulsar's radio pulses are a very good (stable) clock, measurement of the time variation of its Doppler shift yields the orbital parameters to a precision of at least 1 part in 10^6. For example, the eccentricity is $\epsilon = 0.6171308$. This system of two orbiting neutron stars is a laboratory for testing general relativity under conditions of much stronger gravity fields than occur in our solar system. For example, the advance of the orbit in this system is 4.22662 degrees/yr, due entirely to general relativity.

Because the mass components of the system are accelerating, general relativity predicts that energy is lost in the form of gravitational radiation, with the consequence that the orbit period decreases. The observation of this is in excellent agreement with the prediction and provides the first evidence (albeit indirect) of *gravity waves*. The ultimate fate of PSR 1913+16 is that in roughly one billion years the two neutron stars will coalesce (possibly into a black hole), emitting a burst of gravity waves. In this process about 3% of a solar mass will be converted into gravitational radiation.

8.6 Self-Gravitating Bodies: Stars

The sun and other stars are isolated bodies in which self-gravity (the mutual gravitational attraction of their parts) is important. For a detailed description of their internal equilibrium, gravitational force must be everywhere balanced by the pressure gradient. But the virial theorem (derived in §9.3) can give a simple overall view. This theorem states that for an isolated body in mechanical equilibrium (neither collapsing nor expanding) in which gravity is the only important force, the total kinetic energy $K = \sum_A \frac{1}{2} m_A v_A^2$ of its parts and its total gravitational energy $V = -\sum_{\text{pairs}} G m_A m_B / r_{AB}$ are related by

$$K = -\tfrac{1}{2}V \tag{8.76}$$

Its total energy $E = K + V$ can then be expressed as

$$-E = K = -\tfrac{1}{2}V \tag{8.77}$$

Note that $E < 0$. If the material of the body were widely separated and at rest, it would have $V = 0$ and $K = 0$, so the quantity $-E$ is the binding energy of the body. According to (8.77), if the body loses energy Δ (*e.g.*, by radiation), then its internal kinetic energy K increases by an equal amount. These energies are supplied from the potential energy V which changes by -2Δ (by the body shrinking in size).

We next show how (8.76) can be used to give an estimate of the temperature inside a star. In thermal equilibrium, the kinetic energy of a particle is $\frac{3}{2}k_BT$ where k_B is Boltzmann's constant and T is the absolute temperature, so the left-hand side of (8.76) is

$$K \simeq Nk_BT \qquad (8.78)$$

where N is the number of particles composing the star and T now means an average temperature in the star. The factor $\frac{3}{2}$ has been dropped since from now on in this section we make only a semiquantitative discussion.

The number N can be written

$$N = M/m \qquad (8.79)$$

where M is the mass of the star and m is the average mass per particle. An ordinary star like the sun is mostly made of protons and electrons (ionized hydrogen) and so m is the order of the proton mass. On the right-hand side of (8.77) we can write

$$-\tfrac{1}{2}V \simeq GM^2/R \qquad (8.80)$$

where R is the radius of the star. Then (8.76) takes the form

$$Rk_BT \simeq GMm \qquad (8.81)$$

or, written as an equality of dimensionless ratios,

$$\frac{k_BT}{mc^2} \simeq \frac{r_S}{R} \qquad (8.82)$$

where r_S is the Schwarzschild radius of the mass M, defined in (8.56). For the sun $r_S/R \approx 10^{-6}$, and so its mean temperature is given by $k_BT \approx 10^{-6}mc^2 \approx 10^{-6} \times 10^9\,\mathrm{eV} \approx 10^3\,\mathrm{eV}$, *i.e.* $T \approx 10^7$ K. This is much higher than its surface temperature, 6000 K. (According to (8.81), if the 6000 K surface temperature held throughout, the sun would be a thousand times larger than its actual size.) The reason why the mean temperature is higher than the surface temperature is that the heat energy radiated away from the surface must be supplied by conduction (or convection) from within, which means that the temperature must rise inwards from the surface. The energy loss from the surface is supplied by the release of energy by nuclear fusion (hydrogen to helium) in the core of the sun.

The history of the sun (like other stars) is that it started as a gas cloud which lost energy by radiation, and so by (8.77) became continuously hotter (larger K) and denser (smaller R, so that $-V$ is larger) until the core temperature was high enough that the rate of energy release from nuclear fusion balanced the rate of energy loss from the surface. Since that time the sun has been in a quite steady state, as it will continue to be until the hydrogen (protons) in its core gets used up. It will then shrink and heat up, to maintain the fusion rate of hydrogen outside the core. (More precisely, although the bulk of the sun's mass, in the inner $1/10$ or so of its volume, will shrink and heat up, the outer part will expand and cool. This is not easy to explain in simple terms.) Eventually the core temperature will rise to the point where fusion of helium to carbon releases energy at a significant rate. To explain what happens next, and what happens to stars either less or more massive than the sun, a new phenomenon must be included in the virial theorem, namely the so-called degeneracy pressure.

According to the *Pauli principle* in quantum mechanics, each electron must occupy at least a volume in phase space (space volume times momentum volume) of magnitude h^3, where h is Planck's constant. Thus the N electrons in a star, in a space volume $\approx R^3$, must occupy a momentum volume of at least Nh^2/R^3. If the temperature is low, then the electrons will be in the volume of momentum space which has the lowest energy, namely a sphere centered on the origin, with radius p_F (the Fermi momentum) given by

$$p_F^3 \approx Nh^3/R^3 \qquad (8.83)$$

Such a cold gas of electrons is said to be Fermi degenerate. (An example is the conduction electrons in a metal. Another example, though complicated by a spatially varying electrostatic potential, is the electrons in an atom.) Thus the total kinetic energy of the electrons of a cold star is of the order of Np_F^2/m_e. On the other hand, the Pauli principle is unimportant at a high temperature, $k_B T \gg p_F^2/m_e$. Roughly, the kinetic energy of the electrons can be written as simply as the sum

$$K \approx N\left(p_F^2/m_e + k_B T\right) \qquad (8.84)$$

with p_F given by (8.83), which has the right low and high temperature limits. Using this in the virial equation we find that (8.81) is altered to

$$(M/m)^{2/3}\, h^2/(m_e R) + Rk_B T \simeq GMm \tag{8.85}$$

This has the form

$$AR^{-1} + RT = B \tag{8.86}$$

(where A and B are constant for a given star) and so the dependence of T on R is

$$T = BR^{-1} - AR^{-2} \tag{8.87}$$

This says that T has a maximum value

$$T_{\max} = B^2/A, \quad \text{at } R = 2A/B \tag{8.88}$$

Thus as the star loses energy and shrinks it stops heating up when $R \simeq 2A/B$; as it shrinks further it cools. The result is called a white dwarf. According to (8.85) with $T \simeq 0$, its radius *decreases* with increasing mass like $R \propto M^{-1/3}$.

The expression for $T_{\max}$ given in (8.88) can be conveniently written as

$$\frac{k_B T_{\max}}{m_e c^2} \simeq (M/M_0)^{4/3}, \quad \text{where } M_0 = (hc/G)^{3/2} m^{-2} = 6 \times 10^{31}\ \text{kg} \tag{8.89}$$

(It is not a coincidence that M_0 roughly equals the mass of the sun, but we cannot explain this here.) Thus, the more massive a star is, the higher mean temperature it achieves. A sufficiently light star never heats up to the point where nuclear fusion becomes a significant energy source; a proper calculation shows this happens for $M \lesssim 0.01 M_\odot$.

The rough formula (8.85) fails for M too small or too large. If $M \lesssim 0.001 M_\odot$, p_F given by (8.83) is smaller than the mean momentum of electrons in atoms. The consequence is that the white dwarf is made of packed atoms, that is, it is an ordinary liquid or solid. On the other hand, if $M \gtrsim M_\odot$ then p_F is larger than m_e, which means that the electrons are relativistic, that is, their velocities are close to the velocity of light c. If relativistic mechanics is used in the derivation of the virial theorem, the total kinetic energy $K = \sum \frac{1}{2} m v^2 = \frac{1}{2} \sum p^2/m$ gets replaced by $\frac{1}{2} \sum \mathbf{p} \cdot \mathbf{v}$, where $\mathbf{p}$ is particle momentum, $\mathbf{p} = m\gamma \mathbf{v}$ [see (10.57)]. So when $p_F > m_e$

the term p_F^2/m_e in (8.84) should be replaced by $p_F c$, with the result that for sufficiently small R the first term in (8.85) and (8.86) stops growing like R^{-1} and becomes independent of R.

This has two consequences. One is that no $T \simeq 0$ equilibrium is possible if M is too large. A proper calculation concludes that a white dwarf composed of helium, carbon or oxygen cannot have a mass greater than $1.4 M_\odot$; this is called the *Chandrasekhar limit*. The other consequence is that when a sufficiently massive star shrinks, the first term in (8.85) never gets as large as the right-hand side, and so the star's temperature rises without limit as its radius shrinks to zero.

As the temperature of a massive star rises, eventually two energy processes which *absorb* energy from its core become important, namely dissociation of the heavy nuclei back into lighter ones and radiation of neutrinos. The consequence is an essentially free-fall collapse of the core. If the star is not too massive ($M < 15 M_\odot$, it is estimated) the collapse is stopped by the short-range ('hard-core') repulsion between nuclei. Some of the gravitational energy released in the collapse is transferred to the outer part of the star, making a supernova. The collapsed core becomes a neutron star. A neutron star has a radius of the order of thousand times smaller than a white dwarf of the same mass (and has a correspondingly larger binding energy) because the degenerate fermions which support it against gravity are neutrons rather than electrons. [Replace m_e by m_n in (8.85).] It is believed that if the star is too massive, the 'hard-core' repulsion of the nuclei will be unable to stop the collapse of the core and result will be a black hole.

The supernova is one of the most spectacular of all cosmic events and yet it is a natural stage in the evolution of heavy stars. For a few days or weeks a single star's light output rivals the combined output of the ten billion stars of a large galaxy; the energy comes from gravity. The outer part of the progenitor star which is blown off in a supernova event forms a cloud or *nebula*, called a supernova remnant. An example is the "Crab" nebula shown in Fig. 8-16. An important by-product of supernovae explosions is that they are the origin of elements heavier than helium. (Although elements up to iron are made in the cores of the less massive stars which become white dwarfs, most of this material remains buried forever in the white dwarfs.) The oldest stars in our galaxy have little of the heavier elements whereas our solar system condensed more recently from a gas cloud enriched by supernovae.

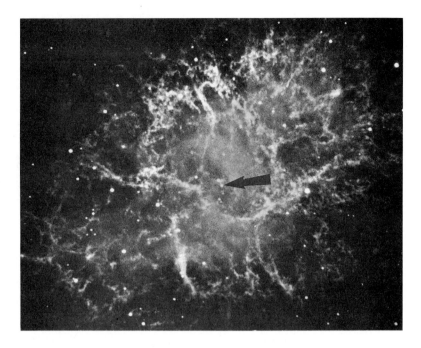

FIGURE 8-16. The Crab Nebula. The star indicated by the arrow became a supernova, which was observed in China in the year 1054. Its outer layers were blown off and a pulsar (spinning neutron star) was left. The Crab pulsar spins 30 times per second. Photo courtesy of Lick Observatory.

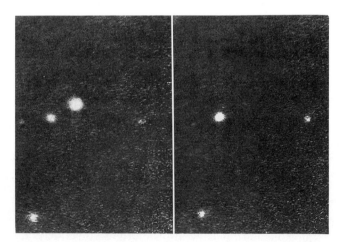

FIGURE 8-17. An optical light picture of the Crab pulsar on and off. Although most pulsars have been discovered by their radio emission the above photo of the Crab pulsar is taken in visible light at maximum and minimum intensity. The Crab pulsar has also been seen via very energetic gamma rays (~ 1 TeV). Photo courtesy of Lick Observatory.

After a supernova explosion much of the magnetic field and angular momentum of the progenitor star remains in the neutron star which has a very large magnetic moment and is spinning very rapidly. The rotation period can be as short as a few milliseconds. The rotating magnetic moment radiates low frequency waves at τ^{-1} Hz, where τ is the rotation period. These waves accelerate electrons in the star's atmosphere to relativistic speeds and these electrons in turn radiate over a broad frequency spectrum. The pulsar in the Crab Nebula is indicated in Fig. 8-17.

PROBLEMS

8.1 Attraction of a Spherical Body: Newton's Theorem

8-1. The density of a spherical planet of radius R with a molten core of radius $\frac{1}{2}R$ is given by ρ for $\frac{1}{2}r < r < R$ and 5ρ for $r < \frac{1}{2}R$, where ρ is a constant. Find:

a) the total mass M in terms of ρ and R,

b) the enclosed mass $M(r)$ in terms of M and R,

c) the force per unit mass inside or outside the planet,

d) the gravitational potential for any distance from the planet's center. Why must the potential match at the boundaries between density changes?

8-2. Find the r dependence of the mass density $\rho(r)$ of a planet for which the gravitational force has constant magnitude throughout its interior.

8-3. If a narrow tunnel were dug through the earth along a diameter, show that the motion of a particle in the tunnel would be simple harmonic. Compare the period to the orbital period of a satellite in a circular orbit close to the earth. Assume that the density of the earth is uniform and neglect the earth's rotation.

8-4. The gravitational attraction due to a nearby mountain range might be expected to cause a plumb bob to hang at an angle slightly different from vertical. If a mountain range could be represented by an infinite half-cylinder of radius a and density ρ_M lying on a flat plane, show that a plumb bob at a distance r_0 from the cylinder axis would be deflected by an angle $\theta \approx \pi a^2 G \rho_M/(r_0 g)$. In actual measurements of this effect, the observed deflection is much smaller. Next assume that the mountain range can be represented by a cylinder of radius a and density ρ_M which is floating in a fluid of density $2\rho_M$, as illustrated. Show that the plumb-bob deflection due to the mountain range is zero in this model. Since the latter result is in much better agreement with observations, it is postulated that mountains, and also continents, are in isostatic equilibrium with the underlying mantle-rock.

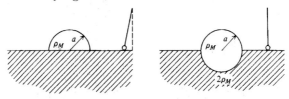

8-5. Show explicitly that the torque due to any gravitational force acting on a spherically symmetric body vanishes. *Hint: an arbitrary gravity field is produced by a superposition of point sources.*

8-6. The *center of gravity* of a system of particles is defined by $\mathbf{N} = \mathbf{R}_G \times \mathbf{F}$ where $\mathbf{F} = \sum m_i \mathbf{g}_i$ is the external forces on the system and $\mathbf{N} = \sum_i m_i \mathbf{r}_i \times \mathbf{g}$ is the torque about the coordinate origin. For a uniform external field $\mathbf{g}_i \equiv \mathbf{g}$ show that the center of gravity and the center of mass are the same point.

8.2 The Tides

8-7. The moon and sun both appear to have nearly the same angular size as viewed from earth. From this fact and the observed tidal maximum ratio what is the implied ratio of average densities? Use the data in the Appendices to check this.

8-8. Pulsars are thought to be rapidly rotating neutron stars. The Crab nebula pulsar has a radius of about 10 km, a mass of about one solar mass, and revolves at a rate of 30 times per second. Find the nearest distance that a man 2 m tall could approach the pulsar without being pulled apart. Assume that his body mass is uniformly distributed along his height, his feet point toward the pulsar, and dismemberment begins when the force that each half of his body exerts on the other exceeds ten times his body weight on earth. What is the period of revolution in a circular orbit about the pulsar at this minimum distance?

8-9. The Crab pulsar mentioned in the previous exercise has a period which increases by 36.526 ns/day. Compute the power loss in rotational kinetic energy in Watts. This power is converted to electromagnetic energy which illuminates the entire nebula.

8.3 Tidal Evolution of a Planet-Moon System

8-10. Find the critical total angular momentum J_c below which corotation does not occur. *Hint: at the critical point the maximum and minimum coalesce so $\partial^2 E/\partial L^2 = 0$.*

8-11. At the present time for the earth-moon system

$$\omega_0 = \frac{2\pi}{1\,\text{day}} = 0.727 \times 10^{-4}\,\text{s}^{-1}$$

$$\Omega_0 = \frac{2\pi}{27.3\,\text{days}} = 2.66 \times 10^{-6}\,\text{s}^{-1}$$

$$L_0 = M_L \Omega_0 r_0^2 = 2.87 \times 10^{34}\,\text{kg m}^2/\text{s}$$

$$S_0 = I\omega_0 = 0.586 \times 10^{34}\,\text{kg m}^2/\text{s}$$

$$r_0 = 1.495 \times 10^8\,\text{km}$$

The moment of inertia of the earth is given by $I \simeq \frac{1}{3} M_E R_E^2$, where the factor of $\frac{1}{3}$ reflects the actual mass distribution within the earth. Show that:

a) The orbital angular velocity is $\frac{\Omega}{\Omega_0} = \left(\frac{r_0}{r}\right)^{3/2}$ and the spin angular velocity is given by $\frac{\omega}{\omega_0} = 5.86 - 4.86 \left(\frac{r}{r_0}\right)^{1/2}$.

b) The present energy is

$$E_0 = 1.75 \times 10^{29}\,\text{J}$$

and the ratio $\frac{E}{E_0}$ is

$$\frac{E}{E_0} = -0.218\frac{r_0}{r} + 29.22\left(1.206 - \sqrt{\frac{r}{r_0}}\right)^2$$

8-12. Repeat the analysis of § 8.3 for the two moons of Mars. The necessary data are

	Mars	Phobos	Deimos
mass	$0.108 M_E$	$1.8 \times 10^{-7} M_L$	$2.4 \times 10^{-8} M_L$
period	1.03 d	0.319 d	1.263 d
radius/distance	$0.52 R_E$	9.4×10^6 m	2.35×10^7 m

Both moons rotate in the same sense as the spin of Mars. Show that the moons are near the unstable corotation solution. What is the eventual fate of each?

8.4 General Relativity: The Theory of Gravity

8-13. Rederive the equivalence principle result for the gravitational frequency shift and light deflection in a uniform gravity field by considering a frame at rest on the earth's surface and a frame in free fall near the earth's surface.

8-14. An object O is lensed by a galaxy G having Schwarzschild radius r_S. The observer, object and galaxy are collinear and at the distances shown. Using a small angle approximation, find an expression for the angular size δ of the Einstein ring in terms of r_S, d_1, and d_2.

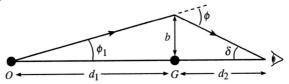

8.6 Self-Gravitating Bodies: Stars

8-15. Show that the gravitational energy $-U$ released in the collapse of a large cloud of mass M into a sphere of radius R and uniform density is

$$U = -\frac{3}{5}\frac{GM^2}{R}$$

where U is the resulting gravitational potential energy. *Hint: bring in a shell of mass dm from infinity and distribute it uniformly on the surface of a sphere of radius r where dm = ρ(4πr²dr). Integrate up to radius R.*

8-16. A gas has n particles per unit volume.

a) Assume the particles move along the x axis with velocity v. Show that the pressure against a wall would be

$$P = nvp$$

where p is the particle's momentum.

b) Show that if the particles obey the Pauli principle the degeneracy pressure is of the order of

$$P_d = \frac{h^2}{m}(n)^{5/3}$$

where m is the mass of a particle and h is Planck's constant.

In the following ignore dimensionless numerical factors.

c) For a fully ionized plasma the electron number density n_e is related to the number of ions by $n_+ = n_e/Z$ where Z is the charge of the ion. Show that the electron degeneracy pressure can be expressed in terms of the radius and mass of a dwarf star by

$$P_d^e = \frac{h^2}{m_e} \left(\frac{Z}{A m_N} \right)^{5/3} \frac{M^{5/3}}{R^5}$$

where A is the number of nucleons in the ion, m_N is the nucleon mass, M the mass of the star and R the radius of the star.

d) The gravitational pressure is roughly

$$P_g = \left(\frac{GM^2}{R^2} \right) \Big/ R^2 = \frac{GM^2}{R^4}$$

For pressure balance in a degenerate dwarf star show that

$$MR^3 = \text{constant}$$

and that the density is proportional to M^2. Thus doubling the mass of such a star causes its volume to reduce by a factor of two.

e) A white dwarf is mostly carbon and oxygen $(A = 2Z)$ when it collapses whereas in a neutron star all the protons convert to neutrons $(A = Z)$. Show that the radius of a neutron star is smaller than a white dwarf of the same mass by a factor

$$\left(\frac{m_e}{m_N} \right) \left(\frac{A}{Z} \right)^{5/3} \sim \frac{1}{600}$$

f) Repeat the arguments leading to part c) except now assume that the degenerate electron gas is relativistic $(v \simeq c)$. Equate this relativistic degeneracy pressure as in d) to obtain a crude estimate of the *Chandrasekhar limit* on the largest possible mass of a white dwarf.

g) Find the ratio of Chandrasekhar limits for neutron and white dwarf stars, neglecting nuclear forces and general relativistic effects.

8-17. Consider a plot of mass density as a function of mass for ordinary matter [density independent of mass], degenerate matter [using part d) of the preceding problem], and an object whose radius is its Schwarzschild radius. As mass increases discuss the evolution on this graph and the inevitability of gravitational collapse. With the aid of this graph, discuss the consequences of increasing the mass. Does a limit to the maximum mass appear?

Chapter 9

NEWTONIAN COSMOLOGY

Newton's laws of motion have been applied in previous chapters to physical problems involving dynamical systems ranging in size from molecules to the solar system. In this chapter we consider the dynamics and the nature of the observable universe. While Einstein's theory of general relativity is needed to fully account for physics at these large scales, Newtonian mechanics still explains many aspects of the phenomena. Also discussed in this chapter is the existence of *dark matter*. There is compelling evidence that most of the mass of the universe is invisible. Its presence is inferred through its gravitational effects on ordinary matter.

9.1 The Expansion of the Universe

A seminal observation was made in 1928 by Hubble and Slipher that the spectrum of the light from distant galaxies is shifted towards the red by an amount proportional to the distance to the galaxy. They interpreted this redshift as a Doppler shift due to the galaxies receding from us with speeds proportional to their distance. If light emitted by an atom at rest is observed to have wavelength λ_0, when the atom is moving away with a velocity v the light will be observed to have a longer wavelength $\lambda = \lambda_0 + \Delta\lambda$, where approximately

$$\frac{\Delta\lambda}{\lambda} \simeq \frac{v}{c} \tag{9.1}$$

if $v \ll c$. According to the empirical Hubble law, the relation between the radial distance and the radial velocity of galaxies is

$$v = H_0 r \quad \text{or} \quad r = v H_0^{-1} \tag{9.2}$$

where H_0 is the *Hubble constant* at the present. (As we shall see, the value of $H = v/r$ had larger values in the past.) The value of H_0 is conventionally expressed as

$$H_0 = 100h \ \frac{\text{km/s}}{\text{Mpc}} \tag{9.3}$$

where h parametrizes the experimental uncertainty; the observational limits on h are

$$0.3 \le h \le 1.0 \tag{9.4}$$

Here Mpc means megaparsec $= 10^6$ pc. The parsec (pc) is a traditional astronomical unit of distance, defined as follows. The position of a comparatively nearby star in the sky will seem to move relative to distant stars (parallax) as the earth moves on its orbit around the sun. A star is said to be at a distance of one parsec if this relative motion has an amplitude equal to one arc second $(4.85 \times 10^{-6}$ rad). Since the radius of the orbit of the earth is $R = 1.5 \times 10^{11}$ m,

$$1 \text{ pc} = 3.09 \times 10^{16} \text{ m} \tag{9.5}$$

Physicists often express astronomical distances in terms of the distance that light travels in a year,

$$1 \text{ light-year} = c(1 \text{ yr}) = (3 \times 10^8 \text{ m/s}) \times (3.16 \times 10^7 \text{ s}) \simeq 10^{13} \text{ km} \tag{9.6}$$

in terms of which

$$1 \text{ pc} \simeq 3.26 \text{ ly} \tag{9.7}$$

The Hubble constant (9.3) can be alternatively expressed as

$$H_0^{-1} = 10^{10} \ h^{-1} \text{ yr} , \quad \text{the } \textit{Hubble time} \tag{9.8}$$

or as

$$cH_0^{-1} = 3000 \ h^{-1} \text{ Mpc} , \quad \text{the } \textit{Hubble distance} \tag{9.9}$$

The size of the observable universe at present is of the order of $cH_0^{-1} \simeq 10^{10}$ ly. The universe may actually be infinite, with infinitely more matter whose light has not yet reached us.

The Hubble law holds for individual galaxies only to a precision of a few hundred km/s, which is the velocity range of galaxies in the same region of space; this is a measure of the 'temperature' of the gas of galaxies that constitutes the visible part of the universe. The law holds much better for the average velocity of the galaxies in a region, that is, it describes the mean flow of the gas of galaxies. A point moving with the mean flow is said to be *comoving*.

The Hubble law has the following interpretation. Suppose that at some time in the past, $t = 0$, all the matter of the universe (or at least of the part of the universe we can see) was at very nearly the same location $\mathbf{r} = 0$ (at very high density) and that from then on all the pieces of matter moved freely, with constant velocities. At a later time t, matter with velocity $\mathbf{v}$ would be at location $\mathbf{r} = \mathbf{v}t$. This has the form of the Hubble law (using a frame in which we are at rest at the origin) but now extended to a vector form

$$\mathbf{v} = H\mathbf{r} \qquad (9.10)$$

with $H^{-1} = t$, the length of time since the density was infinite. It is believed that (9.10) is true; the average 'flow' velocity of galaxies transverse to the line of sight (which is not directly observable) is negligible compared to the line of sight velocity.

It is important to realize that whatever $H(t)$ is, the flow (9.10) is a *uniform expansion* (or compression if H were negative); the same law would be observed from any galaxy. For example, galaxy A at location $\mathbf{r}_A$ has velocity $\mathbf{v}_A = H\mathbf{r}_A$ according to (9.10); subtraction of this equation from (9.10) gives

$$\mathbf{v} - \mathbf{v}_A = H(\mathbf{r} - \mathbf{r}_A) \qquad (9.11)$$

which is just (9.10) again, but now for velocity and position in the frame centered on galaxy A. Another way to see the uniformity is to remark that if at a given time the location of galaxies $1, 2, \ldots$ are $\mathbf{r}_1, \mathbf{r}_2, \ldots$, then according to (9.10) their locations at a time dt later are $(1 + H dt)\mathbf{r}_1, (1 + H dt)\mathbf{r}_2, \ldots$, that is, all change by the same factor $1 + H dt$.

Gravity is the only force which seriously violates the assumption that the pieces of matter of the universe move freely. (Short range forces merely scatter, which does not change an already uniform distribution; electric forces are effectively short range because of charge cancellation, known as shielding.) The universe appears to be uniform in density averaged over large scales (of order 100 Mpc), which gives rise to the *cosmological principle* that the universe is in fact uniform on large scales. We now show that if the mass distribution of the gas of galaxies is uniform at one time, Newtonian gravity leaves uniform both the expansion and the mass density ρ. That is, the flow (9.10) remains of that form; the only effect of gravity is to change the time dependence of H.

We start by choosing the origin of coordinates at an arbitrary galaxy. Assuming that no force except gravity acts on the galaxy, this coordinate frame is an inertial frame. Now consider the gravitational acceleration of some other galaxy with coordinate $\mathbf{r}$. (We will refer to this as the 'test galaxy'.) Since the assumed mass distribution of the universe is uniform, it is spherically symmetric around the origin and we can apply Newton's theorem from § 8.1. The gravitational acceleration of the galaxy at $\mathbf{r}$ is produced by the mass M inside the sphere of radius r

$$M = \tfrac{4}{3}\pi r^3 \rho \tag{9.12}$$

where ρ is the spatially uniform mass density. The equation of motion of the galaxy is

$$\ddot{\mathbf{r}} = -\frac{GM}{r^3}\mathbf{r} = -\frac{G4\pi\rho}{3}\mathbf{r} \tag{9.13}$$

Thus $\ddot{\mathbf{r}}$ is of the form $f(t)\mathbf{r}$; this preserves the Hubble law.

The mass M, Eq. (9.12), is a constant of the motion, because all the galaxies inside the sphere of radius r on which our test galaxy is located stay inside, according to the Hubble flow (uniform expansion). Since the motion is radial ($\mathbf{v} \sim \mathbf{r}$), (9.13) becomes

$$\ddot{r} = -\frac{GM}{r^2}, \qquad M = \text{constant} \tag{9.14}$$

which is identical to the motion of a mass moving radially in the gravitational field of a point mass M at the origin. The energy integral is

$$\tfrac{1}{2}\dot{r}^2 - \frac{GM}{r} = \tfrac{1}{2}C \tag{9.15}$$

The integration constant C has the dimensions of velocity squared; if it is non-negative it has the interpretation that $\dot{r} \to \pm\sqrt{C}$ as $r \to \infty$ (free expansion or contraction).

First we consider expansion in the case $C = 0$, for which

$$\dot{r} = \sqrt{\frac{2GM}{r}} \tag{9.16}$$

Expressing M in terms of ρ,

$$\dot{r} = \sqrt{\frac{8\pi G\rho}{3}}\,r \tag{9.17}$$

This is the Hubble law with H given by

$$H = \sqrt{\frac{8\pi G\rho}{3}} \tag{9.18}$$

Thus the density is proportional to H^2

$$\rho = \frac{3H^2}{8\pi G} \tag{9.19}$$

The time dependence of r can be determined by integrating (9.16)

$$\int_0^r r^{1/2}\,dr = \sqrt{2GM}\int_0^t dt \tag{9.20}$$

giving

$$r = \left(\tfrac{3}{2}\sqrt{2GM}\right)^{2/3} t^{2/3} \tag{9.21}$$

Computing $\dot{r}/r$ from (9.21), the Hubble constant is

$$H = \tfrac{3}{2}t^{-1} \tag{9.22}$$

and the mass density from (9.19) becomes

$$\rho = \frac{1}{6\pi G t^2} \tag{9.23}$$

Note that the final results (9.22) and (9.23) do not depend on r; they are independent of the choice of test galaxy.

In the theory of general relativity this $C = 0$ case corresponds to a spatially flat (Euclidean) space-time and is known as the *Einstein-de Sitter universe*. The mass density of this universe, given by (9.19), is called the *critical mass density* ρ_c. For $C = 0$ the mass density of the universe would be

$$\rho_c = \frac{3H_0^2}{8\pi G} = 1.88 \times 10^{-26}\, h_0^2 \text{ kg/m}^3 \tag{9.24}$$

The mass of a hydrogen atom is 1.67×10^{-27} kg so the critical density is a few hydrogen atoms per cubic meter. From (9.22), if $C = 0$ the present age of the universe would be

$$t_0 = \tfrac{2}{3}H_0^{-1} \approx 6.7h^{-1} \text{ Gyr} \tag{9.25}$$

where $1 \text{ Gyr} = 10^9 \text{ yr}$; for the observationally preferred value $h \simeq 0.5$, the age is $t_0 = 13.7 \text{ Gyr}$. In comparison, the age of the sun and earth is about 5 Gyr and the age of the oldest stars in our galaxy is of the order of 10-15 Gyr.

In the case $C \neq 0$, it is awkward that for a given universe, the value of C depends on the choice of test galaxy. That is, if another is chosen whose distance at a given time is a factor ξ times the original galaxy's distance, $\tilde{r} = \xi r$, then $\tilde{C} = \xi^2 C$ is required in order that $\tilde{r}(t) = \xi r(t)$ for all t. The convenient thing is to choose the test galaxy so that the velocity $\sqrt{|C|}$ is the fundamental velocity c, the velocity of light. Denoting this test galaxy's distance r as R, (9.15) becomes

$$\dot{R}^2 = \frac{2GM}{R} - kc^2 \tag{9.26}$$

where $M = \tfrac{4\pi}{3}R^3\rho$ and $k = +1$ or -1. This is written more simply as

$$\dot{R}^2 = \left(\frac{R_S}{R} - k\right)c^2 \tag{9.27}$$

where

$$R_S = 2GMc^{-2} \tag{9.28}$$

[Compare (8.56).] It is convenient to use (9.26) and (9.27) with $k = 0$ for the $C = 0$ case as well; but it must be remembered that in that case the choice of scale of R is arbitrary.

In general relativity it turns out that R is the radius of curvature of space, positively curved like a sphere (closed, Riemannian space) if $k = +1$, and negatively curved like a potato chip (open, Lobachevskian space) if $k = -1$. For the Euclidean universe with $k = 0$ the radius of curvature is infinite.

The constant of the motion M, or equivalently R_S, has the following significance: when $R \ll R_S$ the term $-kc^2$ in (9.27) is unimportant and the motion of $R(t)$ is the same as if $k = 0$, but it begins to deviate as R (and ct) become comparable to R_S.

The present value of kR/R_S can be measured in two ways. The first is to measure the ratio of the density ρ to the critical density ρ_c of a flat universe

$$\Omega \equiv \frac{\rho}{\rho_c} = \frac{\rho}{(3H^2/8\pi G)} \tag{9.29}$$

This can be related to kR/R_S by using (9.27) to get

$$H^2 = \left(\frac{\dot{R}}{R}\right)^2 = \frac{1}{R^2}\left(\frac{R_S}{R} - k\right)c^2 \tag{9.30}$$

and then (9.12) and (9.28) to get

$$\Omega = \frac{1}{1 - kR/R_S} \tag{9.31}$$

The other is to measure the deceleration $-\ddot{R}$ by finding H at earlier times, by observing more distant galaxies. A convenient dimensionless measure is the deceleration parameter defined by

$$q_0 \equiv -\frac{R\ddot{R}}{\dot{R}^2} \tag{9.32}$$

From (9.26), $R\ddot{R} = -GM/R$ and thus

$$q_0 = \frac{GM/R}{H^2R^2} = \frac{\frac{4\pi}{3}G\rho}{H^2} = \tfrac{1}{2}\Omega \tag{9.33}$$

Consequently Ω and q_0 measure the same thing.

For $k = -1$ (open universe), the solution to (9.27) with $R = 0$ at $t = 0$ can be expressed parametrically as

$$R = \tfrac{1}{2}(\cosh\alpha - 1)R_S \tag{9.34}$$
$$ct = \tfrac{1}{2}(\sinh\alpha - \alpha)R_S \tag{9.35}$$

as can be verified by differentiation, using $\dot{R} = (dr/d\alpha)/(dt/d\alpha)$. (One would like to eliminate the parameter α to get r as a function of t; this cannot be done algebraically but can be done numerically.) The observable quantity Ω, Eqs. (9.29) and (9.31), is given by

$$\Omega = \frac{2}{\cosh \alpha + 1} \tag{9.36}$$

For $k = +1$ (closed universe), the parametric solution to (9.27) is similarly

$$R = \tfrac{1}{2}(1 - \cos \beta) R_S \tag{9.37}$$
$$ct = \tfrac{1}{2}(\beta - \sin \beta) R_S \tag{9.38}$$

and Ω is given by

$$\Omega = \frac{2}{1 + \cos \beta} \tag{9.39}$$

The dependence of R and Ω on time which results from these parametric equations is shown in Figs. 9-1 and 9-2. At small times, $ct \ll R_S$, the radius is small, $R \ll R_S$ and the R_S/R gravity term dominates the right-hand side of (9.27). The expansion is similar to the $k = 0$ case

$$R \sim t^{2/3}, \quad \Omega \simeq 1 \tag{9.40}$$

The nature of the expansion changes for $ct \sim R_S$. For $k = -1$, free expansion occurs when $ct \gg R_S$ and $R \gg R_S$ with

$$R \simeq ct, \quad \Omega \ll 1 \tag{9.41}$$

For $k = +1$, the radius R reaches a maximum at $\beta = \pi$, with

$$R = R_S, \quad ct = \frac{\pi}{2} R_S, \quad \Omega = \infty \tag{9.42}$$

and the radius returns to zero at $ct = \pi R_S$ (the Big Crunch). The time evolution of the universe in the three cases is illustrated in Fig. 9-1 and Fig. 9-2.

The present observational limits on the deceleration parameter and the mass density of the universe are

$$0.1 \leq \Omega \leq 2 \tag{9.43}$$

This is consistent with the value $\Omega = 1$ of the Euclidean universe. Although values of $\Omega \neq 1$ are allowed by observation, they are theoretically unnatural because they require as an initial condition at $t = 0$, the value of the constant of the motion GMc^{-2} to be of the order of 10^{10} years, which is huge compared to the very short natural time scale of the Big Bang.

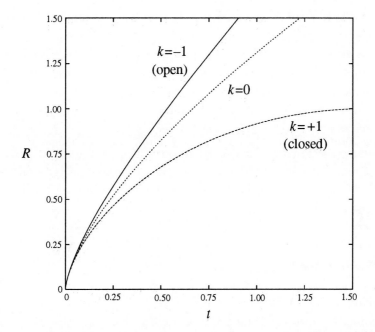

FIGURE 9-1. The radius of the universe versus time (the Big Bang occurs at $t = 0$) for $k = -1$ (open universe), $k = 0$ (Euclidean universe), and $k = +1$ (closed universe); the R scale is in units of R_S and the t scale is in units of R_S/c.

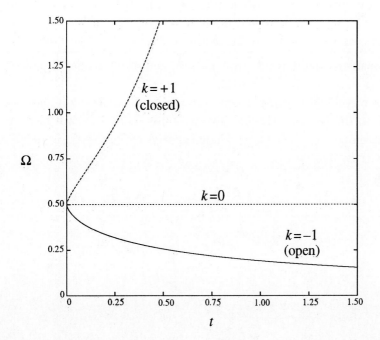

FIGURE 9-2. The time evolution of $\Omega = \rho/\rho_c$ for $k = -1$ (open universe), $k = 0$ (Euclidean universe), and $k = +1$ (closed universe). The units are the same as in Fig. 9-1.

The above cosmology based on Newtonian mechanics is valid during the time that the bulk of the energy content of the universe is in the rest masses of particles, the so-called matter dominated era. At earlier times, in the radiation dominated era, general relativity must be used. The consequence is that (9.26) still holds, but with M equalling the total energy in a comoving volume; if the pressure P is not zero, then M is not constant.

In later sections we discuss the important question of the value of the present mass density of the universe. Here Newtonian dynamics plays a crucial role in determining that most of the matter in the universe is *dark*, that is, it is not detectable by the electromagnetic processes by which we normally observe matter.

9.2 Cosmic Redshift

In looking at distant objects we are looking back in time. One cannot see all the way back to the time $t = 0$ of the Big Bang because at early times the universe, at high temperatures, was opaque. It was a plasma (free charges) so light was strongly scattered. Both radiation and matter cool down as the universe expands according to the *Cosmic Redshift Theorem* which we derive in the next paragraph. The radiation present when the universe became transparent at the plasma-to-gas transition at the age of 300,000 years is observed today as the 2.7° K cosmic black-body radiation.

Consider a light wave propagating past two galaxies separated by a distance r as shown in Fig. 9-3. As seen by galaxy ①, the light passing by has wavelength λ. At a time $\Delta t = r/c$ later this light passes galaxy ②. According to the Hubble law, galaxies ① and ② are moving apart at the velocity $v = Hr$, so the light as seen by galaxy ② is redshifted to wavelength λ', where

$$\frac{\lambda' - \lambda}{\lambda} = \frac{\Delta \lambda}{\lambda} = \frac{v}{c} = H \Delta t \tag{9.44}$$

Since the Hubble constant is related to the expansion by

$$H = \frac{\dot{r}}{r} \tag{9.45}$$

where r is the distance between any two galaxies, (9.44) can be written

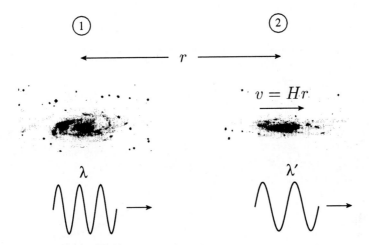

FIGURE 9-3. Cosmic red-shift due to scale increase.

as

$$\frac{\Delta\lambda}{\lambda} = \frac{\Delta r}{r} \qquad (9.46)$$

where $\Delta r = \dot{r}\Delta t$ is the amount that the distance r increased during the time Δt; hence

$$\lambda \propto r \qquad (9.47)$$

The wavelength of freely propagating radiation as seen by comoving observers (*e.g.*, on the galaxies) thus increases in proportion to r, the distance between the two comoving galaxies. As a consequence, λ is proportional to R. The same argument applied to any moving particle says that its momentum, as seen by comoving observers, varies inversely with $R(t)$.

In 1990 the Cosmic Background Explorer (COBE) satellite observed the background radiation very precisely. As illustrated in Fig. 9-4, an excellent description of the intensity of the radiation for wavelengths from 1 cm down to 0.5 mm is given by the Planck blackbody spectrum with temperature $T = 2.73\pm0.006$ K. The radiation coming from various directions is observed to be remarkably uniform. This means that the universe was very uniform at the time that it became transparent, resulting in the uniform and isotropic blackbody radiation. In fact, blackbody radiation defines the Hubble flow much better than do the galaxies. The biggest observed non-uniformity is a small (±0.002) variation in the blackbody temperature depending on direction, which is explainable as the Doppler shift resulting from the motion of the earth (and the entire local group of

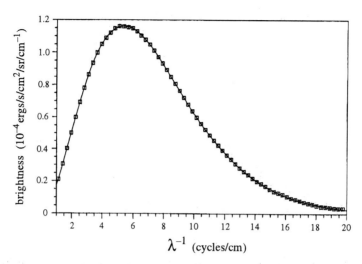

FIGURE 9-4. Measured blackbody spectrum of cosmic background. The curve is the Planck blackbody spectrum for temperature 2.73 K.

galaxies) relative to a uniform Hubble expansion. Much smaller fluctuations on the order of a few parts in 10^5 are observed by COBE and more recent experiments. These tiny temperature fluctuations reflect density fluctuations which are thought to be the "seeds" from which the galaxies condensed.

9.3 Virial Theorem

In chapters 6 and 7 Newton's laws were applied to systems of particles. The most extensive applications concerned the dynamics of rigid body motion. There the constraint that interparticle distances remain constant allowed the behavior of the system to be described by CM motion plus rotations about the CM. For non-rigid systems whose constituent parts interact by a known force we can still draw some general conclusions but only as a statistical average. However a statement about average energies is all that is needed for many astronomical purposes. The virial theorem of Clausius, which we will now derive, concerns time-averaged quantities such as kinetic energy

$$\langle K \rangle_t \equiv \lim_{\tau \to \infty} \frac{1}{\tau} \int_0^\tau K(t)\, dt \tag{9.48}$$

where $\langle \ \rangle_t$ is used to denote the time average.

Consider a system of N particles. The CM is assumed to be fixed and the constituent parts of the system are assumed to be bound or enclosed so that their coordinate positions remain finite. We define the *virial G* as

$$G \equiv \sum_{i=1}^{N} \mathbf{p}_i \cdot \mathbf{r}_i \qquad (9.49)$$

Time differentiation yields

$$\dot{G} = \sum_i \dot{\mathbf{p}}_i \cdot \mathbf{r}_i + \sum_i \mathbf{p}_i \cdot \dot{\mathbf{r}}_i$$
$$= \sum_i \mathbf{F}_i \cdot \mathbf{r}_i + 2K \qquad (9.50)$$

since $\mathbf{F}_i = \dot{\mathbf{p}}_i = m\ddot{\mathbf{r}}_i$. Here $\mathbf{F}_i$ is the force acting on the ith particle. Re-integrating (9.50) from $t = 0$ to τ, dividing the result by τ and taking the limit $\tau \to \infty$ we obtain

$$\lim_{\tau \to \infty} \left[\frac{G(\tau) - G(0)}{\tau} \right] = \left\langle \sum_i \mathbf{F}_i \cdot \mathbf{r}_i \right\rangle_t + 2\langle K \rangle_t \qquad (9.51)$$

By assumption the virial is finite so the factor of $\frac{1}{\tau}$ causes the left-hand side to vanish in the $\tau \to \infty$ limit and thus

$$2\langle K \rangle_t = -\left\langle \sum_i \mathbf{F}_i \cdot \mathbf{r}_i \right\rangle_t \qquad (9.52)$$

This result is known as the *virial theorem*.

If the forces $\mathbf{F}_i$ are due to pair-wise central forces which go like a power of the distance, the right-hand side of (9.52) can be put in a much more useful form. The force on the ith particle due to two-body interactions with the rest of the system is

$$\mathbf{F}_i = \sum_{j \neq i} \mathbf{F}_{ij} \qquad (9.53)$$

The quantity $\sum_i \mathbf{F}_i \cdot \mathbf{r}_i$ appearing in (9.52) can be written as

$$\sum_i \mathbf{F}_i \cdot \mathbf{r}_i = \sum_i \sum_{j \neq i} \mathbf{F}_{ij} \cdot \mathbf{r}_i$$
$$= \sum_i \sum_{j < i} [\mathbf{F}_{ij} \cdot \mathbf{r}_i + \mathbf{F}_{ji} \cdot \mathbf{r}_j] \qquad (9.54)$$

Newton's third law $\mathbf{F}_{ji} = -\mathbf{F}_{ij}$ then leads to the result

$$\sum_i \mathbf{F}_i \cdot \mathbf{r}_i = \sum_i \sum_{j<i} \mathbf{F}_{ij} \cdot \mathbf{r}_{ij} \tag{9.55}$$

where $\mathbf{r}_{ij} \equiv \mathbf{r}_i - \mathbf{r}_j$ and the double summation includes all pairs once.

Now we further assume that the particles interact through an attractive central force of the power-law form

$$\mathbf{F}_{ij} \equiv cr_{ij}^n \hat{\mathbf{r}}_{ij} \tag{9.56}$$

This $\mathbf{F}_{ij}$ can be derived from a potential energy

$$V_{ij} = -\frac{c}{n+1} r_{ij}^{n+1} \tag{9.57}$$

Noting that

$$-(n+1)V_{ij} = \mathbf{F}_{ij} \cdot \mathbf{r}_{ij} \tag{9.58}$$

(9.55) takes the form

$$\sum_i \mathbf{F}_i \cdot \mathbf{r}_i = -(n+1) \sum_{\text{pairs}} V_{ij}$$
$$= -(n+1)V \tag{9.59}$$

where V is the sum of all the pair-wise potential energies and thus is the total potential energy. Using this in (9.52) gives the virial theorem for a self-interacting system

$$2\langle K \rangle_t = (n+1)\langle V \rangle_t \tag{9.60}$$

For harmonic oscillator interactions $n = 1$ and the virial theorem gives $\langle K \rangle_t = \langle V \rangle_t$. For gravitational forces the inverse square law dependence with $n = -2$ gives

$$\langle K \rangle_t = -\frac{1}{2}\langle V \rangle_t \tag{9.61}$$

This relation is used in §9.4B to deduce the presence of dark matter in groups of galaxies.

9.4   Dark Matter

Knowledge about the universe is largely based on the detection of electro-magnetic radiation (radio waves to gamma rays) from *luminous matter* in outer space. The existence of *dark matter* in the universe, which does not significantly emit or absorb electromagnetic radiation, is inferred from its gravitational effects. Possible dark matter candidates include low-mass stars in which thermonuclear reactions barely or never started, black holes, and exotic elementary particles. The presence of an astoundingly large amount of dark matter, at least five to ten times the amount of luminous matter, is inferred from the rotations of spiral galaxies and the dynamics of groups of galaxies.

A. Galactic Rotation Curves

Our sun is one of about 10^{11} stars of the Milky Way galaxy, which is a typical large spiral galaxy. About 10% of the mass of our galaxy is in the form of hydrogen gas. The sun lies about 28 kly ($1\,\mathrm{kly} = 10^3$ light-years) from the center of the galactic disk. About seventy percent of the galaxy's light emanates from within this radius. Our sun moves in a roughly circular orbit with a speed of about $200\,\mathrm{km/s}$; in the 10^{11} years since our galaxy formed, our sun has made about 40 revolutions.

Other spiral galaxies are actually easier to study since then the whole galaxy is often visible and all of the stars in the galaxy are at nearly the same distance from us. Using Doppler shifts of visible light and radio emissions, the circular orbital velocity, or *rotation curve* of galaxies, can be measured. In Fig. 9-5 the rotation velocity v_θ is plotted as a function of distance for several spiral galaxies. The visible light falls off exponentially with increasing radius but by observing atomic radio spectral lines from gas surrounding a galaxy the rotation curve can be measured to large radii. A flat rotation curve is observed extending beyond the luminous region out to the limits of observation. Although the observations at large radii are of radio emission from hydrogen atoms, this hydrogen and the estimated accompanying stars contribute a negligible fraction of the inferred total mass in the outer region. By the following argument Newtonian mechanics implies a preponderance of dark matter in spiral galaxies.

For simplicity we assume a spherical mass distribution, which is the expected distribution of the dominant dark matter component. By New-

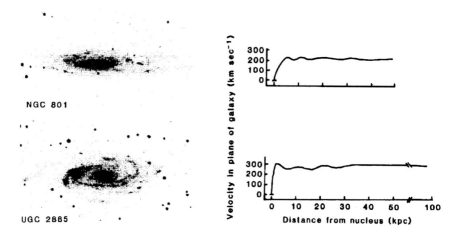

FIGURE 9-5. Photographs and rotation curves for two spiral galaxies. The same horizontal scale applies to the left and right sides of the figure.

ton's theorem of §8.1 and force balance for circular orbits we have

$$\frac{v_\theta^2}{r} = \frac{GM(r)}{r^2} \tag{9.62}$$

or

$$v_\theta = \sqrt{\frac{GM(r)}{r}} \tag{9.63}$$

where $M(r)$ is the total mass within a sphere of radius r. Thus for the inner region of constant density ρ,

$$M(r) = \frac{4}{3}\pi r^3 \rho \propto r^3, \quad \text{so } v_\theta \propto r \tag{9.64}$$

Outside the mass distribution, where $M(r) = $ constant (the so-called Keplerian region), we expect

$$M(r) = \text{constant}, \quad \text{so } v_\theta \propto \frac{1}{\sqrt{r}} \tag{9.65}$$

A constant rotation curve, $v_\theta \simeq$ constant, implies

$$M(r) \propto r, \quad \rho(r) \propto \frac{1}{r^2} \tag{9.66}$$

In the observed rotation curves in Fig. 9-5 the lack of a Keplerian fall-off of v_θ beyond the visible region is evidence for unseen mass. The constant value of v_θ outside the visible part of the galaxy implies that a large dark halo envelopes the galaxy. Representative luminosity and rotation curve observations for a spiral galaxy are shown in Fig. 9-6 and compared with the rotation curve expected from the luminous matter.

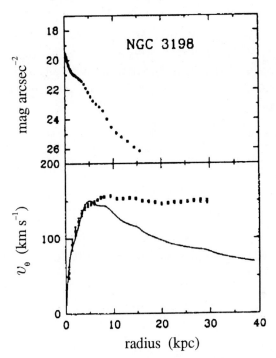

FIGURE 9-6. Light profile and rotation curve of the spiral galaxy NGC 3198. Top: luminosity profile. Bottom: observed rotation curve (dots with error bars) and rotation curve predicted from light and neutral hydrogen gas distributions.

In some cases the rotation curve has been measured about five times further out than the bulk of the visible mass. Since the total mass is proportional to the radius (for v_θ constant), the total mass is at least five times larger than the visible mass.

B. Groups of Galaxies

There are many examples of groups of galaxies bound by gravitational interactions. The "local group" consists of about 30 galaxies of various sizes, dominated by our galaxy and the Andromeda galaxy. The Coma cluster is an example of a group containing thousands of galaxies. The dark matter within groups of galaxies can be determined from an application of the virial theorem.

From (9.61) the virial theorem for a system of galaxies is

$$\sum_i m_i v_i^2 = \sum_{\text{pairs}} \frac{G m_i m_j}{|\mathbf{r}_i - \mathbf{r}_j|} \tag{9.67}$$

where we can drop the time average because of the large number of members of the group. Since we observe only velocities projected onto the line of sight (the direction from us to the astronomical object) and distances projected perpendicular to the line of sight, we need to rewrite (9.67) in terms of the actual observables.

If the group is spherically symmetric, the velocity component u_i of a galaxy along the line-of-sight is related on the average to the full velocity v_i by

$$u_i^2 = \frac{1}{3} v_i^2 \tag{9.68}$$

From Fig. 9-7 the projection of a galactic coordinate $\mathbf{r}$ on the perpendicular plane averaged over solid angle $d\Omega = \sin\theta \, d\theta \, d\phi$ is

$$\frac{1}{R} \equiv \left\langle \frac{1}{|\mathbf{r}|} \right\rangle_\Omega = \frac{1}{4\pi} \int \frac{d\Omega}{r \sin\theta} = \frac{1}{4\pi r} \int_0^\pi d\theta \int_0^{2\pi} d\phi$$
$$= \frac{\pi}{2r} \tag{9.69}$$

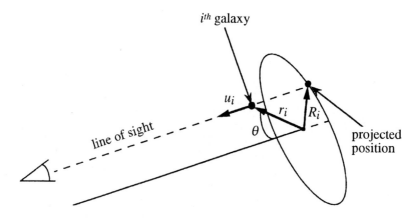

FIGURE 9-7. Line of sight coordinates of a galaxy moving in a group of galaxies.

Using (9.68) and (9.69) the virial theorem of (9.67) becomes

$$3 \sum_i m_i u_i^2 = \frac{2G}{\pi} \sum_{\text{pairs}} \frac{m_i m_j}{R_{ij}} \tag{9.70}$$

Although the galaxies of a group may vary in size one expects that the ratio of mass to luminosity will be about the same for each galaxy. Defining a mass-to-luminosity ratio

$$\Upsilon \equiv \frac{m_i}{L_i} \tag{9.71}$$

and substituting $m_i = \Upsilon L_i$ in (9.70) we find

$$\Upsilon = \frac{3\pi}{2G} \frac{\sum_i L_i v_{L_i}^2}{\sum_{\text{pairs}} L_i L_j / R_{ij}} \tag{9.72}$$

For analyses of several groups of galaxies the value

$$\Upsilon_{\text{group}} \simeq 30 \Upsilon_{\text{local}} \tag{9.73}$$

is deduced, where Υ_{local} is the average value for this ratio for a local portion of our galaxy, taking into account only directly detectable mass. Thus there must exist about thirty times more dark matter than visible matter on the scale of a group of galaxies.

If all the inferred dark matter is included then $\Omega = $ [see (9.29)] exceeds 0.2 but whether enough dark matter is present to realize $\Omega = 1$ for the Einstein-de Sitter universe is not settled. Searches for supernovae in distant galaxies show promise for establishing a more accurate experimental determination of Ω.

PROBLEMS

9.1 The Expansion of the Universe

9-1. Olber's Paradox: Assume a static uniform universe consisting of sun-like stars averaging 5 ly separation.

a) Find the number of stars in a spherical shell of radius r and thickness $\Delta \ll r$.

b) Each star has surface area A. Find the fraction of the shell surface in a) covered by stars.

c) Find the largest such universe for which the fraction in b) does not exceed one. Comment on your result.

9-2. Convert the value of the Hubble constant $H_0 = 50 \frac{\text{km/s}}{\text{Mpc}}$ to a value of H_0^{-1} in years.

9-3.a) Differentiate the Hubble law (9.10) with respect to time and show that the result is consistent with the equation of motion (9.13). Verify that the following equation for time dependence of H is obtained

$$\dot{H} = -H^2 - \frac{4\pi G\rho}{3}$$

With gravity turned off, find the free-expansion solution to this equation.

b) The volume occupied by a group of galaxies changes in a time interval dt by the factor $(1 + H dt)^3$. Using this result show that the mass density ρ of the volume changes at the rate

$$\dot{\rho} = -3H\rho$$

This equation for $\dot{\rho}$ combined with the equation for $\dot{H}$ in part a) are the equations of motion for the expansion. However it is easier to use (9.13) directly.

9-4. Assuming that $\Omega = 1$ and $H_0^{-1} = 2 \times 10^{10}$ yr compute the average mass density at the time of radiation decoupling which occurred at a cosmic time of 300,000 yr. Express your result in hydrogen atoms per cubic meter. Assume Newtonian cosmology is valid for all times.

9-5. Estimate the average visible mass density of

a) the solar system. Assume a radius of 50 AU.

b) the galaxy. Assume 10^{11} sun-like stars and a radius of 5 kpc.

c) the local galactic group. Assume a total visible mass of $2 \times 10^{11} \, M_\odot$ and a radius of 500 kpc.

Compare these densities to the critical density of (9.24). One solar mass is $M_\odot \simeq 2 \times 10^{30}$ kg.

9-6. On dimensional grounds one can argue that the "natural" time scale of the universe is the unit constructed from the fundamental constants G, c and $\hbar$ (reduced Planck's constant). This time unit is known as the *Planck time*. Construct an expression for the Planck

time and evaluate it numerically. This is relevant to the "natural-
ness" argument that $\Omega = 1$ (end of §9.1). Find the corresponding
Planck length.

9.3 Virial Theorem

9-7. Show that the virial theorem of (9.52) is valid if the sum of the
principal moments of inertia of the system increases less rapidly
then quadratically with time. *Hint: express the virial as a time
derivative.*

9.4 Dark Matter

9-8. Assuming the following spherical distribution of mass in a galaxy

visible mass

$$M_V(r) = \begin{cases} M_V(r/r_0)^3, & r < r_0 \\ M_V, & r > r_0 \end{cases}$$

dark mass

$$M_D(r) = \begin{cases} NM_V(r/r_0)^3, & r < r_0 \\ NM_V(r/r_0), & r > r_0 \end{cases}$$

Use the data from Fig. 9-6 to roughly estimate v_θ at large r, the
size of the light matter distribution r_0, and the v_θ due to the visible
matter. Estimate the ratio of dark to visible matter out to 30 kpc.

9-9. For many distant galaxies the distance is determined by using the
Hubble law.

a) If the mass of a spiral galaxy is measured by the rotation curve
out to a given angular radius show that the mass contained within
that radius is proportional to H_0^{-1}.

b) Show that the mass-to-light ratio Υ for a group of galaxies determined by the virial theorem is proportional to H_0. In 1933
F. Zwicky first analyzed the Coma cluster and concluded that
the dark matter was 400 times more massive than the visible
matter. At the time, the Hubble constant was thought to be
560 km s^{-1}Mpc^{-1}. What would he have concluded about the ratio of dark to visible matter using the more current value of (9.3)?
*Hint: remember that the observed luminosity decreases as the in-
verse square of the distance compared to the absolute luminosity.*

Chapter 10

RELATIVITY

Understanding of the physics of space and time was changed forever with the introduction of the special theory of relativity by Albert Einstein in 1905. He dismissed the concept of an ether through which light propagates and postulated that the speed of light is the same in any inertial frame. Among the consequences of this is that the rate of a clock and the length of a ruler depend on their motion. This theory also predicted that mass is a form of energy. For motion with velocity near the speed of light, Newton's laws of classical mechanics must be modified to be consistent with special relativity.

10.1 The Relativity Idea

According to Newton's equations of motion, all inertial coordinate frames are equivalent. This means that the motions following these equations depend only on *relative times* and on the *relative* positions and *relative* velocities of masses. Thus a system of masses following Newton's equations has no behavior which would enable one to determine absolute time, location, orientation or velocity. For example, if one changes from one inertial frame to another which is moving at at different velocity as described by the Galilean transformation of (4.26), Newton's equations remain unchanged.

The situation seemed to change when electromagnetism became part of fundamental physics, that is, when Maxwell's equations were found to describe all laboratory electric and magnetic phenomena. Although Maxwell's equations obey the relativity of time, location and orientation, they do not *seem* to obey the relativity of velocity. They say (in agreement with experiment) that waves of electromagnetic fields ("light") propagate at the velocity $c = (\mu_0 \epsilon_0)^{-1/2} = 3 \times 10^8$ m/s. But then, in another inertial frame with velocity $\mathbf{v}$, according to (4.26), light will propagate at a velocity shifted by $-\mathbf{v}$. That is, Maxwell's equations rewritten in terms of the new coordinates defined by the Galilean transformation (4.26) are *different* equations; these new equations say that the velocity of light varies from $c - v$ to $c + v$, depending on the direction of propagation.

The coordinate frame in which Maxwell's equations hold was historically called the *ether frame* (the rest frame of a hypothetical medium, the *ether*, in which light was considered to propagate). This raises the embarrassing question: why should the 'ether' have the velocity of the earth? What is special about the velocity of the earth? Of course it might just be coincidence that the earth has the magic velocity at which Maxwell's equations hold. But even this can be ruled out, because during a year components of the earth's velocity in the plane of its orbit around the sun vary by ± 30 km/s, and the Michelson-Morley experiment discussed below showed that the velocity of light stayed constant throughout the year to a precision of ≈ 1 km/s.

Subsequent experiments have continued to agree with what Einstein called the *special principle of relativity,* namely that no physical measurement of any sort can establish an absolute time, location, orientation or velocity. As described above, Newton's and Maxwell's equations together do not satisfy the principle. Einstein realized that the way to get agreement with both the principle of special relativity and all existing experimental results was to alter Newton's equations. The clue about how to do this is in the fact that Maxwell's equations by themselves satisfy the principle, that is, they are equally valid in inertial frames differing in location, orientation and velocity, but only if the relation between the coordinates (space and time) of different frames is not the Galilean transformation (4.26) but a different relation, the *Lorentz transformation.* Once Newton's equations are altered into a 'relativistic' form which is valid in all frames related by the Lorentz transformation, then all of the classical equations of motion for dynamics and electromagnetism obey the special principle of relativity. In particular they imply, as desired, that to all observers (that is, in all inertial frames) the speed of light is the same.

10.2 The Michelson-Morley Experiment

In this experiment an incident light beam shown in Fig. 10-1 is split by a glass plate P into two beams which reflect off mirrors M_1 and M_2 and are then compared in phase by this interferometer. The difference of the propagation time of light waves along the two paths can be inferred from a measurement of the phase difference $\Delta\phi = \nu(t_1 - t_2)$ between the waves, where ν is the wave frequency.

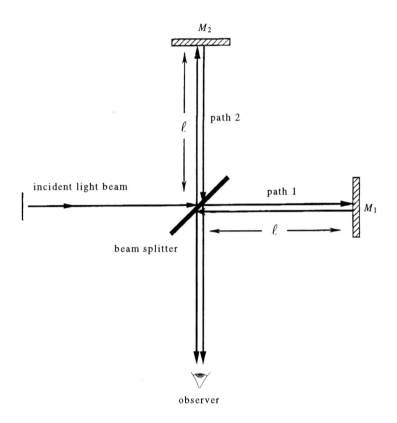

FIGURE 10-1. Schematic representation of Michelson's interferometer. The incident light beam is split and a change in the relative path lengths 1 and 2 is observed by a shift in the interference pattern.

If the apparatus is moving with velocity v with respect to the ether in the direction of path 1 the time taken along path 1 is

$$t_1 = \frac{\ell}{c - v} + \frac{\ell}{c + v} = \frac{2\ell}{c} \frac{1}{1 - v^2/c^2} \tag{10.1}$$

where c is the light velocity in the ether rest frame. For the beam moving along path 2 the time taken is

$$t_2 = \frac{2\ell}{c\sqrt{1 - v^2/c^2}} \tag{10.2}$$

This result for t_2 can be understood from the velocity diagram in Fig. 10-2; the velocity component along path 2 is $\sqrt{c^2 - v^2}$. For the earth's

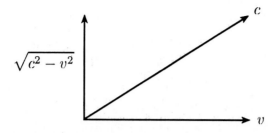

FIGURE 10-2.

motion about the sun $v/c \simeq 10^{-4}$ and so only the lowest order term in v^2/c^2 need be retained giving a phase difference

$$\Delta\phi = \nu(t_1 - t_2) \simeq \nu\ell v^2/c^3 \qquad (10.3)$$

When the apparatus is rotated by 90 degrees from its original orientation in the ether wind the phase difference is $-\Delta\phi$; subtracting $\Delta\phi - (-\Delta\phi)$, an overall "phase shift" of $2\Delta\phi$ is predicted. Since only this change in the phase difference between the paths is important, the two path lengths themselves do not need to be precisely measured. For the Michelson-Morley experiment the ether prediction was $2\Delta\phi \sim \frac{1}{3}$. The experiment was sensitive to a shift forty times smaller, but no phase shift was ever observed.

The predictions of the special theory of relativity have been overwhelmingly confirmed by experiment. The nature of space, time and dynamics when velocities are near the speed of light is radically different from the non-relativistic Newtonian theory.

10.3 Lorentz Transformation

If two observers in inertial frames in uniform relative motion observe the same phenomenon in terms of their respective coordinates one can ask how these coordinates are related. Consider the two inertial coordinate frames S and S' in Fig. 10-3, where the coordinate axes x and x', etc., are taken to be parallel and the origin of S' moves with velocity $\mathbf{v} = v\hat{z}$ relative to S. A given physical point P at a certain time can be specified in terms of either the coordinates $\mathbf{r} = (x, y, z)$ and t or $\mathbf{r}' = (x', y', z')$ and t'. The question we are addressing in this section is how $\mathbf{r}, t$ and $\mathbf{r}', t'$ are related assuming the validity of the principle of special relativity.

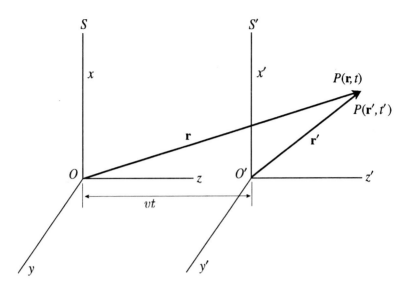

FIGURE 10-3. Two inertial coordinate frames S and S'. The latter frame moves with velocity v relative to the former. The point P can be specified by coordinates in S or S'.

For Newton's laws to be valid in both inertial frames (see § 4.2) the Galilean transformation

$$\mathbf{r} = \mathbf{r}' + \mathbf{v}t \tag{10.4}$$

must hold; in component form

$$
\begin{aligned}
x &= x' \\
y &= y' \\
z &= z' + vt' \\
t &= t'
\end{aligned}
\tag{10.5}
$$

Although this relationship may seem to be obvious it is not consistent with the special relativity principle.

As an electromagnetic example consider the propagation of a light flash which occurs at $t = t' = 0$, when the origins of S and S' are coincident. Observers in S see the light propagate as a spherical shell with radius ct where c is the speed of light. The propagation of the light pulse in S satisfies the equation

$$x^2 + y^2 + z^2 = c^2 t^2 \tag{10.6}$$

An observer using the frame S' also sees the light pulse expand with velocity c as required by the principle of special relativity. The S' observer concludes that the pulse satisfies the equation

$$x'^2 + y'^2 + z'^2 = c^2 t'^2 \tag{10.7}$$

One sees that this does not agree with the Galilean transformation because substitution of (10.5) into the propagation equation (10.6) yields

$$x'^2 + y'^2 + (z' + vt')^2 = c^2 t'^2 \tag{10.8}$$

which is not the same as (10.7); the Galilean transformation is inconsistent with the principle of special relativity.

We now find the coordinate transformation that relates (10.6) and (10.7). First we can restrict consideration to linear transformations since this is the only way to ensure that there are no special points in space or time, *e.g.*, one origin of a coordinate system must be equivalent to any other up to additive constants. The trouble with the Galilean transformation occurred in the z and t transformations and not in the x and y (transverse) parts, so we consider a transformation relating S' and S of the form

$$x = x' \tag{10.9}$$
$$y = y' \tag{10.10}$$
$$z = \gamma(z' + at') \tag{10.11}$$
$$t = \bar{\gamma}(t' + bz') \tag{10.12}$$

where γ, $\bar{\gamma}$, a and b are as yet unknown constants. For small relative frame velocities v, $\gamma \to 1$, $\bar{\gamma} \to 1$, $a \to v$ and $b \to 0$ are needed to reproduce the Galilean transformation (10.5). Actually, $a = v$ just defines the relative frame velocity and must hold for our general transformation. To see this, consider the origin of S as viewed by an observer at the origin of S'. This observer sees S moving to the left with velocity v and hence $z' = -vt'$. Comparing to (10.11) with $z = 0$ requires

$$a = v \tag{10.13}$$

Substitution of the transformation (10.9)–(10.13) into (10.6) gives

$$x'^2 + y'^2 + \gamma^2 \left(z'^2 + 2vz't' + t'^2\right) = c^2 \bar{\gamma}^2 \left(t'^2 + 2bt'z' + b^2 z'^2\right) \tag{10.14}$$

To be consistent with (10.7), the following conditions must be met

$$\gamma^2 - \bar{\gamma}^2 b^2 c^2 = 1 \tag{10.15}$$

$$\gamma^2 v = c^2 b \bar{\gamma}^2 \tag{10.16}$$
$$\bar{\gamma}^2 - \gamma^2 v^2/c^2 = 1 \tag{10.17}$$

One way to solve these three equations for γ, $\bar{\gamma}$ and b is to use (10.16) to eliminate b in (10.15) giving

$$\gamma^2 - \frac{\gamma^4}{\bar{\gamma}^2}\left(\frac{v^2}{c^2}\right) = 1 \tag{10.18}$$

and then use (10.17) to eliminate v^2/c^2. The result is

$$\bar{\gamma}^2 = \gamma^2 \tag{10.19}$$

When (10.19) is substituted back into (10.15)–(10.17) we obtain

$$b = v/c^2$$
$$\gamma^2 = 1/(1 - v^2/c^2) \tag{10.20}$$

Since $\gamma(0) = 1$ we have

$$\gamma(v) = 1 \Big/ \sqrt{1 - v^2/c^2} \tag{10.21}$$

The transformations of (10.9)–(10.12) with a, b, and γ from (10.13), (10.20) and (10.21) are known as the *Lorentz transformation*. The final result is

$$x = x'$$
$$y = y'$$
$$z = \gamma(z' + vt') \tag{10.22}$$
$$t = \gamma(t' + vz'/c^2)$$

where

$$\gamma(v) = \frac{1}{\sqrt{1 - v^2/c^2}} \tag{10.23}$$

The factor γ is always greater or equal to unity and is shown in Fig. 10-4 as a function of velocity v. H. A. Lorentz first proposed this transformation in the context of the form invariance of Maxwell's equations between

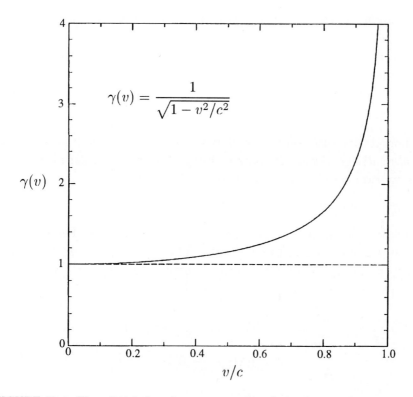

FIGURE 10-4. The relativistic γ factor versus the relative frame velocity v.

two inertial frames. Einstein provided the physical interpretation and demonstrated its universal application.

The inverse transformation for S' coordinates in terms of S frame coordinates is

$$
\begin{aligned}
x' &= x \\
y' &= y \\
z' &= \gamma(z - vt) \\
t' &= \gamma\left(t - vz/c^2\right)
\end{aligned}
\qquad (10.24)
$$

Note that (10.22) and (10.24) are related as follows: if primed and un-primed coordinates are interchanged and the sign of v reversed, one set is obtained from the other.

Only for velocities near that of light or for experiments of extraordinary precision is it necessary to deal with relativistic kinematics. But in many fields of science relativity *is* an everyday experience. For example, at the largest particle accelerators electrons with energy of 50 GeV are common. These electrons travel at a velocity which differs from the velocity of light only in the tenth decimal place. If such an electron were to race light around the earth's equator it would lose by only a few millimeters.

10.4 Consequences of Relativity

The Lorentz transformation contains all of the kinematic information of special relativity. Seemingly paradoxical behavior such as shrinking rulers and slow running clocks are predicted by (10.22).

A. Time Dilation

One relativistic effect concerns the rate of moving clocks. For simplicity consider a clock fixed in the S' frame. The interval between ticks in S' is $(\Delta t)' = t_2' - t_1'$. From (10.22) an observer fixed in frame S perceives these ticks to be separated by an interval

$$\Delta t = \gamma \left[t_2' - t_1' + \frac{v}{c^2}(z_2' - z_1') \right]$$
$$= \gamma \left[(\Delta t)' + \frac{v}{c^2}(0) \right] \qquad (10.25)$$
$$= \gamma(\Delta t)'$$

Since γ is greater than unity, the S-frame time interval Δt between ticks is larger than the S'-frame interval, $(\Delta t)'$, that is, the clock moving at velocity v will tick slower than when it is at rest. Conversely, a clock fixed in S with tick interval Δt will appear to an observer in S' to have a tick interval [using (10.24)]

$$(\Delta t)' = \gamma \left[t_2 - t_1 - \frac{v}{c^2}(z_2 - z_1) \right]$$
$$= \gamma \Delta t \qquad (10.26)$$

Again, the moving clock ticks slower by the factor γ.

In 1971 a direct test of relativistic time dilation was performed [J.C. Hafele and Richard E. Keating, Science **177**, 166 (1972)] in which four cesium-beam atomic clocks were flown on regularly scheduled commercial jet airplanes around the world twice, once eastward and once westward, and then compared to clocks remaining on the ground (see Figure 10-5). The result was that the eastward-flying clock gained -59 ns (the minus sign indicates time loss) and the westward-flying clock gained

+273 ns. There are two effects present here: moving clocks run slow (lose time) and a general relativistic effect that higher clocks run fast (gain time). The height effect should be nearly the same for both flights and cancels the difference of the times lost in the two directions relative to the ground clock.

$$\Delta \equiv t_{\text{EG}} - t_{\text{WG}} = -59 - (+273) = -332 \, \text{ns} \qquad (10.27)$$

Here t_{EG} denotes the trip time registered by the eastward-going clock minus the trip time registered by the ground clock and similarly t_{WG} for the westward-going clock. The uncertainty on the -332 ns value due to the clock inaccuracy is ± 15 ns.

FIGURE 10-5. Direct verification of relativistic clock effects. An airplane containing an atomic clock flies eastward and then westward around the earth. The elapsed times are compared with a clock remaining on the ground.

We shall analyze the flying clock experiment in the approximation that the reference frames on the ground and on the airplane are inertial frames. Consider a reference clock at the north pole, a clock on the ground, and a clock on the airplane. According to (10.25), time intervals recorded on these three clocks are related by

$$t_{\text{ref}} = \gamma_{\text{A}} t_{\text{A}} \qquad (10.28)$$

$$t_{\text{ref}} = \gamma_G t_G \tag{10.29}$$

where $\gamma_A = \left(1 - v_A^2/c^2\right)^{-1/2}$ and $\lambda_G = \left(1 - v_G^2/c^2\right)^{-1/2}$. The velocities v_A and v_G refer to the airborne and ground-based clocks, respectively, relative to the reference frame at the north pole. Thus, by (10.28) and (10.29) t_A and t_G are related by

$$t_A = t_G \frac{\gamma_G}{\gamma_A} = t_G \left(\frac{1 - v_A^2/c^2}{1 - v_G^2/c^2}\right)^{1/2} \tag{10.30}$$

Since $v_A/c \ll 1$ and $v_G/c \ll 1$, we can approximate the square root by the first two terms in the binomial expansion to obtain

$$t_A \simeq t_G \left(1 + \frac{v_G^2 - v_A^2}{2c^2}\right) \tag{10.31}$$

The difference of the time intervals on the air- and ground-based clocks is then

$$t_{AG} = t_A - t_G \simeq t_G \left(\frac{v_G^2 - v_A^2}{2c^2}\right) \tag{10.32}$$

We assume for simplicity that the airplane maintains a constant speed v_P relative to the ground at a particular latitude. Then relative to the inertial reference frame at the north pole, the east-going airplane has speed $v_A = v_G + v_P$. Hence from (10.32) $t_{AG} < 0$ and the airplane clock loses time compared to the ground-based clock. For the west-going airplane the speed is $v_A = v_G - v_P$. Correspondingly $t_{AG} > 0$ and the airplane clock gains time relative to the ground-based clock. The predicted east-west time difference is

$$\Delta = t_{EG} - t_{WG} \simeq \frac{t_G}{2c^2} \left\{ v_G^2 - (v_G + v_P)^2 - \left[v_G^2 + (v_G - v_P)^2\right] \right\}$$
$$= -2 \left(\frac{v_G v_P}{c^2}\right) t_G \tag{10.33}$$

which is negative, in agreement with the sign of the measured Δ in (10.27).

The experimenters used the airplanes' flight records of speeds v_P to accurately compute the predicted time difference Δ. In lieu of such detailed information we make a rough estimate of Δ. The constant flight speed v_P relative to the ground at a particular latitude is

$$v_P = \frac{2\pi r}{t_G} \qquad (10.34)$$

where r is the perpendicular distance to the earth's axis. The ground velocity due to the rotation of the earth is

$$v_G = \frac{2\pi r}{t_{day}} \qquad (10.35)$$

where $t_{day} = 24$ hours. Taking the ratio of the two preceding relations we obtain

$$v_G = v_P \left(\frac{t_G}{t_{day}} \right) \qquad (10.36)$$

The east-west time difference prediction from (10.33) and (10.36) is then

$$\Delta = -2 \left(\frac{t_G}{t_{day}} \right) \left(\frac{v_P}{c} \right)^2 t_G \qquad (10.37)$$

The average trip flying time was $t_G = 45$ hours. For an average jet aircraft speed of 220 m/s we predict a clock difference very close to the measured value in (10.27).

B. Length Contraction

Another kinematic effect of special relativity is an observed shortening of the lengths of moving objects. A stick of rest length L_0, at rest in the S' frame, lying parallel to the z' axis, is measured in the S' frame to have rest length L_0. What is the length of this stick as measured by an observer in S (cf. Fig. 10-3)? By measured we mean that the observer determines the difference in z coordinates of the ends of the stick at a fixed time t. From the inverse Lorentz transformation (10.24) we have

$$z_2' - z_1' \equiv L_0 = \gamma(z_2 - z_1 - v(t_2 - t_1)) = \gamma(z_1 - z_2) = \gamma L \qquad (10.38)$$

The stick's length $L = z_2 - z_1$ as seen from S is thus

$$L = L_0/\gamma \qquad (10.39)$$

The moving stick thus is measured to be shorter by the observer (S) in motion. Similarly, an S' observer will see a contracted length for a stick at rest in frame S. As a historical footnote, length contraction was first proposed by G.F. Fitzgerald and independently by H.A. Lorentz over a decade before relativity theory as the explanation of the null Michelson-Morley experiment. They suggested that if the apparatus arm along the ether wind direction contracted by a factor γ then the time t_1 of (10.1) would be exactly equal to t_2 of (10.2) and no phase difference would be observed.

C. Velocity Addition

The content of the Lorentz transformation can be further illuminated from the inferred relations between a particle's velocity in different frames. The velocity relations follow from the differentials of the Lorentz transformation (10.22)

$$dx = dx'$$
$$dy = dy'$$
$$dz = \gamma(dz' + v\,dt') \tag{10.40}$$
$$dt = \gamma\left(dt' + \frac{v}{c^2}dz'\right)$$

The velocities along the direction of v are

$$u_z \equiv \frac{dz}{dt}$$
$$u'_z \equiv \frac{dz'}{dt'} \tag{10.41}$$

and the velocities transverse to the direction of v are

$$\mathbf{u}_\perp \equiv \frac{dx}{dt}\hat{\mathbf{x}} + \frac{dy}{dt}\hat{\mathbf{y}}$$
$$\mathbf{u}'_\perp \equiv \frac{dx'}{dt'}\hat{\mathbf{x}} + \frac{dy'}{dt'}\hat{\mathbf{y}} \tag{10.42}$$

Using ratios of (10.40) along with (10.41) and (10.42), the following velocity addition formulae are obtained.

$$u_z = \frac{v + u'_z}{1 + vu'_z/c^2} \tag{10.43}$$

$$\mathbf{u}_\perp = \frac{\mathbf{u}'_\perp}{\gamma(v)\,(1 + vu'_z/c^2)} \tag{10.44}$$

When the velocities are small compared to c these approach the Galilean velocity transformation $u_z = v + u'_z$ and $\mathbf{u}_\perp = \mathbf{u}'_\perp$. If we go to the other limit of velocities approaching that of light it can be shown that if v and $\mathbf{u}'$ are less than or equal to c then $\mathbf{u}$ is also bounded by c.

We next express the velocity addition formulae in a way that will be useful for our later discussion of relativistic momentum and energy. The crucial step is the Lorentz transformation of $c^2 dt^2 - d\mathbf{r}^2$. From (10.40) one calculates

$$
\begin{aligned}
c^2 dt^2 - d\mathbf{r}^2 &= c^2 dt^2 - dx^2 - dy^2 - dz^2 \\
&= c^2 \gamma^2 \left(dt' + v dz'/c^2 \right)^2 - dx'^2 - dy'^2 - \gamma^2 \left(dz' + v dt' \right)^2 \\
&= c^2 dt'^2 - d\mathbf{r}'^2
\end{aligned}
$$

$$(10.45)$$

(that is, $c^2 dt^2 - d\mathbf{r}^2$ is an invariant of the transformation). The left-hand side can be rewritten as

$$c^2 dt^2 (1 - \mathbf{u}^2/c^2) = c^2 dt^2/\gamma^2(u) \qquad (10.46)$$

and the right-hand side similarly; thus (10.45) can be written

$$dt^2/\gamma^2(u) = dt'^2/\gamma^2(u') \qquad (10.47)$$

We can take the square root of both sides. The sign is fixed by observing that according to (10.40) $dt = \gamma(1 + vu_z'/c^2)dt'$; hence if both v and u_z' are less than c in magnitude, the ratio dt/dt' is positive. So the result is

$$dt/\gamma(u) = dt'/\gamma(u') \qquad (10.48)$$

This ratio is known as the *proper time* interval and has the same numerical value for all observers. For zero velocity the proper time interval is the time interval in that frame which then yields the time dilation formulae (10.25) and (10.26). If we now divide the left and right sides of each of the equations (10.40) by the left and right side, respectively, of (10.48), we get

$$\gamma(u)\mathbf{u}_\perp = \gamma(u')\mathbf{u}'_\perp \qquad (10.49)$$
$$\gamma(u)u_z = \gamma(v)\left[\gamma(u')u_z' + v\gamma(u')\right] \qquad (10.50)$$
$$\gamma(u) = \gamma(v)\left[\gamma(u') + v\gamma(u')u_z'/c^2\right] \qquad (10.51)$$

According to this, the Lorentz transformation law of the four quantities $\gamma(u)u_x$, $\gamma(u)u_y$, $\gamma(u)u_z$, $\gamma(u)$ is just the same as of dx, dy, dz, dt or of x, y, z, t.

10.5 Relativistic Momentum and Energy

In the absence of external forces the total momentum of a system should remain constant in time. In the S' frame we have

$$\mathbf{P}' = \sum_i \mathbf{p}_i' = \text{(constant vector)}' \tag{10.52}$$

According to the principle of relativity, when this system is observed from a moving frame S the total momentum will also remain fixed in time

$$\mathbf{P} = \sum_i \mathbf{p}_i = \text{(constant vector)} \tag{10.53}$$

For example, the total momentum of a jar of gas in space is zero in some reference frame and remains zero even though the momenta of the constituent molecules are constantly changing due to collisions with each other and the jar. Assuming for the moment that the system components move with non-relativistic velocities, the total momentum in (10.52) is

$$\mathbf{P}' = \sum_i m_i \mathbf{u}_i' \tag{10.54}$$

where the momentum of each mass is $\mathbf{p}_i' = m_i \mathbf{u}_i'$. An observer in S can also calculate the total momentum. If the relative frame velocity $v \ll c$ the Galilean velocity law is valid then

$$\mathbf{u}_i = \mathbf{v} + \mathbf{u}_i' \tag{10.55}$$

and the momentum as seen in S is

$$\mathbf{P} = \sum_i m_i \mathbf{u}_i = M\mathbf{v} + \mathbf{P}' \tag{10.56}$$

where $M \equiv \sum_i m_i$ is the total mass. We see that if the total momentum is constant in one inertial coordinate frame then the momentum as observed from a different inertial frame is also constant, in accord with the principle of relativity.

If the particle and/or the relative frame velocities are large, the Galilean velocity transformation formula is inaccurate. Using the relativistic velocity addition formulae and (10.53) and (10.54), a constant $\mathbf{P}'$ no longer implies a constant $\mathbf{P}$. In order to retain the principle of relativity, we must give up Newton's expression for momentum $\mathbf{p}_i = m_i \mathbf{u}_i$.

We will show that the correct relativistic alteration of $\mathbf{p}_i = m_i\mathbf{u}_i$ is

$$\mathbf{p}_i = m_i\mathbf{u}_i\gamma(u_i) \tag{10.57}$$

This $\mathbf{p}_i$ reduces to $m_i\mathbf{u}_i$ in the non-relativistic limit since

$$\gamma(0) = 1 \tag{10.58}$$

Consider the transverse components of the system's momentum as observed in frames S and S'

$$\mathbf{P}'_\perp = \sum_i [m\mathbf{u}'_\perp\gamma(u')]_i \tag{10.59}$$

$$\mathbf{P}_\perp = \sum_i [m\mathbf{u}_\perp\gamma(u)]_i \tag{10.60}$$

Using the velocity transformation (10.49) the S frame transverse momentum of (10.60) can be written as

$$\mathbf{P}_\perp = \sum_i [m\mathbf{u}'_\perp\gamma(u')]_i \tag{10.61}$$

We see immediately that $\mathbf{P}_\perp$ and $\mathbf{P}'_\perp$ have the same form and value. Thus the assumed form of relativistic momentum (10.57) assures that if $\mathbf{P}_\perp$ is conserved as viewed from one reference frame that it will be conserved for any observer. Moreover all observers will agree on its value.

Turning to the z-component of the system's momentum, as viewed in S'

$$P'_z = \sum_i [mu'_z\gamma(u')]_i \tag{10.62}$$

while viewed in S

$$P_z = \sum_i [mu_z\gamma(u)]_i \tag{10.63}$$

Using the velocity transformation of (10.50), (10.63) becomes

$$P_z = \gamma(v) \sum_i [m(u'_z + v)\gamma(u')]_i \tag{10.64}$$

Thus we obtain

$$P_z = \gamma(v) \left[P'_z + \frac{v}{c^2}E' \right] \tag{10.65}$$

where we have defined

$$E' = \sum_i \left[mc^2\gamma(u')\right]_i \qquad (10.66)$$

If for an observer in S' the quantity E' is constant, as well as the momentum component P'_z, the identity (10.65) implies that the corresponding momentum component P_z will be constant as observed in frame S. We thus obtain relativistic momentum conservation in all inertial frames but only if E' is also conserved.

This shows that for an observer in S', if the quantity E', as well as P'_z, is constant, then the momentum component P_z is constant as observed in S. Then also E is constant as observed in all frames, as shown explicitly by its transformation law for E, derived in the same way as (10.65)

$$E = \gamma(v)\left[E' + vP'_z\right] \qquad (10.67)$$

So in special relativity the constancy of $\mathbf{P}$ in all frames requires also the constancy of the new quantity E, in all frames. Einstein considered this the most significant result of his theory of special relativity.

To interpret E, we take the non-relativistic limit for a single particle

$$E = mc^2\gamma(u) = \frac{mc^2}{\sqrt{1 - \frac{u^2}{c^2}}} \simeq mc^2\left[1 + \frac{u^2}{2c^2} + \cdots\right]$$
$$E \simeq mc^2 + \frac{1}{2}mu^2 + \cdots \qquad (10.68)$$

We recognize that the term $\frac{1}{2}mu^2$ is the *non-relativistic kinetic energy* of the particle. Consequently we associate E with the *relativistic energy* of a particle. The value of E for zero velocity is the *rest energy* of the particle

$$E(u = 0) = mc^2 \qquad (10.69)$$

The *relativistic kinetic energy* K is defined in general as the difference of the total energy and the rest energy

$$K = E - mc^2 \qquad (10.70)$$

To see how relativistic particle energy conservation works we consider the collision of two slowly moving putty balls each of mass m, as illustrated in Fig. 10-6. In the usual Newtonian analysis carried out in the center-of-mass system the initial kinetic energy is mv^2 and the final kinetic energy is zero so the mechanical energy is converted into heat energy

$$Q = mv^2 \qquad (10.71)$$

which warms up the final putty ball. If we conserve relativistic free particle energy we find

$$E_{\text{initial}} \simeq 2\left(mc^2 + \tfrac{1}{2}mv^2\right) = E_{\text{final}} \equiv Mc^2 \qquad (10.72)$$

Thus the final putty ball mass is

$$M = 2m + \frac{mv^2}{c^2} = 2m + \frac{Q}{c^2} \qquad (10.73)$$

The heat energy is equivalent to an increase of mass of Q/c^2. This additional mass Q/c^2 adds to the inertia of the ball and in all respects is equivalent to a "real" mass. It is in the realm of nuclear and elementary particle physics where the concept of relativistic particle energy conservation becomes of crucial importance.

before collision

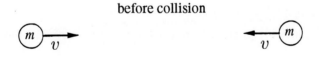

after collision

FIGURE 10-6. Collision of two non-relativistic putty balls in the CM system. The resulting single ball is warmer than before. It also is more massive than the sum of the original putty masses at rest.

Consider the important fusion reaction where two isotopes of hydrogen convert to helium and a neutron

deuterium	tritium	Helium	neutron
atom	atom	atom	

$$\text{H}^2 \quad + \quad \text{H}^3 \quad \rightarrow \quad \text{He}^4 \quad + \quad n$$

This reaction plays a central role in the energy production in stars, controlled fusion and the hydrogen bomb. In the CM system the initial energy is

$$E_i = (M_{\text{H}^2} + M_{\text{H}^3})\, c^2 + K_i \tag{10.74}$$

The final energy is

$$E_f = (M_{\text{He}^4} + M_n)c^2 + K_f \tag{10.75}$$

By energy conservation $E_i = E_f$, and the change in kinetic energy is

$$
\begin{aligned}
K_f - K_i &= (M_{\text{H}^2} + M_{\text{H}^3} - M_{\text{He}^4} - n)c^2 \\
&= \left[2.014102 + 3.016050 - 4.002603 - 1.008665\right] (931.5 \text{ MeV}) \\
&= 17.6 \text{ MeV}
\end{aligned}
\tag{10.76}
$$

Thus this reaction releases 17.6 million electron volts of kinetic energy.

We summarize the transformation laws of energy and momentum

$$
\begin{aligned}
\mathbf{P}_\perp &= \mathbf{P}'_\perp \\
P_z &= \gamma \left(P'_z + \frac{v}{c^2} E' \right) \\
E &= \gamma \left(E' + v P'_z \right)
\end{aligned}
\tag{10.77}
$$

Comparing with the coordinate-time Lorentz transformation (10.22) we see that $\mathbf{P}c$ transforms like $\mathbf{r}$ and E/c transforms like time. From the transformation (10.77) it follows directly that

$$E^2 - (Pc)^2 = E'^2 - (P'c)^2 \tag{10.78}$$

That is, this combination is frame independent. This is the momentum-energy analog of $r^2 - (ct)^2 = r'^2 - (ct')^2$. For a system consisting of a single particle we have

$$E^2 - (pc)^2 = \left(mc^2\gamma(u)\right)^2 - \left(mu\gamma(u)c\right)^2 = \left(mc^2\right)^2$$

or

$$E = \sqrt{(pc)^2 + \left(mc^2\right)^2} \tag{10.79}$$

The rest energy of the particle is mc^2.

As an example of the use of the Lorentz invariance relation (10.78) we consider the collision of a positron (e^+) with an electron (e^-) resulting in a pair of equal mass pi-mesons π^+ and π^-

$$e^+ + e^- \rightarrow \pi^+ + \pi^- \tag{10.80}$$

The pi-mesons each have a mass $m_\pi = 273.1 m_e$ and hence the e^+e^- collision must be quite energetic to cause this reaction to occur. One way to do this experiment is to collide the e^+ on an electron at rest. In the lab system $E = E_+ + m_e c^2$ and $P = p_+$. The left-hand side of (10.78) is

$$\begin{aligned} E^2 - (pc)^2 &= E_+^2 + 2E_+ m_e c^2 + (m_e c^2)^2 - (p_+ c)^2 \\ &= 2E_+ m_e c^2 + 2(m_e c^2)^2 \end{aligned} \tag{10.81}$$

where we have used (10.79) for the positron. In the CM system $\mathbf{P'} = 0$ and the invariant is just the relativistic energy squared. Relativistic energy is conserved in a collision and the minimum energy of the two resulting pi-mesons is their rest energy $E' = 2m_\pi c^2$; from (10.81) the threshold energy is

$$E_+ = \frac{2m_\pi^2 c^2 - m_e^2 c^2}{m_e} \tag{10.82}$$

Numerically $m_\pi c^2 = 139.57\,\text{MeV}$ and $m_e c^2 = 0.511\,\text{MeV}$ so the threshold lab positron energy is $E_+ = 76.24\,\text{GeV}$. This experiment has been carried out at the Fermilab and CERN particle accelerators to probe the electric charge distribution of the pi-meson.

10.6 Relativistic Dynamics

The generalization of Newton's law of motion to relativistic velocities is

$$\frac{d\mathbf{p}}{dt} = \mathbf{F} \tag{10.83}$$

where

$$\mathbf{p} = m\mathbf{v}\gamma(v) \tag{10.84}$$

For a charged particle moving in an electromagnetic field $\mathbf{F}$ is the Lorentz force

$$\mathbf{F} = q(\mathbf{E} + \mathbf{v} \times \mathbf{B}) \tag{10.85}$$

where q is the charge of the particle. The predicted motion has been accurately tested in particle accelerators and beams, electron lenses and numerous other ways.

As an illustration consider the motion of a charge q released from rest at $t = 0$ in a constant electric field $\mathbf{E} = E\hat{\mathbf{z}}$. The equation of motion is

$$\frac{d}{dt}\left(\frac{m\dot{z}}{\sqrt{1 - \dot{z}^2/c^2}}\right) = qE \tag{10.86}$$

We can directly integrate and impose the boundary value $\dot{z}(t = 0) = 0$ to obtain

$$\frac{\dot{z}/c}{\sqrt{1 - \dot{z}^2/c^2}} = \frac{t}{t_0} \tag{10.87}$$

where

$$t_0 = \frac{mc}{qE} \tag{10.88}$$

Solving for $\dot{z}$ gives

$$\frac{\dot{z}}{c} = \frac{t/t_0}{\sqrt{1 + (t/t_0)^2}} \tag{10.89}$$

Here the sign of the square root was chosen to agree with (10.87). The velocity increases from zero and rises asymptotically to the speed of light.

After a time $t = t_0$, the velocity is $\dot{z}(t_0) = c/\sqrt{2}$. For an electron in an electric field of $10^6 \, V/m$ this time is

$$t_0 = \frac{(9.1 \times 10^{-31})(3 \times 10^8)}{(1.6 \times 10^{-19})(10^6)} = 1.7 \times 10^{-9} \, \text{s} \qquad (10.90)$$

A final integration of (10.89) yields

$$z = c \, t_0 \int_0^{t/t_0} d\tau \, \frac{\tau}{\sqrt{1+\tau^2}} = c \, t_0 \left[\sqrt{1 + (t/t_0)^2} - 1 \right] \qquad (10.91)$$

The distance covered in time t_0 is

$$z = c \, t_0 \left(\sqrt{2} - 1 \right) = 0.21 \, \text{m} \qquad (10.92)$$

The electron energy after this distance will be

$$\begin{aligned} E &= (0.511 \, \text{MeV})\gamma \\ &= (0.511)\sqrt{2} = 0.72 \, \text{MeV} \end{aligned} \qquad (10.93)$$

and its kinetic energy is

$$K = E - m_e c^2 = 0.21 \, \text{MeV} \qquad (10.94)$$

It is thus relatively easy to use electric fields to accelerate electrons to relativistic energies. The deflection of relativistic electrons by magnetic and electric fields provided the earliest tests of the correctness of relativistic dynamics.

Finally we should emphasize that a better name for the speed of light would be "limiting velocity". Any body with rest mass, if accelerated continually, approaches c arbitrarily closely. Particles having zero rest mass travel exactly with the speed of light c.

PROBLEMS

10.3 Lorentz Transformation

10-1. Invert the Lorentz transformation (10.22) to solve for S' coordinates in terms of S coordinates as given in (10.24). Show that the operation $x' \to x$, $y' \to y$, etc. and $v \to -v$ give the same result.

10-2. Show that the Lorentz transformation can be represented as a rotation in the z, ict plane through an imaginary angle.

10.4 Consequences of Relativity

10-3. From § 8.4 the rates of a clock at a height h (S' frame) as seen from the ground (S frame) is

$$\nu = \nu'(1 + gh/c^2)$$

Consider the flying clocks of § 10.4A

a) The sum of the east- and west-going clock differences is

$$\Sigma \equiv t_{\mathrm{EG}} + t_{\mathrm{WG}} = \Sigma_v + \Sigma_h$$

where Σ_v is due to clock motion and Σ_h is due to the height of the clock. Show that

$$\Sigma_v = -t_{\mathrm{G}} \left(\frac{v_{\mathrm{P}}}{c}\right)^2$$
$$\Sigma_h = 2t_{\mathrm{G}} gh/c^2$$

b) Using the plane velocity of (10.37) as determined from Δ compute Σ_v. Assuming the plane flies at a height of 10 km estimate Σ_h and compare with the measured value of $\Sigma = 214$ ns.

10-4. Two spaceships are traveling at speeds of $0.9c$ but in opposite directions. What is the speed of one as seen by the other?

10-5. The *pseudorapidity* of a particle moving in frame S with z-component of velocity u_z is defined by $\eta_z \equiv \tanh^{-1}(u_z/c)$. Its pseudorapidity in frame S' is $\tanh^{-1}(u_z'/c)$. Show that $\eta_z = \eta + \eta_{z'}$ where $\eta = \tanh^{-1}(v/c)$ and v is the velocity of frame S' relative to S. Use this result to demonstrate that if $v < c$ and $u_z' < c$ then $u_z < c$.

10.5 Relativistic Momentum and Energy

10-6. For a body of mass m and velocity $\mathbf{u}$ show that $\mathbf{p}/E = \mathbf{u}/c^2$.

10-7. The Tevatron at Fermilab can produce electrons of energy 0.5 TeV (10^{12} eV). Recall that the electron's rest energy is 0.511×10^6 eV. Suppose a race were held from the earth to the moon between one of these electrons and a photon. What distance would the electron lose by? *Hint: Use approximate formulae accurate for electron velocities nearly equal to the speed of light.*

10-8. A hyperon particle Λ has a rest mass of 1.116 GeV/c^2 and a mean lifetime of 2.6×10^{-10} s. A beam of Λ's can be produced at Fermilab with energy 30 GeV.

   a) Find an approximate formula for the difference between the Λ's velocity and c in terms of γ. Find γ and v in the present case.

   b) What is the mean distance that the Λ goes before decay?

   c) What would the mean decay distance be without relativistic lifetime dilation?

10-9. A relativistic particle of momentum p_1 and mass m_1 collides with a particle of mass m_2 at rest.

   a) Find the center of momentum (CM) velocity. *Hint: use the inverse momentum/energy Lorentz transformation of* (10.77).

   b) Find the relativistically correct relation between the CM and lab angles of particle 1 after an elastic collision.

10-10. In order to travel to the stars a spaceship must reach a significant fraction of the speed of light. The only practical way of doing this is to use matter-antimatter annihilation as fuel. Suppose that a mass dm is so converted into a massless reaction product having momentum $(dm)\, c$,

   a) Following an analogous argument to that in §4.1, show that the rocket equation in free space is

$$m\frac{dv}{dm} + c\left(1 - \frac{v^2}{c^2}\right) = 0$$

where the momentum-energy Lorentz transformation provides the exhaust momentum in the rest system to be $dm\,\gamma(c - v)$.

b) If $m = m_0$ when $v = 0$, integrate the rocket equation and show that

$$v = c \left[1 - \left(\frac{m}{m_0} \right)^2 \right] / \left[1 + \left(\frac{m}{m_0} \right)^2 \right]$$

How much fuel must be burned to reach a speed of $\frac{3}{5}c$?

c) Demonstrate that if $\Delta m = m_0 - m \ll m_0$ the result reduces to the non-relativistic force-free rocket for an exhaust velocity of the speed of light $(u = c)$.

10-11. To gain a perspective on the feasibility of interstellar travel suppose we wish to travel to a star 10 light years distant in a time of about 100 years. Assume the spacecraft has an antimatter drive of high efficiency and that mass of the fuel required is much less than the mass of the spacecraft.

a) The spacecraft accelerates at 10 m/s² to a cruising velocity of $\frac{1}{10}c$, coasts for most of the trip and then decelerates at the same rate. Find the time of acceleration.

b) What fraction of the total mass must be fuel?

c) If 1 gigawatt (10^9 W) electric generating facility were dedicated to producing anti-hydrogen at perfect efficiency for ten years, how massive a spacecraft could be sent?

10.6 Relativistic Dynamics

10-12. Show that the single-particle Lagrangian

$$L = -mc^2 \sqrt{1 - v^2/c^2} - V(\mathbf{r})$$

yields the equation of motion (10.83). Find the corresponding Hamiltonian and compute Hamilton's equations.

10-13. Starting with (10.83) in one dimension compute the work done $W = \int F\, dx = \int \frac{d}{dt}(mv\gamma)v\, dt$. Derive the relativistic work-energy theorem. Interpret the meaning of the resulting energy.

Chapter 11

NON-LINEAR MECHANICS: APPROACH TO CHAOS

There are two major reasons for studying non-linear mechanics. The first and most basic is that the equations of motion of almost all systems are non-linear. The second reason is that even a relatively simple system which obeys a non-linear equation of motion can exhibit unusual and surprisingly complex behavior for certain ranges of the system parameters. In a variety of non-linear systems the same features show up. Much of the knowledge of non-linear behavior has been obtained from numerical solutions. The traditional methods of classical mechanics, which lead to analytic (explicit) expressions for motion, fail for most problems in non-linear mechanics. Numerical integration of the equations of motion is usually necessary.

From the time of Newton until the twentieth century, physicists and philosophers viewed the universe as a sort of enormous clock which, once wound up, behaves in a predictable manner. This conception was badly shaken by the discovery of quantum mechanics and its Heisenberg uncertainty principle. However even in classical mechanics the motion of a system obeying Newton's equations is sometimes very difficult to predict; the system is then said to exhibit chaos.

The essential difference between a chaotic system and a non-chaotic one is the degree of predictability of the motion given the initial conditions to some level of accuracy. We recall some familiar examples. Suppose that two identical linear damped oscillators with the same sinusoidal driving force are started off with different initial conditions. After a time long enough for transients to have died out, the oscillators will end up moving in exact synchronism. Thus the final motion here is independent of the initial conditions. As a second example consider a point mass falling near the earth's surface. For initial height z_0 and velocity v_0, the height is given by the solution to Newton's laws to be $z(t) = z_0 + v_0 t - \frac{1}{2}gt^2$. If the initial conditions were instead $z_0 + \delta z_0$ and $v_0 + \delta v_0$ the two trajectories would differ by $\delta z(t) = \delta z_0 + (\delta v_0)t$. Thus the difference in height changes at most linearly in time. The dependence of the difference of two solutions on t to some power is what is normally encountered when an analytic solution to Newton's law exists.

One of the best known examples of poor predictability is the weather. At one time it was believed that with a large number of atmospheric measurements and powerful computers to integrate the fluid mechanics equations it would be possible to make long term weather predictions. It is now realized that this was a naive hope and that the solutions of the weather equations are exponentially sensitive to the initial conditions. One can imagine that the perturbation due to a butterfly flapping its wings in Africa may grow exponentially into a great weather front in North America (the *butterfly effect*).

Even though the motion of a complex system cannot be precisely predicted certain features can often still be relied upon. For example, the exact path of a given molecule of water down a white-water stretch of the Colorado River is certainly not predictable. We can say with high probability, however, that the molecule will traverse the rapids and remain within the canyon walls. It is a real challenge to deduce such a 'robust feature' of the solutions of a given nonlinear equation.

11.1   The Anharmonic Oscillator

The driven one-dimensional non-linear oscillator provides a good illustration of the fundamental ideas in non-linear mechanics and chaotic motion; it also has immediate applicability to many actual physical problems. We shall consider here an oscillator called the *anharmonic oscillator*, which is a mass m subject to a conservative force

$$F(x) = -k(x + \alpha x^3) \qquad (11.1)$$

(The coordinate x of the mass will often be referred to as the *displacement*, since $x = 0$ is a point of static equilibrium.) If k (the spring constant at zero displacement) is positive and the coefficient α of the cubic term vanishes, $F(x)$ is a Hooke's law spring force, and the motion $x(t)$ will be simple harmonic. If the anharmonic coefficient α is instead positive, the restoring spring force grows faster than linear in x and the spring is said to be *hard*. If α is negative the spring is called *soft*.

The potential energy corresponding to the force (11.1) is

$$V(x) = -\int_0^x F(x)\,dx = k\int_0^x dx\,(x + \alpha x^3) = k\left(\tfrac{1}{2}x^2 + \tfrac{1}{4}\alpha x^4\right) \qquad (11.2)$$

Figure 11-1 illustrates the potential energy for hard and soft springs ($k >$ 0), the double-well ($k < 0$, $\alpha < 0$) and inverted-well ($k < 0, \alpha > 0$) cases. For the double-well there are three equilibrium points ($F = 0$) at

$$x = 0, \qquad x = \pm\sqrt{\frac{1}{|\alpha|}} \tag{11.3}$$

The effective spring constants $k_{\text{eff}} = -d^2V/dx^2$ at these points are

$$k_{\text{eff}} = -k, \qquad k_{\text{eff}} = 2k \tag{11.4}$$

The equilibrium is unstable at $x = 0$ and stable at the other two points; see Fig. 11-1(c).

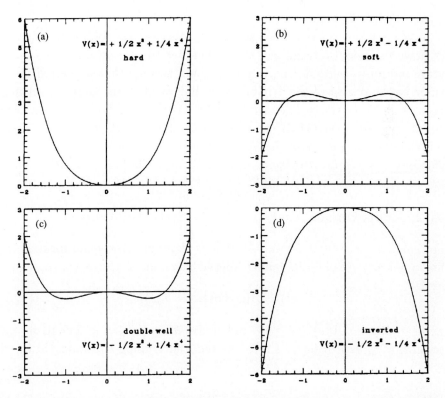

FIGURE 11-1.   Potential energies of (11.2) with $k = \pm 1$ and $\alpha = \pm 1$. The free (undriven) motion is confined in (a) and (c), unconfined in (d), and may be either confined or unconfined in (b). We will confine our discussion almost exclusively to the hard spring (a).

In addition to the anharmonic spring force (11.1) we assume that the mass m is subject to a linear damping force $F_{\text{damp}} = -2\gamma m\dot{x}$ and a simple harmonic (sinusoidal) time-dependent driving force $F_{\text{drive}} = mf\cos\omega t$. The resulting equation of motion is

$$\ddot{x} + 2\gamma\dot{x} + \omega_0^2 x + \beta x^3 = f\cos\omega t \tag{11.5}$$

where as usual $\omega_0^2 = k/m$. This driven, damped, anharmonic equation of motion is known as *Duffing's equation*. It is convenient to reduce the number of parameters in the equation from five to three by choosing the scales of t and x. By choice of the scale of time the small-amplitude frequency ω_0 can be made 1, and similarly by rescaling x the coefficient of the anharmonic (cubic) term can be made ± 1. The equation becomes the *reduced Duffing equation*

$$\ddot{x} + 2\gamma\dot{x} + x \pm x^3 = f\cos\omega t \tag{11.6}$$

We shall restrict our attention to the steady-state solutions of (11.6). A powerful analytic technique to find these is to assume that the steady-state motion is periodic (naively one would assume that its period is the same as the period of the driving force, but we shall see later that it may be a multiple of this.) Then $x(t)$ can be written as a Fourier series. The equation of motion (11.6) then provides relations which determine the amplitudes and phases of the Fourier terms. In the next section we work this out in the approximations of keeping only the first term or the first two terms of the Fourier series.

11.2 Approximate Analytic Steady-State Solutions

For small f we assume a trial solution of the same functional form as the linear spring case of § 1.9, namely simple harmonic at the driver frequency

$$x(t) = A(\omega)\cos\left[\omega t - \theta(\omega)\right] \tag{11.7}$$

with amplitude $A(\omega)$ ($A \geq 0$) and phase $\theta(\omega)$ ($0 \leq \theta < 2\pi$) to be determined. Substituting this into the reduced Duffing equation (11.6) we obtain

$$A(1-\omega^2)\cos(\omega t-\theta)-2\gamma\omega A\sin(\omega t-\theta)\pm A^3\cos^3(\omega t-\theta) = f\cos\omega t \tag{11.8}$$

This is to hold for all values of t. If the four functions of t in (11.8) (namely $\cos(\omega t - \theta)$, $\sin(\omega t - \theta)$, $\cos^3(\omega t - \theta)$ and $\cos\omega t$) were linearly independent, then their coefficients would all have to vanish. In fact they are not independent. The systematic procedure is to write them in terms of $\cos n\psi$, $n = 0, 1, 2, \ldots$, and $\sin n\psi$, $n = 1, 2, \ldots$, where $\psi = \omega t - \theta$ (it would be less convenient to choose $\psi = \omega t$) because we know that these are a complete and independent set of functions of t of period $2\pi/\omega$. We use the following trigonometric identities

$$\cos^3(\omega t - \theta) = \tfrac{3}{4}\cos(\omega t - \theta) + \tfrac{1}{4}\cos 3(\omega t - \theta)$$
$$\cos \omega t = \cos\theta\cos(\omega t - \theta) - \sin\theta\sin(\omega t - \theta) \qquad (11.9)$$

which can be established using the complex exponential forms of cosine and sine. Then (11.8) becomes

$$\begin{aligned}
&\left[A\left(1 - \omega^2 \pm \tfrac{3}{4}A^2\right) - f\cos\theta\right]\cos(\omega t - \theta) \\
&+ \left[-2\gamma\omega A + f\sin\theta\right]\sin(\omega t - \theta) \qquad (11.10) \\
&\pm \tfrac{1}{4}A^3\cos 3(\omega t - \theta) = 0
\end{aligned}$$

This has the form $a\cos\psi + b\sin\psi + c\cos 3\psi = 0$, which is equivalent to the three equations $a = 0$, $b = 0$, $c = 0$. It is easy to see that these conditions cannot be satisfied by any values of A and θ (if $f \neq 0$) which merely tells us that $x(t)$ of the simple form (11.7) cannot be an exact solution. But the first two of these three equations imply $A \propto f$, and so if $|f| \ll 1$ the third equation, $A^3 = 0$, is nearly satisfied. An alert reader might complain that if the last term of (11.10) is going to be neglected on the grounds that A^3 is very small, then the third term, $\pm\tfrac{3}{4}A^3\cos(\omega t - \theta)$, can be dropped as well, since it has the same order of magnitude. However, in the next approximation, Eq. (11.15), the contribution of the term $B\cos 3(\omega t - \phi)$ cancels the last term of (11.10), while contributing only higher order terms to the first two terms. So for small f (11.10) is nearly satisfied if

$$f\cos\theta = A\left(1 - \omega^2 \pm \tfrac{3}{4}A^2\right) \qquad (11.11)$$
$$f\sin\theta = 2\gamma\omega A \qquad (11.12)$$

hold. Eliminating θ by summing the squares of these two conditions gives

$$f^2 = A^2\left[\left(1 - \omega^2 \pm \tfrac{3}{4}A^2\right)^2 + (2\gamma\omega)^2\right] \qquad (11.13)$$

After solving this for A, (11.11) and (11.12) then determine θ. Note that in the limit $A \to 0$ the results reduce to the forced harmonic oscillator solution of (1.123) and (1.125), as one would expect, since the Duffing equation (11.6) reduces to the driven simple harmonic (linear) oscillator equation when $|x(t)| \ll 1$.

For a given frequency ω, (11.13) is cubic in A^2 and gives one or three real values for A^2. Alternatively, by regarding the equation as determining ω^2 for a given A, a quadratic equation results which is easier to solve. In Fig. 11-2 we plot A versus ω for the case of a hard spring with a damping constant $\gamma = 1/10$ and a force constant $f = 1/2$. The dashed curve is the amplitude-frequency relation for an undamped free oscillation, that is, the limit $f = \gamma = 0$ of (11.13)

$$\omega = \sqrt{1 \pm \tfrac{3}{4}A^2} \quad \text{or} \quad A(\omega) = \left[\pm \tfrac{4}{3}\left(\omega^2 - 1\right)\right]^{1/2} \qquad (11.14)$$

This curve is called the *spine* of the resonance; it is the locus of the resonance peak as f is varied, in the $\gamma \to 0$ limit. The anharmonic cubic term causes the resonance amplitude curve to lean over and if f is sufficiently large A and θ become triple valued over an interval in ω. (The corresponding resonance for a soft spring leans to the left.) The middle values in fact correspond to an *unstable* steady motion, and we will see later that numerical integration never gives this for the steady-state motion at large t. The phase angle $\theta(\omega)$ calculated from (11.11) and (11.12) is depicted in Fig. 11-3; again the phase is compared to the linear solution.

The next approximation beyond (11.7) is to keep the first two terms of the Fourier expansion of $x(t)$,

$$x(t) = A \cos(\omega t - \theta) + B \cos 3(\omega t - \phi) \qquad (11.15)$$

The second term is $\cos 3(\omega t - \phi)$, not $\cos 2(\omega t - \phi)$ because the first term when cubed gives a $3\omega t$ cosine, but not a $2\omega t$ cosine. More generally, it is consistent that $x(t)$ have only odd harmonics because the spring force $x + x^3$ will likewise. If this trial solution is substituted into Duffing's equation (11.6) the values of A, B, θ, and ϕ can be chosen to satisfy all harmonics through $3\omega t$. Figure 11-4 shows the resulting amplitude as a function of ω, for the same values of γ and f as in Fig. 11-2. The numerically calculated result is also shown (see § 11.3 below); the agreement is good. A new feature, not seen in the lowest approximation (Fig. 11-2), is a small resonance peak near $\omega \simeq 0.4$. This is called the third-harmonic resonance. The coefficient B of the third harmonic ($3\omega t$ term) of $x(t)$ peaks there

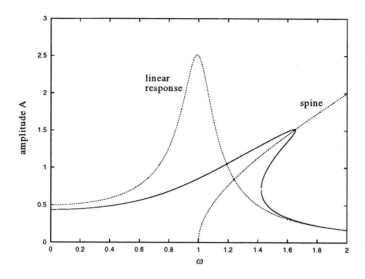

FIGURE 11-2. Hard spring approximate analytic amplitude (11.13) for $\gamma = 0.1$ and $f = 0.5$. The linear response is shown for reference. The other dashed curve is the "spine" of the resonance, (11.14).

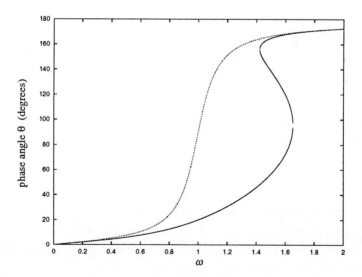

FIGURE 11-3. Hard spring approximate analytic phase angle θ from (11.11) and (11.12) for the same parameters as the preceding figure. The corresponding linear oscillator phase angle is shown by the dashed curve.

as a consequence of 3ω being a little larger than the natural frequency 1. Similarly because of higher odd harmonics one sees in Fig. 11-5 a series of harmonic resonances at frequencies $1/N$ where N is odd.

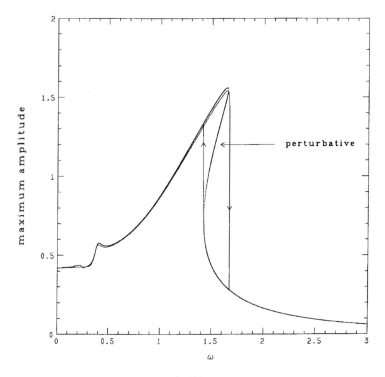

FIGURE 11-4. Steady-state amplitude $|x(t)|$ during a period for $\gamma = 0.1$ and $f = 0.5$. Hysteresis is seen at the primary resonance and one harmonic resonance is seen. The two-term approximate analytic prediction of (11.15) is compared with the numerical solution.

11.3   Numerical Solutions of Duffing's Equation

The differential equation of motion can be numerically integrated to any desired accuracy. The numerical algorithm we have used is the *fourth order Runge-Kutta* method, which for a given desired accuracy is considerably more efficient than the numerical method discussed in § 1.6. In the numerical work in this section we examine solutions to Duffing's equation (11.6) with γ fixed at the value 1/10 and for various values of the driving frequency ω and the driving amplitude f.

An important feature of the numerical result shown in Fig. 11-4 is *mechanical hysteresis*. In the frequency range $1.4 < \omega < 1.7$, where the steady-state amplitude is a triple-valued function of ω, initial conditions (for example, the values of x and $\dot{x}$ at $t = 0$) determine which of these three motions is the actual steady state reached at large t. A practical equivalent to choosing initial conditions is *sweeping in frequency*. One starts the oscillator off at some initial conditions and waits until the transient motion has damped out and the motion has become steady. The frequency ω is then changed slightly and one waits until the motion damps to a steady state at the new frequency.

In the present case, if one starts at a low frequency, where the steady state is unique, and sweeps up in ω, one finds that in the triple valued region the steady motion remains the one with the highest amplitude. At the top end of the region at $\omega = 1.7$, the amplitude drops abruptly as shown by the vertical line with downward-pointing arrow. On the other hand, if one starts at a high frequency and sweeps down in ω, the amplitude remains the lowest, and at the bottom end of the region at $\omega = 1.4$ it jumps up abruptly. This dependence of the steady-state motion on the direction of sweeping is called mechanical hysteresis. The middle-amplitude steady motion is never found as the steady motion at large t; it is in fact unstable. (Warning: not all stable steady-state motions can be found by sweeping in ω starting with a given steady state.)

Figure 11-5 shows the result of numerical calculation (and also the simplest analytic approximation) for a larger driving force, $f = 3$. In addition to the third-harmonic resonance there are seen other odd-harmonic resonances: 3,5,7... There is also seen something qualitatively new, an even-harmonic resonance in which the steady state $x(t)$ has a non-vanishing $2\omega t$ term. If $x_A(t)$ is a solution of the Duffing equation (11.6), then so is $x_B(t) \equiv -x_A(t + \pi/\omega)$. This is because the equation is unchanged if x is replaced by $-x$ and simultaneously t is shifted by a half period, π/ω. For the motions with $f = 0.5$ these two solutions were the same, that is, the motions had the property $x(t + \pi/\omega) = -x(t)$, which is equivalent to saying that $x(t)$ had only odd harmonics. In the present case with $f = 3$, when $\omega_{2-} < \omega < \omega_{2+}$ where $\omega_{2-} = 0.88$ and $\omega_{2+} = 1.05$, solution x_B is different from x_A, that is to say, the steady-state motion has even harmonics. This shows itself in the figure by the amplitude being double valued; the maximum (positive) values of x_A and x_B are different. (Actually there is always a steady-state motion which has only odd harmonics, but in the range $\omega_{2-} < \omega < \omega_{2+}$ it is unstable.)

Yet more complicated motions occur for larger values of the driving force. The numerical integration on a computer of the steady state equation of motion is equally straightforward (once you have a program) for any values of the parameters, but an intelligible description of the resulting steady-state motions is a challenge. A very useful concept is the *Poincaré section*. The motion is sampled periodically, at the period of the driving force $(2\pi/\omega)$, and the values of x and $\dot{x}$ at those times are

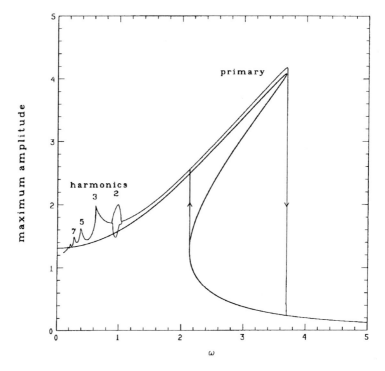

FIGURE 11-5. Steady-state amplitude $|x(t)|$ during a period for an attractor with $f = 3$ and $\gamma = 0.1$. Compared to Fig. 11-4 with $f = 0.5$ the emergence of new harmonic resonances should be noted, especially the second harmonic resonance. A smooth sweep was made up and down in frequency. The simplest approximate analytic calculation is compared with the numerical solution.

plotted as dots with coordinates x and $\dot{x}$. That is, it is a 'stroboscopic' picture of the motion of the oscillator in the $x, \dot{x}$ plane (*phase space*). Thus the nth dot has coordinates

$$x_n = x(t_0 + 2\pi n/\omega)$$
$$y_n = \dot{x}(t_0 + 2\pi n/\omega)$$
(11.16)

where t_0 determines the fixed phase of the driving force at which sampling occurs. The basic property of this is the following: a given pair of values x_n, y_n, *i.e.*, the values of x and $\dot{x}$ at the time $t_0 + 2\pi n/\omega$, determines (by way of the equation of motion) the motion, and therefore, the values of x_{n+1}, y_{n+1} (that is, x and $\dot{x}$ at the time one driving period later, $t_0 + 2\pi(n+1)/\omega$). That is to say

$$x_{n+1} = f(x_n, y_n)$$
$$y_{n+1} = g(x_n, y_n)$$
(11.17)

where the functions f and g *do not depend* on n. That is, the values of x_{n+1}, y_{n+1} determine the value of x_{n+2}, y_{n+2} exactly the same way as x_n, y_n determined x_{n+1}, y_{n+1}, because the equation of motion (11.6) is periodic; it is the same at t and at $t + 2\pi/\omega$. (The functions f and g do depend on the choice of t_0, that is, on the choice of the phase of the driving force at which the "stroboscope flashes".)

The pair of equations (11.17) is a sort of equation of motion; given x and $\dot{x}$ at one time in the sequence $t_0 + 2\pi n/\omega$, it yields x and $\dot{x}$ at the next time. In mathematical terminology, (11.17) describes a *mapping* of the $x, \dot{x}$ plane into itself.

A *fixed point* of the mapping is a point which maps to itself, *i.e.*

$$x = f(x, y)$$
$$y = g(x, y) \tag{11.18}$$

This corresponds to a steady state. If there are points which, after more and more repetitions of the mapping (*i.e.*, time steps) approach closer and closer to a fixed point, the fixed point is called an *attractor*.

In Figures 11-6 and 11-8 a simplified version of the attractors of the Poincaré section, namely just the x coordinate (called the Poincaré displacement), is plotted versus ω. The "stroboscopic" phase has been chosen to be zero, *i.e.*, at maximum driving force.

11.4   Transition to Chaos: Bifurcations and Strange Attractors

A new feature appears for $f = 20$, Fig. 11-6, namely *period doubling*. Whereas in previous figures the two values of the amplitude or of the Poincaré displacement seen in some ranges of ω corresponded to two attractors; here the two values seen in the range $1.2 < \omega < 1.4$ correspond to a single attractor, called a *two-cycle*, consisting of two points. That is, the attractor is a pair of points of period 2; calling the two points x_A, y_A and x_B, y_B, the mapping sends $x_A, y_A \to x_B, y_B \to x_A, y_A \to \cdots$. In other words, this steady-state motion has twice the period of the driving force. This is sometimes called a subharmonic motion, since the fundamental frequency of the motion is a subharmonic (a fraction) of the driving frequency.

What happens as ω increases past the critical value 1.2, or decreases past 1.4, is that the single-point (fixed-point) "simple" attractor turns into an unstable fixed point (not seen in the figures) and a two-point (period 2) attractor. One says the attractor *bifurcates*.

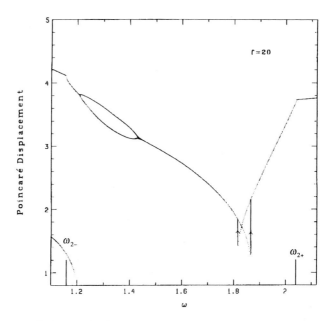

FIGURE 11-6. Poincaré displacement for $f = 20$. A period doubling bifurcation appears between $1.2 < \omega < 1.4$. Hysteresis at $\omega = 1.8$ has become prominent. Only one attractor is shown. As in Fig. 11-5, the other branch is on the other side of the same orbit.

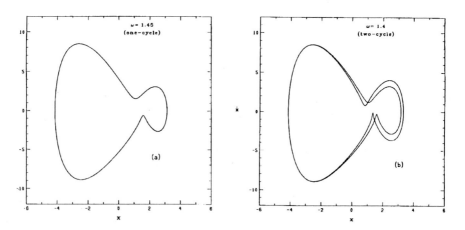

FIGURE 11-7. Two examples of phase space orbits in the frequency range of Fig. 11-6. In (a) at $\omega = 1.45$ a simple attractor is seen. In (b) at $\omega = 1.4$ the attractor has period 2.

In Fig. 11-7 we see what happens to the orbit in phase space in a period-doubling bifurcation. Referring back to Fig. 11-6 we note that at $\omega = 1.45$ the attractor is simple. The corresponding phase space orbit is shown in Fig. 11-7(a). Sweeping down in frequency to $\omega = 1.4$, the attractor bifurcates to a two-point attractor. The corresponding orbit is shown in Fig. 11-7(b). The orbit is similar to the previous one except now two driver periods are needed before it closes.

In Fig. 11-8 is shown the attractor (Poincaré displacement) for $f = 25$. One sees that as ω varies, period-doubling (bifurcation) occurs repeatedly (*cascade of bifurcation*), and the period of the attractor rises through the values 1,2,4,8,16... to ∞ at $\omega = \omega_\infty \approx 1.29$. This figure is a detailed view of the initial part of the bifurcation region showing the cascade of period doublings. The similarity to the well known *logistic* or *quadratic iterative map* is striking, as discussed in § 11.5.

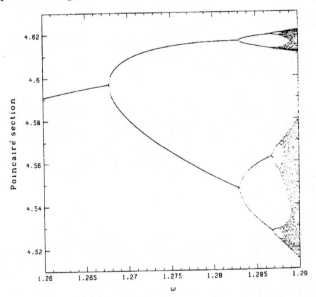

FIGURE 11-8. Poincaré displacement for $f = 25$. In this plot the amplitude is recorded each time the driving force is maximum. At the lowest driving frequencies a single point is found corresponding to a simple attractor. A cascade of period-doubling bifurcations occurs and by $\omega = 1.29$ the motion has become quite chaotic.

The full Poincaré section is shown in Fig. 11-9 at $\omega = 1.2902$, which is beyond $\omega = \omega_\infty$; the attractor is an infinite number of points. The "steady-state" motion of the oscillator at this f and ω is thus not periodic at all; the motion is *chaotic*. An attractor of this sort is known as a *strange attractor*. Its infinitude of points are arranged in a strange self-similar (fractal) manner. An expanded view of the portion of the attractor within the rectangle in Fig. 11-9 is shown in Fig. 11-10. In principle this magnification can be continued but numerical limitations soon intercede.

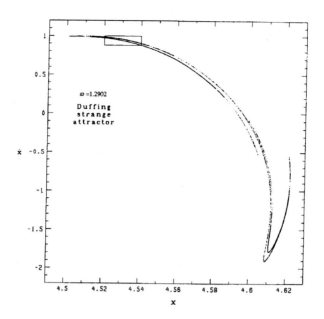

FIGURE 11-9. The Duffing strange attractor at $\omega = 1.2902$. The plot is a Poincaré section at maximum driving force. The plot contains ten thousand points, each from one driving period.

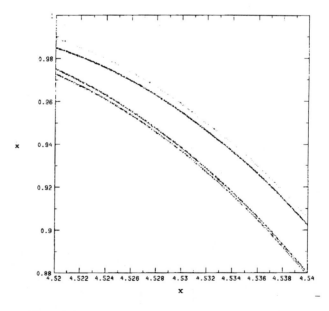

FIGURE 11-10. Magnified portion within the box of Fig. 11-9. The Poincaré section shown is based on 30,000 driving periods and shows some of the detail present in a strange attractor.

11.5   Aspects of Chaotic Behavior

We conclude this chapter by touching upon some general properties common to a wide range of chaotic motion and transitions to chaos. The first topic is a deceptively simple mapping which exhibits many of the aspects encountered in nonlinear differential equations. We then define the Hausdorf or fractal dimension which characterizes the geometry of the strange attractor and illustrate with the two dimensional Henon map. Finally we briefly discuss the Lyapunov exponent which is a measure of sensitivity to initial conditions and characterizes chaotic motion.

A. The Quadratic (or Logistic) Map

In (11.17) we defined a *mapping*, in which the mapping functions f and g were determined by the Duffing equation but not known explicitly. We now consider a much simpler mapping, in fact the simplest nontrivial (nonlinear) mapping for one variable, namely the quadratic map, also known as the logistic equation. This map is simple enough to explain to an elementary school child and to analyze on a pocket calculator yet subtle enough to capture the essence of a wide class of real world nonlinear phenomena. The map is

$$x_{n+1} = \lambda x_n (1 - x_n) \tag{11.19}$$

If

$$0 \leq \lambda \leq 4 \tag{11.20}$$

then $0 \leq x_n \leq 1$ implies $0 \leq x_{n+1} \leq 1$, so we can assume that x_n is always in the interval 0 to 1. The quadratic map function $\lambda x(1 - x)$ is illustrated for two values of λ in Fig. 11-11. The name 'logistic' refers to its origin as a simple population model.

If $\lambda < 1$ we see from (11.19) or Fig. 11-11 that $x_{n+1} < x_n$ for all x_n. The ultimate result of repeated iterations is thus inevitably $x = 0$. Thus when $\lambda < 1$ the mapping has one fixed point, which is an attractor. It is easy to find the fixed points of the mapping for any λ. By (11.19) the fixed-point condition is

$$x = \lambda x(1 - x) \tag{11.21}$$

with solutions

$$x = 0 \tag{11.22}$$

$$x = 1 - \lambda^{-1} \tag{11.23}$$

Geometrically the fixed point is the intersection of the quadratic map function with the line $x_{n+1} = x_n$; see Fig. 11-11. Note that the fixed point $x = 1 - \lambda^{-1}$ is in the interval 0 to 1 only when $\lambda \geq 1$.

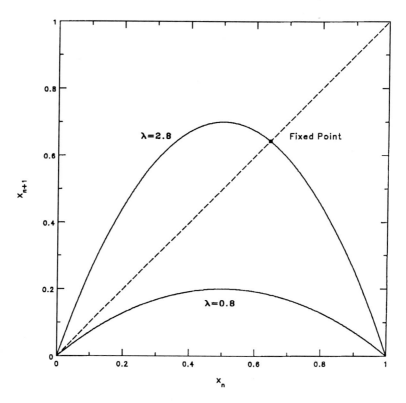

FIGURE 11-11. The quadratic map (or logistic equation). The mapping of (11.19) with two values of λ is shown.

The next question is whether the fixed points of (11.22)–(11.23) are stable (*i.e.*, are attractors). To settle this we start with a point near the fixed point and see if the result of repeated mapping converges to the fixed point. We write x_n as

$$x_n \equiv \bar{x} + \delta_n \tag{11.24}$$

where $\bar{x}$ is a fixed point and δ_n is (at least initially) small in magnitude. Substituting (11.24) into the map equation (11.19), and retaining only terms linear in δ_n, we find that

$$\frac{\delta_{n+1}}{\delta_n} = \lambda(1 - 2\bar{x}) \tag{11.25}$$

In obtaining this result we have used $\bar{x}[\lambda(1 - \bar{x}) - 1] = 0$ which is the condition (11.21) that $\bar{x}$ is a fixed point. If $|\delta_{n+1}| < |\delta_n|$ then with repeated mappings the point (11.24) moves closer and closer to $\bar{x}$ with increasing n and so this fixed point is called stable or attracting; on the other hand if $|\delta_{n+1}| > |\delta_n|$ the point moves away from $\bar{x}$ and the fixed point is called unstable or repelling. For a general map $x_{n+1} = F(x_n)$ it is easy to show that $\bar{x}$ is an attractor if

$$\left| \frac{dF}{dx} \right|_{x=\bar{x}} < 1 \tag{11.26}$$

A simple but informative geometrical construction of the iteration process is shown in Fig. 11-12.

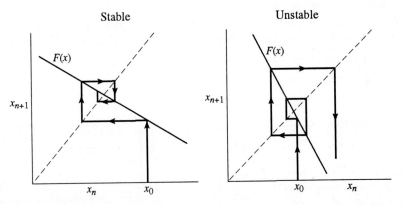

FIGURE 11-12. Stable and unstable fixed points. The iteration $x_{n+1} = F(x_n)$ can be graphically constructed as shown. If the slope of the mapping function $F(x)$ has an absolute value less than unity the fixed point is stable.

According to (11.25) the fixed point $\bar{x} = 0$ is stable (it is an attractor) when $0 < \lambda < 1$, while the fixed point $\bar{x} = 1 - \lambda^{-1}$ is stable when $1 < \lambda < 3$.

Now that we have established the stability of the fixed points when $0 < \lambda < 3$ we venture into the region $3 < \lambda < 4$. Since we found that there are no stable fixed points (simple attractors) when $\lambda > 3$, we consider *periodic points*, that is, points which return to their original value after some number of mappings. For instance, period-2 points satisfy $x_{n+2} = x_n$; they are fixed points of the once iterated mapping

$$
\begin{aligned}
x_{n+2} &= \lambda x_{n+1}(1 - x_{n+1}) \\
&= \lambda^2 x_n (1 - x_n) - \lambda^3 x_n^2 (1 - x_n)^2
\end{aligned}
\tag{11.27}
$$

This double map function is plotted in Fig. 11-13 for two values of λ. For the lower value of $\lambda = 2.8$ the single fixed point is stable and is in fact the fixed point of the single mapping discussed just above. For $\lambda = 3.2$ there are three fixed points. The middle one is the unstable fixed point of the single mapping at $x = 1 - (3.2)^{-1} = 0.6875$. We can establish (see problems 11-6 and 11-7) that the two remaining fixed points of the double mapping are stable in the range $3 < \lambda < 3.449 \ldots$ Note that these two points are a single *pair* of period-2 points; calling them x_A and x_B, the mapping takes one into the other: $x_B = F(x_A)$ and $x_A = F(x_B)$. This transition, as the value of λ is raised past a critical value (3 in this case), from one stable fixed point to a pair of stable period-2 points, is known as a *bifurcation* or *period doubling*. As λ is raised past 3.499 a second bifurcation occurs, that is, the pair of stable period-2 points turns into a quartet of period-4 points. Such bifurcations occur faster and faster until an infinite number of bifurcations occur at $\lambda \simeq 3.56994 \ldots$ This portion of the mapping is shown in Fig. 11-14.

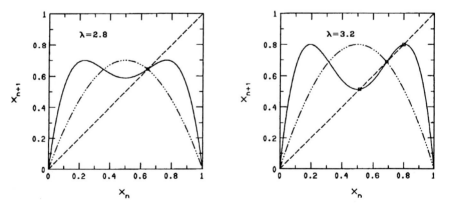

FIGURE 11-13. Iterated map for $\lambda = 2.8$ and $\lambda = 3.2$. At the larger λ there are three fixed points; the extreme points are stable while the middle one is unstable. The dot-dash curves are the single maps.

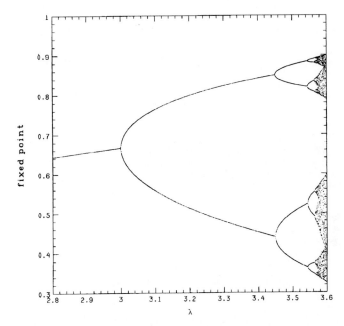

FIGURE 11-14. Quadratic map fixed points for $2.8 < \lambda < 3.6$. A cascade of bifurcations is seen leading to a chaotic mapping. Comparison to the Poincaré section of Fig. 11-8 shows the universal nature of bifurcation cascades.

Denoting by λ_k the critical value of λ at which the bifurcation from a stable period-k set of points to a stable period-$(k + 1)$ set occurs, it is found that

$$\lim_{k \to \infty} \frac{\lambda_k - \lambda_{k-1}}{\lambda_{k+1} - \lambda_k} = 4.669201\ldots \tag{11.28}$$

known as the *Feigenbaum number*. This ratio turns out to be universal for any map with a quadratic maximum and is seen in a wide range of physical problems. Indications of this ratio appeared in the Poincaré section plots for the Duffing equation attractors in Fig. 11-8. One of the conclusions one can draw from the existence of the Feigenbaum number is that each bifurcation looks similar up to a magnification factor. This *scale invariance* or *self similarity* plays an important role in the transition to chaos and, as we will see shortly, in the structure of the strange attractor.

The quadratic map for $2.8 < \lambda < 4$ is shown in Fig. 11-15. Above $\lambda_c = 3.56994\ldots$ the attractor set for many (but not all) values of λ shows no periodicity at all. For these values of λ the quadratic map exhibits chaos and is a strange attractor. In the region $\lambda_c < \lambda < 4$ there are also "windows" where attractors of small period reappear.

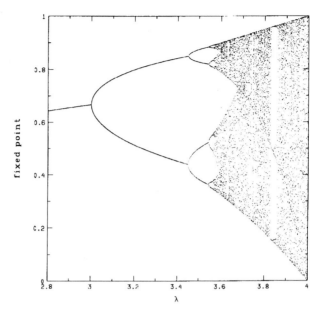

FIGURE 11-15. Fixed points of the quadratic map in the parameter region $2.8 < \lambda < 4$.

B. Fractal dimension, scaling, and the Henon strange attractor

One of the very interesting properties of the strange attractor is its dimensionality. The Duffing strange attractor of Fig. 11-10 is not a simple curve nor does it densely fill any region of phase space. One can think of it as a non-integral dimensional curve with a dimension somewhere between one and two. To describe such attractors we need to generalize our usual conception of dimension.

The Hausdorf (or fractal) dimension d of a set of points in p dimensional space is defined as

$$d = \lim_{\epsilon \to 0} \frac{\ln N(\epsilon)}{\ln(1/\epsilon)} \tag{11.29}$$

where $N(\epsilon)$ is the number of p-dimensional cubes of side ϵ needed to cover the set. Some examples which reduce to previous results are the sets consisting of

1. <u>a single point</u>. Only one "cube" is required so $N(\epsilon) = 1$ and $d = 0$

2. <u>a line of length ℓ in a plane</u>. The number of cubes of side ϵ (=line

segments of length ϵ) required is $N(\epsilon) = \ell/\epsilon$ so

$$d = \lim_{\epsilon \to 0} \frac{\ln \ell/\epsilon}{\ln 1/\epsilon} = \lim_{\epsilon \to 0} \left(1 + \frac{\ln \ell}{\ln 1/\epsilon} \right) = 1$$

3. <u>a square of area ℓ^2</u>. Here $N(\epsilon) = \ell^2/\epsilon^2$ and hence

$$d = \lim_{\epsilon \to 0} \frac{\ln \ell^2/\epsilon^2}{\ln 1/\epsilon} = 2$$

The point of the above is to be able to define the dimension of something less obvious. A nice example is provided by

4. <u>the Cantor Set</u>. As shown in Fig. 11-16 we start with a line of unit length. Then the middle third is removed leaving two line segments. The process is continued by removing at each step the middle third of each segment. To cover this set at steps $k \geq K$, $N(K) = 2^K$ cubes of side $\epsilon = (1/3)^K$ are needed and the resulting Hausdorf dimension of the Cantor set is

$$d = \lim_{K \to \infty} \frac{2^K}{3^K} = \frac{\ln 2}{\ln 3} = 0.631 \ldots$$

An important aspect of this set is its scale invariance. Whatever the k level of the set it is similar to the other levels up to a scale factor.

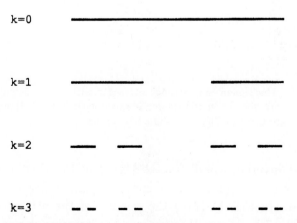

k=0

k=1

k=2

k=3

FIGURE 11-16. The Cantor Set. At each iteration the middle third of each segment is removed.

This scale invariance appears in another context when we consider the two-dimensional *Henon mapping*

$$x_{n+1} = 1 - cx_n^2 + y_n$$
$$y_{n+1} = \beta x_n$$

$$(11.30)$$

This is the simplest nonlinear two-dimensional map. By setting $\beta = 0$ (and rescaling x) the Henon map reduces to the quadratic map. In Fig. 11-17 the attractor is shown for $c = 1.4$ and $\beta = 0.3$.

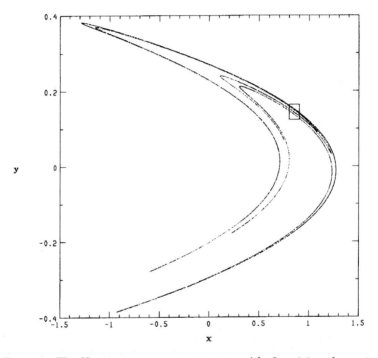

FIGURE 11-17. The Henon map strange attractor with $\beta = 0.3$ and $c = 1.4$. Ten thousand points are shown on this map. The similarity of the Henon map to the Duffing strange attractor of Fig. 11-9 should be noted.

The scale invariance and fractal nature of the Henon attractor becomes evident in Fig. 11-18 where successive magnifications of the map are examined. In Fig. 11-18(a) the portion of the attractor within the box in Fig. 11-17 is shown. In each case the iteration is run long enough that there are about 10,000 points in the figure. The Hausdorf dimension of this attractor is $d \simeq 1.264$.

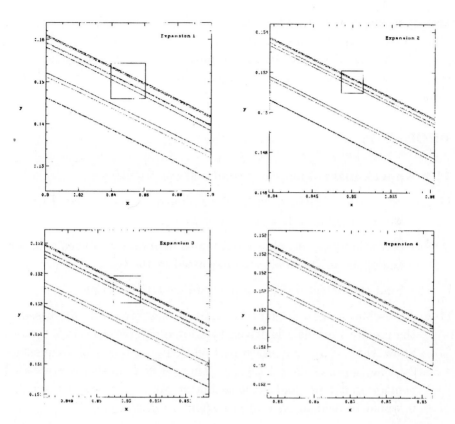

FIGURE 11-18. Henon attractor with $\beta = 0.3$ and $c = 1.4$. Part a is a magnification of the contents within the box in Fig. 11-17. Part (b) magnifies the portion in the box in part (a) and so forth. The scale invariance of the strange attractor is seen and the fractal nature of the curve is evident.

C. The Lyapunov exponent

An important property of chaotic motion is extreme sensitivity to initial conditions. To express this sensitivity quantitatively we introduce the Lyapunov exponent. Consider two points in phase space separated by a distance d_0 at time $t = 0$. If the motion is regular (non-chaotic) these two points will remain relatively close, separating at most according to a power of time. In chaotic motion the two points separate exponentially with time according to

$$d(t) = d_0 e^{\lambda_L t} \qquad (11.31)$$

The parameter λ_L is the Lyapunov exponent. If λ_L is positive the motion is chaotic. A zero or negative coefficient indicates non-chaotic motion. The exponent can vary somewhat from point to point on the attractor and often an average exponent is computed. If the phase space is two dimensional there will be two Lyapunov exponents. If one is positive and the other is negative, a small ball of phase space will be stretched into a spaghetti-like structure as the system evolves. This is typical chaotic behavior.

PROBLEMS

11.2 Approximate Analytic Steady-State Solutions

11-1. A damped oscillator with a non-linear spring force $F(x) = -k(x + \alpha x^2)$ has a driving force $mf \cos \omega t$.

a) Scale the space and time coordinates to obtain a reduced equation analogous to the cubic case discussed in the text.

b) Find the lowest order approximate analytic solution.

11-2. An oscillator with a quadratic anharmonic force term discussed in the previous problem is driven by a force $mf_1 \cos \omega_1 t + mf_2 \cos \omega_2 t$. Show that the steady-state motion contains terms of frequencies $2\omega_1, 2\omega_2, \omega_1 \pm \omega_2, 2\omega_1 \pm \omega_2 \ldots$. These new frequencies are called *combination tones* and if the oscillator were a high-fidelity speaker it would be said to have *intermodulation distortion*.

11-3. Show that the undamped Duffing equation $\ddot{x} + x + x^3 = f \cos \omega t$ can have the exact solution $x(t) = A_0 \cos \frac{\omega t}{3}$. Find the conditions under which such simple *subharmonic* solutions exist.

11.3 Numerical Solutions of Duffing's Equation

11-4. Find an expression for the Poincaré section at $\omega t = 2n\pi$, $n = 0, 1, 2, \ldots$ for a linear oscillator (Eq. (1.115) with $\beta = 0$) attractor.

11-5. For the lowest harmonic solution to the damped and forced Duffing equation, (11.11) and (11.12), find analytic expressions for the maximum amplitude. Evaluate for the parameters of Fig. 11-4 and compare with the numerical result.

11.5 Aspects of Chaotic Behavior

11-6. For any mapping function $x_{n+1} = F(x_n)$ show that a fixed point $\bar{x}$ will be stable if $|F'(\bar{x})| < 1$, where $F'(\bar{x}) \equiv dF/dx|_{\bar{x}}$.

For the iterated quadratic map $x_{n+2} = F(F(x_n)) \equiv F_2(x_n)$ show that its fixed points satisfy

$$\bar{x}^3 - 2\bar{x}^2 + \frac{(\lambda+1)}{\lambda}\bar{x} - \frac{(\lambda^2-1)}{\lambda} = 0\,.$$

One of the fixed points $\bar{x} = 1 - \lambda^{-1}$ is the single cycle fixed point. Factor this root out and show that the two remaining roots are

$$\bar{x} = \frac{1+\lambda}{2\lambda}\left[1 \pm \sqrt{\frac{\lambda-3}{\lambda+1}}\right].$$

11-7. The stability condition for the iterated quadratic map at a fixed point is $|F_2'(\bar{x})| \leq 1$, where $F_2(x) = F(F(x))$. Using the values of the preceding problem for $\bar{x}$ show that the period-2 fixed points are stable when

$$|4 + 2\lambda - \lambda^2| \leq 1$$

and that λ which satisfies this is $\lambda = 3.449\ldots$

11-8. If in the quadratic map $x_{n+1} = \lambda x_n(1 - x_n)$ we assume that x_n changes slowly with n we can approximate $x(n+1) \simeq x(n) + \frac{dx}{dn}$. Show that the differential equation limit of the quadratic map takes the form

$$\frac{dy}{dt} = y(1-y)$$

where y is proportional to $x(n)$ and t is proportional to n. Solve the differential equation and discuss the relationship of the solution to the properties of the iterative map.

11-9. In the limit $\beta = 0$ show that the Henon map of section 11.5B reduces to the one-dimensional quadratic map. Relate the parameter c to λ. For $c = 1.4$ what is the significance of the corresponding λ? *Hint: define a new variable by* $x_n = b_1 + b_2 x_n'$.

11-10. Consider the Henon map

$$x_{n+1} = 1 - cx_n^2 + y_n$$
$$y_{n+1} = \beta x_n$$

If β is small, then y is small compared to x so y_n can be neglected in the upper equation. For $\beta = 0.3$ and $c = 1.4$ compare this approximation to the Henon attractor given in Fig. 11-7.

Appendix A

TABLES OF UNITS, CONSTANTS AND DATA

TABLE A-1. Abbreviations for Units

Length:	centimeter	cm
	meter	m
Mass:	gram	g
	kilogram	kg
Time:	second	s
	hour	h
	year	y
Velocity:	meters per second	m/s
	kilometers per second	km/s
	kilometers per hour	km/h
Astronomical distance:	astromical unit	AU
	light year	ly
	parsec	pc
Angular velocity:	radians per second	rad/s
Energy:	electron volt	eV
	million electron volts	MeV
Force:	newton	N
Charge:	coulomb	C

TABLE A-2. Conversion Factors

	Multiply	By	To Obtain
Distance:	feet	0.3048	meters
	meters	3.281	feet
	kilometers	3281	feet
	feet	3.048×10^{-4}	kilometers
	kilometers	0.6214	miles
	miles	1.609	kilometers
	miles	5280	feet
	meters	100	centimeters
	centimeters	10^{-2}	meters
	kilometers	1000	meters
	centimeters	0.3937	inches
	inches	2.540	centimeters
	astronomical unit	1.495×10^{8}	kilometers
Velocity:	feet/second	0.3048	meters/second
	meters/second	3.281	feet/second
	meters/second	2.237	miles/hour
	miles/hour	0.4470	meters/second
	feet/second	0.6818	miles/hour
	miles/hour	1.609	kilometers/hour
	kilometers/hour	0.6214	miles/hour
	kilometers/second	2237	miles/hour
	miles/hour	4.470×10^{-4}	kilometers/second
Mass, weight and force:	pounds (weight)	0.4536	kilograms (mass)
	kilograms (mass)	2.205	pounds (weight)
	newtons	1	kg m/s^2
	newtons	10^{5}	dynes
	pounds	4.448	newtons
	newtons	0.2248	pounds
Liquid measure:	gallons	3.785	liters
	liters	0.2642	gallons
	liters	10^{-3}	cubic meters
Volume and pressure:	cubic feet	0.02832	cubic meters
	cubic meters	35.31	cubic feet
	pounds/square inch	68950	dynes/square cm
	dynes/square cm	1.450×10^{-5}	pounds/square inch
Energy and Power:	newton meter	1	joule
	dyne centimeter	1	erg
	joule	10^{7}	ergs
	electron volts	1.602×10^{-19}	joule
	joule	6.242×10^{18}	electron volts
	electron volts	10^{-6}	million electron volts
	joule	0.7376	foot pound
	horsepower	746	watts
	watt	1	joule/second
Time:	mean solar day	8.640×10^{4}	seconds
	year	3.156×10^{7}	seconds
	hour	3600	seconds
	r/min	0.1047	radians/second
	radians/second	9.549	revolutions/min

TABLE A-3. Some Physical Constants

Gravitational constant:
  $G = 6.673 \times 10^{-11}\,\mathrm{N\,m^2/kg^2}$
Electron charge:
  $e = 1.602 \times 10^{-19}\,\mathrm{C}$
Proton mass:
  $m_p = 1.6725 \times 10^{-27}\,\mathrm{kg} = 938.3\,\mathrm{MeV}$
Neutron mass:
  $m_n = 1.6748 \times 10^{-27}\,\mathrm{kg} = 939.6\,\mathrm{MeV}$
Electron mass:
  $m_e = 9.1096 \times 10^{-31}\,\mathrm{kg} = 0.511\,\mathrm{MeV}$
α particle (He^{++}) mass:
  $m_\alpha = 6.644 \times 10^{-27}\,\mathrm{kg} = 3727.4\,\mathrm{MeV}$
Velocity of light
  $c = 2.998 \times 10^8\,\mathrm{m/s}$
Planck's constant
  $\hbar(= h/2\pi) = 1.05457266 \times 10^{-34}\,\mathrm{J\,s}$

TABLE A-4. Some Numerical Constants

$$\pi = 3.1415927$$
$$e = 2.7182818$$
$$\ln 2 = 0.69314718$$
$$1\,\mathrm{rad} = 57.2957795^\circ$$

TABLE A-5. Vector Identities

$\mathbf{A} \times \mathbf{B} = -\mathbf{B} \times \mathbf{A}$
$\mathbf{A} \times (\mathbf{B} \times \mathbf{C}) = (\mathbf{A} \cdot \mathbf{C})\mathbf{B} - (\mathbf{A} \cdot \mathbf{B})\mathbf{C}$
Differential Forms:
  $\boldsymbol{\nabla}(st) = s\boldsymbol{\nabla}t + t\boldsymbol{\nabla}s$
  $\boldsymbol{\nabla} \cdot (s\mathbf{A}) = s\boldsymbol{\nabla} \cdot \mathbf{A} + \mathbf{A} \cdot \boldsymbol{\nabla}s$
  $\boldsymbol{\nabla} \times (s\mathbf{A}) = s(\boldsymbol{\nabla} \times \mathbf{A}) - \mathbf{A} \times (\boldsymbol{\nabla}s)$
  $\boldsymbol{\nabla} \times \boldsymbol{\nabla} = 0$
  $\boldsymbol{\nabla} \cdot \boldsymbol{\nabla} \times \mathbf{A} = 0$
  $\boldsymbol{\nabla} \times (\boldsymbol{\nabla} \times \mathbf{A}) = \boldsymbol{\nabla}(\boldsymbol{\nabla} \cdot \mathbf{A}) - \nabla^2\mathbf{A}$
  $\boldsymbol{\nabla} \cdot (\mathbf{A} \times \mathbf{B}) = \mathbf{B} \cdot \boldsymbol{\nabla} \times \mathbf{A} - \mathbf{A} \cdot \boldsymbol{\nabla} \times \mathbf{B}$
  $\boldsymbol{\nabla} \times (\mathbf{A} \times \mathbf{B}) = (\mathbf{B} \cdot \boldsymbol{\nabla})\mathbf{A} - \mathbf{B}(\boldsymbol{\nabla} \cdot \mathbf{A}) + \mathbf{A}(\boldsymbol{\nabla} \cdot \mathbf{B}) - (\mathbf{A} \cdot \boldsymbol{\nabla})\mathbf{B}$

TABLE A-6. Sun and Earth Data

Mean distance from sun to earth	1.495×10^8 km
Mass of sun	$M_\odot = 1.987 \times 10^{30}$ kg
Mass of earth	$M_E = 5.97 \times 10^{24}$ kg
Sun-to earth mass ratio	$M_\odot/M_E = 332{,}946$
Mean radius of earth	$R_E = 6371$ km
Mean radius of sun	$R_\odot = 696{,}000$ km
Mean gravity on earth	$g = 9.8064$ m/s^2
Equatorial earth gravity	$g_E = 9.7805$ m/s^2
Polar earth gravity	$g_P = 9.8322$ m/s^2

TABLE A-7. Moon Data

Semimajor axis of orbit	3.84×10^5 km
Eccentricity of orbit	0.055
Sidereal period about the earth	27.32 days
Inclination of orbit to ecliptic	5.15°
Radius	$R_L = 1{,}741$ km
Mean density	3.33 g/cm^3
Mass	$M_L = M_E/81.56$
Surface Gravity	$0.165g$
Escape velocity	2.4 km/s
Orbital velocity about the earth	1.0 km/s

TABLE A-8. Properties of the Planets

	Mercury	Venus	Earth	Mars	Jupiter	Saturn	Uranus	Neptune	Pluto
Mean distance from sun, AU*	0.39	0.72	1.00	1.52	5.2	9.54	19.2	30.1	39.4
Eccentricity of orbit	0.206	0.007	0.017	0.093	0.048	0.056	0.047	0.009	0.250
Sidereal period, years	0.24	0.615	1.00	1.88	11.9	29.5	84.0	164.8	247.7
Inclination of orbit to ecliptic	7.00°	3.39°	0.0°	1.85°	1.31°	2.50°	0.77°	1.78°	17.15°
Equatorial radius (earth= 1)†	0.38	0.95	1.0	0.53	11.2	9.41	4.01	3.88	0.18
Mean density, gm/cm³	5.43	5.24	5.52	3.94	1.33	0.70	1.30	1.64	2.0
Mass (earth= 1)**	0.055	0.185	1.0	0.107	317.8	95.2	14.5	17.1	0.002
Rotation period, days	59	−243	1.0	1.03	0.41	0.44	0.72	0.67	0.39
$\epsilon = (R_E - R_P)/R_E$			1/298	1/193	1/15	1/9.2	1/45		
Surface gravity (earth= 1)‡	0.38	0.91	1.0	0.38	2.54	1.08	0.91	1.18	0.05
Escape velocity, km/s	4.3	10.4	11.2	5.0	59.6	35.6	21.3	23.8	1.2
Orbital velocity, km/s	48	34.9	29.8	24.1	13.0	9.6	6.8	5.4	4.7

*Where earth = 1.495×10^8 km
†Where earth = 6378 km
**Where earth = 5.975×10^{24} kg
‡Where earth = 9.8 m/s²

Appendix B

ANSWERS TO SELECTED PROBLEMS

Chapter 1

1-1. $R = \dfrac{2v_0}{g}(v_0 \cos\theta + v_r)\sin\theta = 97.2\,\text{m}$

1-2. $m = 42.0$ kg

1-3. $\sin^2\theta = \dfrac{1}{2}\left(\dfrac{1}{1 + \frac{gh}{v_0^2}}\right) = 0.455 \qquad \theta = 42.4°$

1-5. $x = x_0 \cosh\left(\sqrt{\dfrac{g}{\ell}}\,t\right)$

1-6. $v(t) = \dfrac{a}{bm}\left(1 - e^{-bt}\right) \qquad x(t) = x_0 + \dfrac{a}{bm}\left[t - \dfrac{1}{b}\left(1 - e^{-bt}\right)\right]$

1-7. $F(v) = -2mcv^{1/2}$

1-8.a) $v = v_0 - \dfrac{bx}{m}$; at $v = 0$, $x_f = mv_0/b$.

b) $v = v_0 e^{-cx/m} \qquad x = \dfrac{m}{c}\ln\left(\dfrac{cv_0 t}{m} + 1\right)$

1-9. $x = \dfrac{m}{2c}\ln\dfrac{v_t^2}{v_t^2 - v^2}\bigg|_{v=\frac{2}{3}v_t} = \dfrac{m}{2c}\ln\dfrac{9}{5} = 87.5\,\text{meters}$

1-10.a) $v_0 = \sqrt{2gx} = 14\,\text{m/s} \qquad t_0 = \sqrt{2x/g} = 1.43\,\text{s}$

b) $v = v_0 e^{-\frac{c}{m}x}$

c) $x = \dfrac{1}{0.4}\ln 10 = 5.8\,\text{m}$

d) $t = 1.15\,\text{s}$

1-12. $v_0^3 = (v + w)^2 v$

$v = 11.87\,\text{m/s for head wind}$

$v = 18.51\,\text{m/s for tail wind}$

1 13. $\dfrac{v_t}{2g\mu}\ln\left(\dfrac{v_t + v}{v_t - v}\right) = t \qquad \dfrac{v_t^2}{2g\mu}\ln\left(\dfrac{1}{1 - (v/v_t)^2}\right) = x$

$v_t = 123.6$ m/s $\qquad \mu = 4.22$

1-14.a) new height $= \ell - \dfrac{mg}{k}$

b) $x(t) = \ell - \dfrac{mg}{k} - c\cos\omega_0 t$

c) $c_{\text{crit}} = \dfrac{mg}{k}$

1-15.a) $y = \frac{1}{2}gt^2 = 2.9\,\mathrm{m}$

b) $\sin 2\alpha = \frac{gx}{v_0^2} = 0.116$ $\alpha = 3.33°$

c) $x_{\max} = \frac{v_0^2}{g} = 431\,\mathrm{m}$

d) $x = (v_0 \cos\alpha)t = 69.4\,\mathrm{m}$

1-19. 955 rpm

1-20. $x_0 = f\left(\omega_0^2 - \omega^2\right)/r^2$ $v_0 = 2\gamma\omega^2 f/r^2$

1-21. $x = \frac{f}{r}\,Im\,e^{i(\omega t-\theta)} = \frac{f}{r}\sin(\omega t - \theta)$,
where r and θ are given in Eqs. (1.122) and (1.123).

1-22.a) $L = 84.4\,\mu\mathrm{H}$

b) $\gamma = 2.97 \times 10^4$

c) $V_{\mathrm{cap}}^{\mathrm{max}} = 0.106$ Volts

d) Ratio of amplifications $= 0.23$

1-23.a) $x = Ae^t$, $A = \frac{1}{4}f$

b) $x = At^2 e^{-t}$, $A = \frac{1}{2}f$

c) $x(t) = \frac{1}{4}\left(t^2 - t - \frac{1}{2}\right)fe^{-t} + \frac{1}{8}fe^t$

1-24. $x(t) = C\cos(\omega_0 t + \alpha) + \dfrac{a/me^{-bt}}{\omega_0^2 + b^2}$

$\tan\alpha = \dfrac{b}{\omega_0}$ $C = \dfrac{a}{m\omega_0\sqrt{\omega_0^2 + b^2}}$

1-25. $\langle P \rangle = \dfrac{-m\gamma f^2 \omega^2}{r^2}$

1-26. $\Omega_n \equiv (2n + 1)\omega$

$r_n^2 = \left(\omega_0^2 - \Omega_n^2\right)^2 + (2\gamma\Omega_n)^2$

$\tan\theta_n = \dfrac{2\gamma\Omega_n}{\omega_0^2 - \Omega_n^2}$

$x(t) = \sum_{n=0}^{\infty} x_n(t) = \sum_{n=0}^{\infty} \frac{1}{(2n+1)^2}\frac{\cos(\Omega_n t - \theta_n)}{r_n}$

Chapter 2

2-1. $V = \dfrac{1}{12}\rho(bh)^2 g = 2.4 \times 10^{12}\mathrm{N\text{-}m}$

$\Delta V = 1.1 \times 10^8 \mathrm{N\text{-}m}$ $\dfrac{V}{\Delta V} = 22000$ person years

2-2.a) $R = \dfrac{v_0^2}{g} = \dfrac{\pi F_{\max} d}{2mg} = 1188\,\text{m}$

b) The range of the Turkish bow is larger by a factor of $\pi/2$ than that of a bow that acts like a linear spring.

2-3. $v_{\text{esc}}(\text{moon}) = 0.21 v_{\text{esc}}(\text{earth}) = 2.37\,\text{km/s}$

2-4. $v = \sqrt{\dfrac{2GM_L}{R_L}} \left[1 - \dfrac{R_L}{x} - \dfrac{M_E}{M_L}\left(\dfrac{R_L}{d-x} - \dfrac{R_L}{d-R_L}\right) \right]^{1/2}$

$v = 0.956 v_{\text{esc}}(\text{moon}) = 2.27\,\text{km/s}$

2-5. $\dfrac{\text{KE}}{\text{Fe atom}} = 35\dfrac{\text{eV}}{\text{Fe atom}}$

2-6. $k = aF_0 \qquad \omega = \sqrt{\dfrac{aF_0}{m}}$

2-7. $x_0 = +a$ is a stable equilibrium point with $\omega = \sqrt{\dfrac{2V_0}{ma^2}}$

$x_0 = -a$ is an unstable equilibrium point.

2-8. Approximate $V(r)$ near $r = 0.74\,\text{Å}$ by $V(r) = \frac{1}{2}k(r - 0.74)^2 - 4.52\,\text{eV}$ with $k \approx 47\,\text{eV/Å}^2$.

$\nu_{\text{vib}} = \dfrac{1}{2\pi}\sqrt{2k/m_{\text{proton}}} = 1.5 \times 10^4\,\text{Hz}$

2-9.a) $A = \sqrt{29} \qquad B = \sqrt{17}$

b) $\mathbf{A} \cdot \mathbf{B} = 4 \qquad \theta = 79.6° \text{ or } 280.4°$

c) $|\mathbf{A} \times \mathbf{B}| = 21.84 \qquad \theta = 79.6° \text{ or } 100.4°$.
Consistent solution from b) and c) is $79.6°$

2-11.b) $V(r) = \dfrac{K}{3}(x - z)^3$

2-12. $\dfrac{\partial F_i}{\partial x_j} - \dfrac{\partial F_j}{\partial x_i} = (x_i x_j - x_j x_i)\dfrac{1}{r}\dfrac{d}{dr}\left(\dfrac{F(r)}{r}\right) = 0$

2-13. $\nabla \times \mathbf{F} = -2\mathbf{a}$. Not conservative. $\oint = \mathbf{F} \cdot d\mathbf{r} = -2\pi a_z R^2$

2-16. $\omega = \sqrt{g/\ell \cos\theta}$

2-17. $\cos\theta = \frac{2}{3} \qquad \theta = 47.1°$ measured from the bottom

2-18. $T = mg(-3\cos\alpha + 2\cos\alpha_0)$
$T = 0$ for $\cos\alpha = \frac{2}{3}\cos\alpha_0$

2-19. For $\theta \le 90°$, $\quad 0 \le E \le mg\ell$
For a complete circular arc the string is taut if $E > \frac{5}{2}mg\ell$.

2-20. $W_f = \frac{8}{3}\theta_0^3 \ell^2 gc$
$c = \frac{1}{2}C_D(\pi R^2)\rho$
$W_f = -\Delta V = -(\Delta h)mg$

$\Delta h = -0.81$ cm.

Distance from release point is $d = 0.81/\sin(36.9°) = 1.35$ cm

2-21. $2m\ddot{x}_1 = -2kx_1 + k(x_2 - x_1)$

$m\ddot{x}_2 = -k(x_2 - x_1)$

$\omega_+ = \sqrt{k/2m}$ for the mode $x_1 + x_2$

$\omega_- = \sqrt{2k/m}$ for the mode $2x_1 - x_2$

2-22. $x_1 = A\cos\omega t$, $x_2 = B\cos\omega t$

$$A = \frac{f(\omega_2^2 - \omega^2)}{(\omega_1^2 + \omega_2^2 - \omega^2)(\omega_2^2 - \omega^2) - \omega_2^4}$$

$$B = \frac{f\omega_2^2}{(\omega_1^2 + \omega_2^2 - \omega^2)(\omega_2^2 - \omega^2) - \omega_2^4}$$

When $\omega = \omega_2$, m_1 remains fixed.

Chapter 3

3-1. $m\ddot{x}_1 - k(x_2 - x_1) = 0$

$m\ddot{x}_2 + k(x_2 - x_1) = 0$

$\omega_0 = \sqrt{2k/m}$

3-2. $L = m\dot{x}^2 - \frac{1}{2}kx^2 + mgx$

$2m\ddot{x} - mg + kx = 0$

$$x(t) = \frac{mg}{k}\left(1 - \cos\sqrt{\frac{k}{2m}}\,t\right)$$

3-3. $L = \frac{1}{2}(2m\dot{x}_1^2) + \frac{1}{2}m\dot{x}_2^2 - \frac{1}{2}(2k)x_1^2 - \frac{1}{2}k(x_2 - x_1)^2$

3-4. $L = K - V = \frac{1}{2}m(\dot{x}_1^2 + \dot{x}_2^2) - \frac{1}{2}m\omega_1^2 x_1^2 - \frac{1}{2}m\omega_2^2(x_2 - x_1)^2$

$Q_1(t) = mf\cos\omega t$ $Q_2(t) = 0$

3-5.a) $L = K - V = \frac{1}{2}m_1\dot{x}_1^2 + \frac{1}{2}m_2\dot{x}_2^2 - \frac{1}{2}k(x_2 - x_1)^2$

b) $\omega = 0$ for the mode $mx_1 + mx_2$

$\omega = \sqrt{k/\mu}$ for the mode $x_1 - x_2$, where $\mu = \dfrac{m_1 m_2}{m_1 + m_2}$

c) $m_1 x_1 + m_2 x_2 = A + Bt$

$x_1 - x_2 = C\cos\omega t + D\sin\omega t$

d) $x_1 = \dfrac{m_1}{m_1 + m_2}v_0 t + \dfrac{m_2}{m_1 + m_2}\dfrac{v_0}{\omega}\sin\omega t$

$x_2 = \dfrac{m_1}{m_1 + m_2}v_0 t - \dfrac{m_1}{m_1 + m_2}\dfrac{v_0}{\omega}\sin\omega t$

e) $x_{CM} = \dfrac{m_1 x_1 + m_2(x_2 + \ell)}{m_1 + m_2} = \dfrac{m_2\ell + m_1 v_0 t}{m_1 + m_2}$

$x_1 - x_2 = \dfrac{v_0}{\omega}\sin\omega t$

3-6.a) $L = \frac{1}{2}m\omega^2 R^2 \sin^2\theta + \frac{1}{2}mR^2\dot{\theta}^2 + mgR\cos\theta$

$\ddot{\theta} + \left(\frac{g}{R} - \omega^2\cos\theta\right)\sin\theta = 0$

b) $\Omega = \sqrt{g/R} \equiv \omega_0$

c) Stable equilibrium at $\theta = \theta_0$ where $\cos\theta_0 = \omega_0^2/\omega^2$.

3-7.a) $L = \frac{1}{2}\ell^2(m_1 + m_2)\dot{\theta}^2 + \frac{1}{2}\ell^2 m_2\dot{\phi}^2 + \ell^2 m_2\dot{\phi}\dot{\theta}\cos(\theta - \phi)$
$\quad + (m_1 + m_2)g\ell\cos\theta + m_2 g\ell\cos\phi$

$\ddot{\theta} + \frac{m_2}{m_1 + m_2}\left[\dot{\phi}^2\sin(\theta - \phi) + \ddot{\phi}\cos(\theta - \phi)\right] + \frac{g}{\ell}\sin\theta = 0$

$\ddot{\phi} + \left[-\dot{\theta}^2\sin(\theta - \phi) + \ddot{\theta}\cos(\theta - \phi)\right] + \frac{g}{\ell}\sin\phi = 0$

b) $\omega_{\pm}^2 = \dfrac{g/\ell}{1 \pm \sqrt{m_2/(m_1 + m_2)}}$

c) $m_1 \gg m_2$: $\omega_+ \approx \omega_- = \sqrt{g/\ell}$
The system becomes two weakly coupled degenerate pendula each of length ℓ. m_2 is light and does not affect the motion of m_1.

$m_1 \ll m_2$: $\omega = \sqrt{g/2\ell}$ or ∞
The heavy mass m_2 oscillates like a simple pendulum of length 2ℓ. In the limit $m_2/m_1 \to \infty$ the tension in the rod due to m_2 is infinitely large compared to $m_1 g$, causing m_1 to oscillate with infinite frequency.

3-8.a) $L = \frac{1}{2}mR^2\left(\dot{\theta}_1^2 + \dot{\theta}_2^2 + \dot{\theta}_3^2\right)$
$\quad - \frac{1}{2}m\omega_0^2 R^2\left[(\theta_2 - \theta_1)^2 + (\theta_3 - \theta_1)^2 + (\theta_2 - \theta_3)^2\right]$

b) $\theta_1 + \theta_2 + \theta_3 = 3\Omega t + \theta_0$
$p_1 + p_2 + p_3 = mR^2(\dot{\theta}_1 + \dot{\theta}_2 + \dot{\theta}_3) = 3mR^2\Omega$

c) $\ddot{\theta}_1 + 3\omega_0^2\theta_1 = 0$
$\ddot{\theta}_2 + 3\omega_0^2\theta_2 = 0$

3-10. $L = \frac{1}{2}m(\dot{x}^2 + \dot{y}^2) - mgy$
$L = \frac{1}{2}m\dot{x}^2(1 + 4a^2 x^2) - mgax^2$
$\ddot{x}(1 + 4ax^2) + 4a\dot{x}^2 x + 2gax = 0$
$\ddot{x} + (2ga)x \simeq 0$

3-14. $H = \dfrac{p_r^2}{2m} + \dfrac{p_\theta^2}{2mr^2} + V(r)$

3-15. $\dfrac{\partial \dot{q}_j}{\partial q_j} + \dfrac{\partial \dot{p}_j}{\partial p_j} = \dfrac{\partial^2 H}{\partial p_j \partial q_j} - \dfrac{\partial^2 H}{\partial q_j \partial p_j} = 0$

Chapter 4

4-2. $v = v_0 e^{-\sigma t/m_0}$

4-3. $\dfrac{v}{v_0} = \dfrac{1}{1 + (\sigma/m_0)t}$, where σ is the rate of mass of water pickup

$v(t = 10\,\text{s}) = 182\,\text{km/hr}$

$m(t)\dfrac{dv}{dt} = -(b + \sigma)v$

$v = v_0 \left[m_0/m(t) \right]^{1+(b/\sigma)}$, where $m(t) = m_0 + \sigma t$

4-4. For maximum momentum, $m = m_i/e$
Velocity at this point is $v = u$

For maximum kinetic energy, $m = m_i/e^2$
Velocity at this point is $v = 2u$

4-5. $1.451 \times 10^5\,\text{N}$

4-8. For first stage of Apollo rocket, $h = 329\,\text{km}$
For $\alpha \to \infty$, $h = 691\,\text{km}$

4-9. $K_1 = 3\,\text{MeV}$
$\theta_2 = -60°$
$K_2 = 1\,\text{MeV}$

4-10. $M = 3m_\alpha$

4-11. $M/m = 3$, where M is the stationary mass after the collision.

4-13. $v_S = v_0 \left(\dfrac{3M - m}{M + m} \right)$

4-15.a) $h = v_0^2/(8g)$

   b) $h = v_0^2/(2g)$

4-17. $v_f = 1.75 v_0$ $\theta = 4.1°$

4-18. Dissociation energy $= 2.6\,\text{eV}$

4-19. Maximum compression $s = v\sqrt{m/6k}$

4-21. Energy lost $= F\ell$

Chapter 5

5-1. $\mathbf{L} = \mathbf{L}_0 e^{-\frac{b}{m}t}$

5-2. $\ddot{r}_0 = 0$ $\dfrac{L^2}{mr_0^3} = \lambda c r_0^{-\lambda-1}$

   $r = r_0 + \delta$ $\omega_r^2 = \dfrac{\lambda c}{m} r_0^{-\lambda-2}(2 - \lambda)$

   $\ddot{\delta} + \omega_r^2 \delta = 0$ $\omega_\theta^2 = \dfrac{L^2}{m^2 r_0^4}$

5-3.a) $V(r) = -\dfrac{2c^2}{\sqrt{r}}$

b) Closed orbit for $E < 0$

Unbounded motion for $E \geq 0$

c) $r_0 = \left(\dfrac{L^2}{mc^2}\right)^{2/3}$

$\tau = \dfrac{2\sqrt{m}\, r_0^{5/4}}{c}$

d) $\omega_r^2 = \dfrac{3}{2}\omega_\theta^2$

5-4. $F(r) = -\dfrac{6CL^2}{mr^4} - \dfrac{L^2}{mr^3}$

5-5. $K' = K$ $V' = \frac{1}{2}V$ $E' = 0$

5-6.a) $v = 1016$ m/s $\tau = 27.4$ days

b) $v_{sE} = 7786$ m/s $v_{sL} = 1587$ m/s

5-7. $v_{esc} = 4.23$ m/s

$v_{max}(\text{Earth}) = 3.13$ m/s

5-9. 129 days

5-11. On the equator in the direction of rotation, $r = 6.6R_E$

5-12.a) $a = 17.9$ A.U.

b) $r_{max} = 35.3$ A.U.

c) $\epsilon = 0.967$

5-13. $d_{min} = 1.85$ cm

5-14. $\dfrac{dE}{dt} = \dfrac{dK}{dt} = m\mathbf{v} \cdot \dfrac{d\mathbf{v}}{dt}$

5-16. $a = 6995$ km

$\tau = 1.62$ hr

5-17. $v^2 = (v_{esc}^E)^2 + v_1^2$

$v_1 = v_{esc}^\odot - v_{orb}$

$v = 16.7$ km/s

5-19.a) $\epsilon = 0.360$ $a = 0.735$ A.U.

$\Delta v = -6$ km/s

b) $\epsilon = 0.225$ $a = 0.817$ A.U.

$\Delta v = -3.55$ km/s

5-20. $\sigma = \pi R_L^2 \left[1 + \left(v_{esc}^L/v_0\right)^2\right]$

5-21. $\sigma = \dfrac{4\pi}{v_0}\sqrt{\dfrac{2C}{m}}$

5-22. $\theta_s^{\max} = 133$ degrees

$\quad\quad \theta_s^{\min} = 0.006$ degrees

5-24.a) $u = 14.7$ km/s $\quad\quad \theta = 19.8°$

$\quad\quad$ b) $\epsilon = 1.222$

$\quad\quad$ c) $r_{\min} = 1.85R_J$

Chapter 6

6-1. $r_{CM} = 4651$ km, where r_{CM} is distance of CM from center of the earth.

6-3. $\dfrac{16}{7}$ m.

6-4. $x = 6$ cm, $\quad v = \sqrt{125\,f/M}$

6-5. $\mathbf{F}_{L\odot} = 2.18\mathbf{F}_{LE}$

6-6.a) $r_0 = \left(\dfrac{a}{b}\right)^2$, $\quad \omega = \dfrac{b^6}{a^5\sqrt{\mu}}$, $\quad E = \dfrac{b^8}{12a^6}$

$\quad\quad$ b) $L_{\max} = \dfrac{\sqrt{\mu}b^2}{2a}$, $\quad r_{\max} = 2(a/b)^2$

$\quad\quad$ c) $v_1' = \sqrt{\mu}\,\dfrac{b^4}{\sqrt{96m_1a^3}}$, $\quad v_2' = \sqrt{\mu}\,\dfrac{b^4}{\sqrt{96m_2a^3}}$

6-7.a) $r_0 = \ell$, $\quad \omega = \sqrt{k/\ell}$

$\quad\quad$ b) $r_0 = \ell + \dfrac{L^2}{\mu r_0^3 k}$, $\quad \omega = \sqrt{\dfrac{k}{\mu}\left(1 + \dfrac{3L^2}{\mu k r_0^4}\right)}$

6-8.a) $K' = \dfrac{1}{2}\dfrac{m_1 m_2}{m_1 + m_2}v_1^2$, $\quad L = 0$

$\quad\quad$ c) $r_{\min} = \sqrt{\dfrac{2V_0 a^2}{\mu v_1^2}}$

$\quad\quad$ d) $\mathbf{v}_{1f} = \dfrac{m_1 - m_2}{m_1 + m_2}\mathbf{v}_1$

6-9.a) $E' = \frac{1}{2}\mu v_1^2$

$\quad\quad L' = \mu v_1 b$

$\quad\quad$ b) $r_0^2 = \dfrac{b^2}{2} + \dfrac{1}{2}\sqrt{b^4 + \dfrac{8V_0}{\mu v_1^2}}$

$\quad\quad$ c) $\cos\beta = \dfrac{(m_1 - m_2)\sqrt{v_1^2 - v_{1f}^2}}{2\sqrt{m_1 m_2}\,v_{1f}}$

6-12. $m_V = 3$ kg located on a line between the legs supporting 2 and 3 kg and 2/3 of the way toward the 3 kg leg. Table tips if $m_V > 3$ kg located on edge midway between two legs.

6-13.a) right

b) $\cos \theta = R_1/R_2$

c) $T_{\text{max}} = \mu M g \left/ \left(\dfrac{R_1}{R_2} + \mu \sqrt{1 - \left(\dfrac{R_1}{R_2}\right)^2} \right) \right.$

6-14. $x = \dfrac{1}{3} h$

6-16. $a = \mu g$. Using the notation of Problem 6-15,
$$N_1 = \frac{M g(b_2 + \mu h)}{b_1 + b_2}, \qquad N_2 = \frac{M g(b_1 - \mu h)}{b_1 + b_2}$$

6-17. $\omega_p = M g \ell / I \omega$

6-19. $v_f = R\omega_0/3$

6-20. $\dfrac{1}{32}$ rpm

6-21.a) For one century $\Delta t = -5.87$ s
For 3000 years $\Delta t = -14.7$ h

b) 0.6×10^{13} W

6-22. $T = \dfrac{1}{4} M g$

6-23. $\ell = \sqrt{I/M}$ for minimum period

6-24.a) $I = m_a a^2 + m_b b^2$

c) $\omega^2 = \dfrac{m_b b - m_a a}{m_b b^2 + m_a a^2} g$

d) $\tau = \dfrac{2\sqrt{2}}{\omega} \displaystyle\int_0^{\theta_{\text{max}}} \dfrac{d\theta}{\sqrt{\cos \theta_0 - \cos \theta_{\text{max}}}}$

e) $\dot{\theta}_0 = 2\omega$

6-25. $T = \dfrac{Mg}{1 + (2r^2/R^2)}$. Same descending or ascending

6-26. $I_{xx} = M(Y^2 + Z^2) + I_{xx}^{\text{CM}}$
$I_{xy} = -MXY + I_{xy}^{\text{CM}}$
$I_{xz} = -MXZ + I_{xz}^{\text{CM}}$

$I_{yx} = I_{xy}$
$I_{yy} = M(X^2 + Z^2) + I_{yy}^{\text{CM}}$
$I_{yz} = -MYZ + I_{yz}^{\text{CM}}$

$I_{zx} = I_{xz}$
$I_{zy} = I_{yz}$
$I_{zz} = M(X^2 + Y^2) + I_{zz}^{\text{CM}}$

6-27.a) $\tan 2\phi_0 = \dfrac{2I_{xy}}{I_{yy} - I_{xx}}$

d) $2I_{x'x'} = I_{xx} + I_{yy} + D$
$2I_{y'y'} = I_{xx} + I_{yy} - D$
$D^2 = (I_{yy} - I_{xx})^2 + (2I_{xy})^2$

6-28. $x = \dfrac{I}{m}t$ $\theta = \dfrac{6I}{m\ell}t$

6-29.a) $x = \frac{3}{4}\ell$

b) $I = \frac{4}{3}m\ell^2$

c) $I_{CM} = \frac{5}{24}m\ell^2$

d) $I_{CM}\dot{\theta}_f = \left(\frac{3}{4}\ell - h\right)I$

e) $h = \frac{8}{9}\ell$

6-30. $v_{min} = \dfrac{\sqrt{\frac{10}{7}gh}}{1 - \frac{5}{7}h/a}$

6-33. Changes sign

Chapter 7

7-1. $v_0 = -\omega_0 r_0$

7-2.b) 8.54 rpm

7-3. $N = 41.9$ dyne cm

7-5.a) $m\ddot{r} - mr\omega^2 = 0$

b) $2mr\dot{r}\omega = Q'_\theta = rF'_\theta$

7-6.a) $\ddot{\theta} + (\omega_0^2 - \omega^2\cos\theta)\sin\theta = 0$, where $\omega_0^2 = g/R$

b) $\Omega = \omega_0$

c) $\cos\theta_e = (\omega_0/\omega)^2$

7-7. $g_{eff} = g\sqrt{1 - 0.36\sin^2\theta}$
$v = 0.949v_{esc}$

7-8. $m\dfrac{\delta^2 z}{\delta t^2} = -mg$ $m\dfrac{\delta^2 \rho}{\delta t^2} = m\omega^2\rho$

7-9. $R = v/2\Omega$

7-14. $F_B = Ma\omega^2/10\sqrt{5}$

7-15.a) $I_{xx} = my_0^2$
$I_{xy} = -mx_0y_0$
$I_{yy} = mx_0^2$
$I_{xz} = I_{yz} = I_{zz} = 0$

b) $I = 0$ and $I = I_0 \equiv m(x_0^2 + y_0^2)$

7-18.a) $t_{up} = 1$ s

b) $\omega_2(0) = 30.3$ rad/s

c) $t = 0.19$ s

7-20. $\omega_3/\omega_1 = 1/2$

7-24. $L = K - V$

$K = \frac{1}{2}I_{CM}(\dot{\theta}^2 + \dot{\phi}^2 \sin^2 \theta) + \frac{1}{2}M\ell^2\dot{\theta}^2 \sin^2 \theta + \frac{1}{2}I_3(\dot{\psi} + \dot{\phi}\cos \theta)^2$

$V = Mg\ell \cos \theta$

7-25.a) $\omega_p = \dfrac{Mgz_0}{I_3\omega_3}$

b) $z_{CM} = \dfrac{2}{3}\ell \cos \alpha$

Chapter 8

8-1.a) $M_{core} = \dfrac{4\pi R^3}{3}\left(\dfrac{5}{8}\rho\right)$

$M_{outer} = \dfrac{4\pi R^3}{3}\left(\dfrac{7}{8}\rho\right)$

$M_{tot} = \dfrac{4\pi R^3}{3}\left(\dfrac{3\rho}{2}\right)$

b) For $r < \frac{1}{2}R$, $M(r) = \frac{10}{3}M\,(r/R)^3$

For $\frac{1}{2}R < r < R$, $M(r) = \frac{2}{3}M\left[\frac{1}{2} + (r/R)^3\right]$

For $r > R$, $M(r) = M$

c) $\dfrac{F}{m} = -GM\begin{cases} \frac{10}{3}rR^{-3} & 0 < r < \frac{1}{2}R \\ \frac{1}{3}r^{-2} + \frac{2}{3}rR^{-3} & \frac{1}{2}R < r < R \\ r^{-2} & r > R \end{cases}$

d) $\Phi = GM\begin{cases} \frac{5}{3}r^2R^{-3} - 2R^{-1} & 0 < r < \frac{1}{2}R \\ -\frac{1}{3}r^{-1} + \frac{1}{3}r^2R^{-3} - R^{-1} & \frac{1}{2}R < r < R \\ -r^{-1} & r > R \end{cases}$

8-2. $\rho(r) \propto 1/r$

8-3. $\tau = 2\pi\sqrt{R^3/GM}$, which is the same as the period of a close orbit

8-7. $\rho_{moon}/\rho_{sun} = 2.36$

8-8. $R = 872$ km, $\tau = 0.44$ s

8-9. Power $= \dot{K} = 3.7 \times 10^{31}$ W

8-10. $J_c = \frac{4}{3}(3m\alpha^2 I)^{1/4}$

8-14. $\delta = \sqrt{\dfrac{2r_sd_1}{d_2(d_1 + d_2)}}$

Chapter 9

9-1.a) $N = (4\pi r^2 \Delta) \left(\dfrac{1}{d^3}\right)$

b) Fraction covered in shell $= \left(\dfrac{\Delta A}{d^3}\right)$

c) $R = \dfrac{d^3}{A} = 8.1 \times 10^{15}$ ly

9-2. $H_0^{-1} = 2 \times 10^{10}$ yr

9-4. $\rho = 0.55 \times 10^{10}$ atoms/m^3

9-5.a) 1.13×10^{-9} kg/m^3

b) 1.29×10^{-20} kg/m^3

c) 2.6×10^{-22} kg/m^3

9-6. $t_P = \left(G\hbar/c^5\right)^{1/2} = 0.54 \times 10^{-43}$ s,
$L_P = ct_P = 1.62 \times 10^{-35}$ m

9-8. $\dfrac{M_D}{M_V} \simeq 2.25$

Chapter 10

10-3.b) $\Sigma_v = -177$ ns, $\Sigma_n = 353$ ns

10-4. $0.9945c$

10-7. 0.2 mm

10-8.a) $v \simeq c - \dfrac{c}{2\gamma^2}$, $\gamma = 26.88$

b) 2.1 m

c) 7.8 cm

10-9.a) $v = \dfrac{c^2 p_{1i}}{E_1 + m_2 c^2}$

b) $\tan\theta = \dfrac{\sin\theta'}{\gamma\left(1 + \gamma v^2/c^2\right)\cos\theta' + \gamma v/u'_{1f}}$

10-11.a) $\dfrac{1}{10}$ yr

b) 1% must be fuel

c) 300 kg

10-12. $H = mc^2 \sqrt{1 + \left(\dfrac{p}{mc}\right)^2} + V(\mathbf{r})$

10-13. $W = mc^2(\gamma - 1)$
$W = K_f - K_i$, where $K = mc^2(\gamma - 1)$

Chapter 11

11-1.a) $\ddot{x} + 2\gamma\dot{x} + x + x^2 = f\cos\omega t$

b) $x = a\cos(\omega t - \alpha) + C, \quad C = -\dfrac{1}{2}a^2$

$\tan\alpha = \dfrac{2\gamma\omega}{1 - \omega^2 - a^2}$

$f^2 = a^2\left[\left(1 - \omega^2 - a^2\right)^2 + (2\gamma\omega)^2\right]$

11-3. $A_0 = 2\left(f/2\right)^{1/3}, \quad \omega = 3\sqrt{1 + 3(f/2)^{2/3}}$

11-4. Poincaré coordinates $(x_P, \dot{x}_P)$

$x_P = \dfrac{f(\omega_0^2 - \omega^2)}{r^2}, \quad \dot{x}_P = -\dfrac{2\gamma f}{r^2}$

$r^2 = (\omega_0^2 - \omega^2) + (2\gamma\omega)^2$

11-5. $3A_{\max}^2 = \sqrt{4\left(1 - \gamma^2\right)^2 + \dfrac{3f^2}{\gamma^2}} - 2(1 - \gamma^2)$

$\gamma = 0.1, \quad f = 0.5, \quad A_{\max} = 1.517$

Index

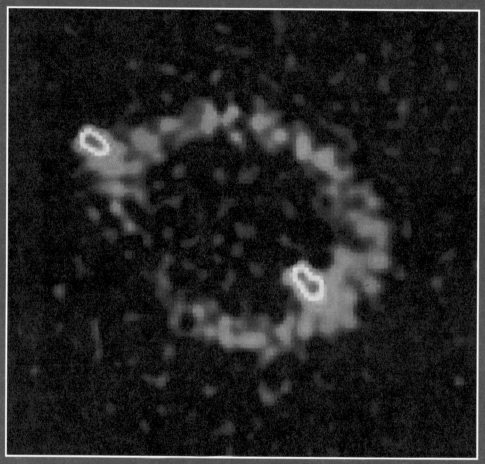